Green Energy and Technology

More information about this series at http://www.springer.com/series/8059

Alan J. Sangster

Electromagnetic Foundations of Solar Radiation Collection

A Technology for Sustainability

Springer

Alan J. Sangster
School of Engineering and Physical Science
Heriot-Watt University
Edinburgh
UK

ISSN 1865-3529 ISSN 1865-3537 (electronic)
ISBN 978-3-319-08511-1 ISBN 978-3-319-08512-8 (eBook)
DOI 10.1007/978-3-319-08512-8

Library of Congress Control Number: 2014945259

Springer Cham Heidelberg New York Dordrecht London

Printed on acid-free paper

Springer is part of Springer Science+Business Media (www.springer.com)

To my Grandchildren
Modern consumer-driven capitalism has become a major mistake perpetrated on the planet by my generation. It is the uncontrollable force driving our ecological crisis. Hopefully, your generation will have the skill to untangle the mess.

Preface

This book is intended to be a technology resource for students of electrical science and for electrical engineering departments in universities and colleges with an interest in developing courses focused on the rapidly burgeoning topic of solar radiation collection. This development has been awakened by a growing concern of the impending dangers, for future generations, of climate change. The author hopes that this text will contribute, in some small way, to the evolution of a technological route out of our self-inflicted predicament caused by an unsustainable addiction to fossil fuels.

During the first decade of the twenty-first century, it is probably now fair to say that the vast majority of reputable scientists with an interest in anthropogenic global warming would have accepted that the 'canary in the mine' providing the warning for its arrival would be the state of summer sea ice in the Arctic. Should it 'expire', this would herald 'real evidence' for the dubious non-scientific world of dangerous man-made climate change. In the summer of 2012 the canary fell off its perch! Arctic Sea ice cover in late summer of 2013 almost disappeared.

The reaction to this cataclysmic event in the media and other electronic outlets which generally claim to be representative of public opinion was almost precisely nothing. In 2013, *Homo sapiens*' collective head was still stuck firmly in the ground despite the seemingly endless breeching of weather records around the world, particularly in the United States, where the 'bread-basket' southern states are suffering 'dust-bowl' conditions as global warming brings desertification. The record breaking hurricane Sandy which struck the East Coast of North America in the autumn of 2012—dubbed the Frankenstorm in the US media and in blogs on the Internet—had a devastating impact on New York, and some reports and reactions suggested that it may have been the 'light bulb switch on' moment in the consciousness of the US public. At the very least, it is perhaps valid to suggest that the 'Denial Lobby', which for 25 years has strenuously and vociferously dismissed the notion of anthropogenic climate change, has finally been defeated. The battle now, is over whether society should adapt to the inevitable ravages wreaked by global warming, or should it adopt the obvious fundamental solution to the problem which entails abandoning the fossil fuels which by combustion are 'poisoning' the atmosphere? These alternative futures are well described in Jorgen Randers' book entitled simply "2052". The dilemma for mankind is succinctly put in this quotation from that book (the additional observations in parentheses are mine):-

> Thus the main challenge in our global future is not to solve the problems we are facing (these are do-able), but to reach an agreement to do so (almost impossible). The real challenge is to have people and capital owners accept short-term sacrifice, roll up their sleeves, and do the heavy lifting. The agreement to act will arise, sooner or later, but it will come late in the day, and the resulting solution even later. As a consequence, humanity will have to live with the unsolved problem (of climate change) longer than if the action had been started at once. Waiting for the 'market' (as we are doing) to give the start signal will lengthen the temporary period of forced sacrifice. Forward-looking political leadership (almost extinct) could kick start the societal response but may be kept from doing so by the democratic majority of voters with a short term perspective.

The science of climate change and the dangers it poses for mankind has been reiterated five times by climate scientists on the International Panel on Climate Change (IPCC5), and with increasing forcefulness. The latest warning has very recently (Spring 2014) been spelt out, comprehensively and with all the relevant evidence, in the 5th Report sponsored by the United Nations.

A feature of renewable power sources such as wind, wave and solar, which is raised repeatedly in debates about their capacity to replace fossil fuel powered electricity generators, is intermittency of supply. However, at the global, or continental level (Europe, say), the variability of renewables can be addressed more easily. When the wind is not blowing in Scotland, or the sun is not shining in Germany, the former will likely be gusting in Portugal, while the latter will be sizzling in Spain! Under the auspices of the European Community, several reports have been generated to assess the feasibility of a direct current (DC) super-grid connecting geothermal power stations in central Europe, solar power stations in southern Europe and North Africa, wind farms in Western Europe, wave/tidal systems in Scandinavia and Portugal, and hydroelectric stations in Northern Europe. This system would be backed up by massive storage facilities based on pumped hydro-storage in reservoirs or artificial lagoons, on compressed gas and hot water thermal storage using cathedral sized underground caverns, on massive flywheel farms, on battery storage barns the size of football pitches and on huge super-cooled magnetic storage devices. Prototype examples of all of these technologies already exist, and undersea power lines from Scotland to the continent of Europe are seriously being evaluated.

Clearly the technologies already largely exist to make a Europe-wide electrical power supply system a reality. In fact, it should be emphasized that almost all of the renewable technologies listed above are relatively conventional. In principle, therefore, sustainable power systems based on these technologies could be executed very quickly if drive, leadership, determination, enthusiasm and cooperation can be imbued in the international community, to recruit and deploy significant human and capital resources towards implementing the task. But where, at 'short notice', would the scientists and engineers required to implement the paradigm shift to renewables come from, and how could the required unprecedented expansion of manufacturing capability be achieved? The major components of renewable power stations, such as turbines, gear trains, generators, propeller blades, nacelles, control electronics, management systems, metering, mirrors, etc., are, in engineering terms, not unlike what is currently manufactured in considerable volume by the automobile and aeronautic industries. Consequently, the answer to the above question is not too difficult to find if we accept that the future has to be oil-free. We must shift the manufacturing emphasis of these major factories, away from building, soon-to-be-redundant vehicles and aircraft, towards providing the infrastructure for renewable power plants, and we must use the capabilities of other fossil fuel dependent industries, such as those involved in chemicals and plastics, to develop storage systems and materials for a superconducting grid.

The book seeks to provide coherent and wide ranging instructional material on electromagnetic solar power collection techniques by collating all of the currently available developments in this technology sector embellished with enough mathematical detail and discussion to enable the reader to fully comprehend the basic physics. As far as the author is aware, the full range of solar power technologies has not previously been presented in a single textbook, which seeks to illuminate and explain through electromagnetism the technological challenges associated with collecting direct radiation from the sun, the primary source of almost all renewable energy, including wind, wave and biomass.

An introductory chapter establishes the 'technological route' that mankind needs to pursue in order to transition away from fossil fuels towards renewables. It also introduces the range of solar techniques available to assist in this endeavour. Subsequently, the content of the book divides naturally into two sections. The first section, Chaps. 2–6, provides the mathematical and conceptual tools which are required to develop fully analysed and comprehensive treatments of the primary solar radiation collection systems as expounded in Chaps. 7–10.

In the early chapters (Chaps. 2–5) which provide the basic electromagnetic theory, the electrical science and the mathematical tools to support the chapters on solar power collection technologies, the author's conscious choice has been to present this material through the agency of classical electromagnetism and waves as opposed to the quantum electro-dynamic approach which emphasises the exchange of particles or photons in the treatment of fields. This preference is justified in Chap. 6 which examines the wave/particle duality issue in some detail.

Naturally, all views, assertions, claims, calculations and items of factual information contained in this book have been selected or generated by myself, and any errors therein are my responsibility. However, the book would not have seen the light of day without numerous personal interactions (too many to identify) with family, with friends, and with colleagues at the Heriot-Watt University, on the topic of global warming and solar energy. So if I have talked to you on this topic, I thank you for your contribution, and the stimulus it may have provided for the creation of this book. I would also particularly like to thank my son Iain for his assistance with image manipulation and the members of staff at the Heriot-Watt University library, who have been very helpful in ensuring that I was able to access a wide range of written material, the contents of some of which have been germane to the realisation of this project.

Contents

1 **Energy from Ancient and Modern Sunshine** 1
1.1 Fossil Fuels—the 'Fruit' of Ancient Sunshine. 1
1.2 Conservation of Energy for Earth 3
1.3 Harnessing Radiant Solar Power 7
1.3.1 Solar Geometry 9
1.3.2 Thermal Solar Conversion 13
1.3.3 Concentrated Solar Power 16
1.3.4 Solar Photovoltaics 18
1.3.5 Orbital Collection 21
1.3.6 Nantennas 23
References. 24

2 **Electromagnetic Waves** 27
2.1 Electromagnetic Spectrum 28
2.2 Electromagnetic Theory and Maxwell's Equations. 28
2.2.1 Flux and Circulation 30
2.2.2 Boundary Conditions. 33
2.3 Plane Wave Solution. 34
2.3.1 Second-Order Differential Equation. 35
2.3.2 General Solution. 37
2.3.3 Snell's Laws 39
2.3.4 Wave Guiding 46
References. 50

3 **Classical Radiation Theory** 51
3.1 Radiation Mechanism 52
3.2 Maxwell's Equations: Source Form. 54
3.3 Auxiliary Potential Functions 58
3.3.1 Electrostatics Analogy. 60
3.3.2 Magnetostatics Analogy. 62

3.4 Hertzian Dipole . . . 65
References. . . . 72

4 Aperture Antennas for Solar Systems . . . 73
4.1 Introduction . . . 73
4.2 Auxiliary Potentials. . . . 77
4.3 Radiation From a Linear Aperture. . . . 79
4.3.1 Huygen's Principle and Equivalent Sources . . . 80
4.3.2 Plane Wave Spectrum . . . 81
4.4 Spectrum Functions. . . . 84
4.4.1 Pattern Sidelobes . . . 86
4.4.2 Mainlobe Beamwidth . . . 86
4.4.3 Pattern Gain. . . . 87
4.5 Rectangular Aperture. . . . 88
4.5.1 Uniformly Illuminated Rectangular Aperture . . . 90
4.5.2 Directivity . . . 94
References. . . . 95

5 Array Antennas for Solar Systems. . . . 97
5.1 Introduction . . . 97
5.2 Uniform Linear Array of Isotropic Elements. . . . 98
5.2.1 Radiation Patterns. . . . 100
5.2.2 Broadside Array . . . 101
5.2.3 End-Fire Array. . . . 103
5.2.4 Scanned Array . . . 105
5.3 Array Design Using Theory of Polynomials . . . 106
5.3.1 Optimum Element Spacing. . . . 112
5.3.2 The Binomial Array . . . 114
5.3.3 Supergain Array . . . 114
5.4 Radiation Pattern Synthesis . . . 115
5.4.1 Tschebyscheff Technique. . . . 116
5.4.2 Fourier Series Method. . . . 118
References. . . . 120

6 Solar Radiation and Scattering: Waves or Particles? . . . 121
6.1 Introduction: What Is Really Being Collected? . . . 122
6.2 Classical Electromagnetism . . . 123
6.2.1 Influence of QED . . . 124
6.3 Photon Scattering . . . 125
6.3.1 Compton Effect . . . 126
6.3.2 Young's Experiment . . . 127
6.3.3 Photons and Interference . . . 130
6.3.4 Coherence . . . 132

6.4 Electron Waves in a Superconducting Ring 134
6.5 Electromagnetic Ring Resonator 138
6.6 EM Waves and QED. 141
References. 144

7 Solar Photovoltaics 145
7.1 Introduction—Photovoltaic Collectors 145
7.1.1 Solar Cell Electronics 146
7.1.2 PN Junction Basic Equations 149
7.1.3 Photovoltaic Action. 153
7.2 PV Array Simulation. 154
7.2.1 Newton Iteration Procedure 157
7.2.2 Solar Cell Conductance Method 159
7.3 Cells, Modules and Arrays. 161
7.3.1 Electrical Circuit Representation 163
7.3.2 Efficiency 166
7.3.3 Array Sizing, Monitoring and Optimisation 168
7.3.4 State-of-the-Art Cell Fabrication 170
References. 171

8 Concentrated Solar Power. 173
8.1 Introduction 173
8.2 Solar Collectors as Antennas 174
8.2.1 Huygen's Principle and Rays 177
8.2.2 Geometrical Optics 180
8.2.3 Theoretically Optimum CSP Collector. 184
8.3 Concentrator Thermodynamics 189
8.4 Architecture of CSP Systems 192
8.4.1 Parabolic Trough Collector System 195
8.4.2 Linear Fresnel Reflector System 197
8.4.3 Heliostat Field System. 199
8.4.4 Parabolic Dish System. 201
8.4.5 Concentrated Photovoltaic System. 202
References. 206

9 Solar Power Satellites (SPS) 207
9.1 Introduction 207
9.2 Space-Based Photovoltaic Array 211
9.3 Microwave Power Generation. 211
9.3.1 Klystron 212
9.3.2 Magnetron 214
9.3.3 Solid-State Microwave Source 216

9.4 Microwave Array Antennas 218
9.4.1 Waveguide Slot Arrays 221
9.4.2 Waveguide Phased Array 227
9.4.3 Retro-directive Array Techniques 229
9.4.4 Micro-strip Patch Array 234
9.5 Rectenna-Based Receiver Arrays 237
References 239

10 Optical Antennas (Nantennas) 241
10.1 Introduction 241
10.2 Antenna Efficiency at Nanoscale 243
10.2.1 Conventional Dipole 244
10.2.2 Efficiency Anomaly 246
10.2.3 Modal Attenuation 249
10.3 Impedance and Conductivity Issues 251
10.4 Radiation Efficiency of a Filamentary Dipole 255
10.5 Superconduction Techniques 258
References 261

Index 263

Abbreviations

AC	Alternating current
AU	Astronautical unit
BARITT	Barrier injected transit time
BWFN	Beamwidth for nulls
CEM	Classical electromagnetism
CPC	Compound parabolic concentrator
CPV	Concentrated photovoltaic
CSP	Concentrated solar power
CW	Continuous wave
CO_2	Carbon dioxide
°C	Degree centigrade
DC	Direct current
EM	Electromagnetism
EMW	Electromagnetic wave
FET	Field effect transistor
GEO	Geostationary
HPBW	Half power beamwidth
H_2O	Water
IMPATT	Impact avalanche and transit time
IR	Infra-red
LSA	Limited space charge accumulation
MF	Medium frequency
MPP	Maximum power point
NASA	National aeronautics and space agency
N-	Negative
P-	Positive
PCB	Printed circuit board
PCC	Phase conjugating circuit
PIN	Positive-insulator-negative
PSD	Passive solar design

PPMV	Parts per million by volume
PV	Photovoltaic
QED	Quantum electrodynamics
SPS	Solar power satellite
TE	Transverse electric
TED	Transferred electron diode
TEM	Transverse electromagnetic
TM	Transverse magnetic
TRAPATT	Trapped plasma and avalanche transit-time
UV	Ultraviolet
VHF	Very high frequency
VLF	Very low frequency
YBCO	Yttrium barium copper oxide

Chapter 1
Energy from Ancient and Modern Sunshine

I have no doubt that we will be successful in harnessing the sun's energy... If sunbeams were weapons of war, we would have had solar energy centuries ago.

Sir George Porter

The human race must finally utilise direct sun power or revert to barbarism.

Frank Shuman

Abstract The importance of harnessing direct sunshine as a route to providing energy to sustain our modern sophisticated societies, in the not too distant future, is addressed in this chapter. The need to transition away from fossil fuels, because by their combustion in the atmosphere, we are triggering dangerous climate change, is explained in a simple engineering friendly manner, emphasising as it does the thermodynamics and well-established electromagnetic wave propagation principles which underlie the science. That a sustainable, fossil fuel-free future for mankind is entirely possible is also reinforced. The energy in sunshine is garnered naturally by photosynthesis, but this is too inefficient for modern requirements. The range of artificial methods that provide the promise of the collection of very large levels of power from solar rays is broached here, and the various modes identified are examined in detail in later chapters. These come in the form of electricity-generating solar farms ranging from those employing photovoltaic panels located on the ground and in space, to those employing sophisticated optical reflector techniques to gather the incident rays. Optical antennas with their potential for high-efficiency light collection are also considered.

1.1 Fossil Fuels—the 'Fruit' of Ancient Sunshine

The consumer-driven global market system which underpins modern economic activity is lubricated (pun intended) by ridiculously cheap fossil fuels. These fossil fuels, a bounty donated by millions of years of photosynthesising ancient sunshine,

A.J. Sangster, *Electromagnetic Foundations of Solar Radiation Collection*,
Green Energy and Technology, DOI 10.1007/978-3-319-08512-8_1

have lain undisturbed and harmless for millions of years, until Homo sapiens discovered their abundance in accessible locations about 200 years ago. Today, even inaccessible and difficult sites are being exploited with ferocious determination, despite our now certain knowledge that continued burning of these materials is harming the planet.

Plant photosynthesis is nature's method of gathering energy from solar flux. The chemistry of photosynthesis is very well understood, although perhaps not by electrical engineers, notwithstanding the fundamental role played in this natural process by electromagnetism, a topic we shall pursue further in Chap. 2. Essentially, photosynthesis encapsulates the mechanism by which light bombardment (electromagnetic waves) on the leaves of a plant enables it to chemically produce glucose (a form of sugar), and hence leaf tissue, from carbon dioxide (CO_2) and water (H_2O) [1]. The stoichiometric equation describing the process has the form:

$$6CO_2 + 6H_2O \Rightarrow C_6H_{12}O_6 + 6O_2 \tag{1.1}$$

The first term on the right-hand side is of course glucose (plant tissue), and helpfully for animals, the process also generates oxygen (O_2). In words, the equation expresses the fact that within the cellular spaces of the leaf, and facilitated by energy extracted from the solar radiation incident upon it, six molecules of carbon dioxide and six molecules of water can be made to combine to generate one molecule of glucose and six of oxygen. Photosynthesis is a thermodynamically driven chemical process whereby plant cells generate glucose from carbon dioxide absorbed through the stomata in the foliage, and water taken in through the roots. The chemical reaction is propelled by electromagnetic radiation incident upon the leaf which acts as a 'gathering antenna'. Thus, plant growth provides a 'sink' for atmospheric carbon.

As intimated above, fossil fuels, namely coal oil and gas, are the product of photosynthetic processes in terrestrial and marine flora, but energised by 'ancient sunshine' which illuminated the earth over 100 million years ago. After millions of years of dying, decaying and sinking into the earth's crust, this rich plant life from long-gone inter-glacial eras was compressed into the coal seams and the oil wells and natural gas reservoirs, through the agency of geological activity driving the carbon-rich material deep into the crust where high pressure and high temperature has done the rest. Today, mankind is successfully exploiting this gift. Given the vast timescales involved in this particular process of transforming sunlight into an energy source accessible to mankind, not surprisingly, the conversion efficiency from sunlight to directly useful energy (coal, oil, gas) is actually extremely low, as Table 1.1 conveys. There are several other conversion routes. Interestingly, energy from renewable sources created by today's sunlight is acquired very much more efficiently as we shall see [2]. It is also clear from the table that the fossil fuel formation modes, having taken many millions of years to establish the energetic bounty, which we are reaping today, are not going to be repeated during mankind's geologically brief sojourn on earth. Therefore, it is safe to say that fossil fuels are a

Table 1.1 Efficiency of conversion of solar power to useful energy for humans [2]

Energy source	Time to harvest in years	Solar conversion efficiency (%)
Coal	>150,000,000	<0.001
Oil, gas	>100,000,000	<0.001
Wood	1–30	0.1–1.0
Biomass	0.1–1	<1
Reservoirs	0.01–1	<1
Wind	Continuous	0.2–2.0
Optically collected sunlight	Continuous	6–25

strictly limited resource, and if current rates of combustion are allowed to continue, they could be exhausted by early next century.

Notwithstanding the inefficiency of their formation, it is fairly obvious from the recent history of the human race that fossil fuels have been little short of a ‘windfall’ for economic and cultural advancement. As we well know, it has been exploited extremely vigorously over the past 200 years. This has meant that over and above the serious inconvenience of inevitable resource exhaustion, an atmospheric ‘side effect’ accompanies fossil fuel combustion. Climate science has discovered that there is an unexpectedly severe penalty [3, 4] for mankind if it persists in unnaturally returning to the biosphere, carbon previously extracted from it, over a period of millions of years in the ancient past. Evidence is accumulating that recent carbon dioxide build-up in the atmosphere, by combusting fossil fuels, is upsetting the thermal balance in the biosphere between incoming solar radiation and outgoing ‘black body’ radiation. The consequence is climate instability on the earth as the biosphere warms to correct the imbalance. This outcome has been long predicted by the laws of thermodynamics, and in particular the first law [4] which counsels the inviolability of energy conservation in natural processes.

1.2 Conservation of Energy for Earth

Without delving too deeply into all the mechanisms involved, the thermodynamics of a large lump of rock in empty space orbiting, and warmed by, a nearby star, can be deduced by asking the question: why does planet Earth in inter-glacial eras such as the Holocene, the geological era which we are currently experiencing, display a relatively invariant mean temperature of approximately +14 °C? As we know this current era is highly favourable to biodiverse life. It is a question which Jean Baptiste Fourier (1768–1830) [5] posed to himself over 200 years ago. Armed with a rudimentary knowledge of solar radiation (approximately 1,400 W/m^2 at the top of the atmosphere), solar transmissivity through the atmosphere, and approximate solar reflection levels at ice, terrestrial and marine surfaces, he determined that the earth should be an ice-encased orb with a mean temperature of about −15 °C. This

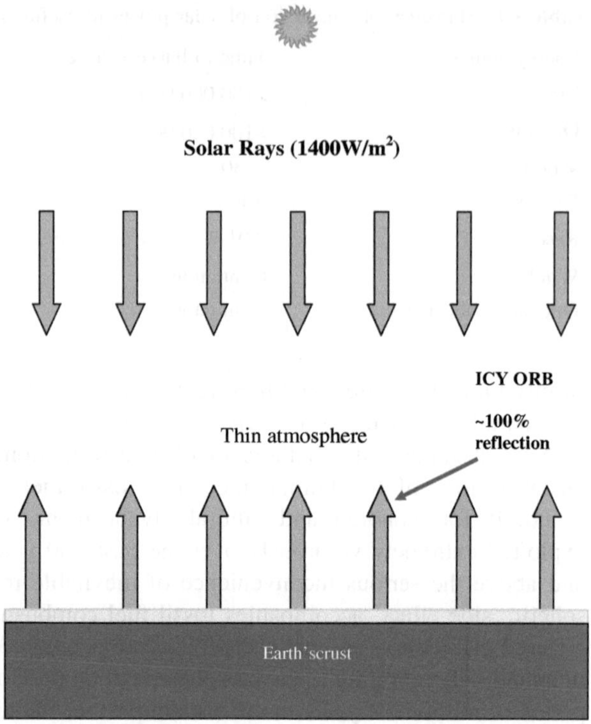

Fig. 1.1 Ice-covered earth—no atmospheric attenuation

is illustrated in Fig. 1.1 where the down arrows, representing incoming solar power density, are matched by the up arrows representing ground reflections. His calculations seemed to suggest that the ice coverage from the preceding ice age would not retreat as the earth entered an 'inter-glacial' orbit promising a strengthening of solar radiation and potentially an enhancement of warming influences. Reflected light from the surface seemed to balance the incoming radiation with no net heating. Fourier soon realised that his assumption that reradiated waves from the earth could simply pass unattenuated through the atmosphere back into space was unsustainable. In this he was correct although he did not manage to unravel the mechanism of radiation trapping.

When solar radiation impinges on ice-free terrestrial surfaces, it is partially absorbed rather than predominantly reflected, warming the surface. The warmed surface then acts as a 'black body', which radiates, from the agitated molecules, electromagnetic waves at infrared and even lower frequencies, into the atmosphere [6]. At these lower frequencies, several atmospheric gases become absorbers of electromagnetic waves. This is a quantum mechanical manifestation which results in wave attenuation as the 'black body' radiation passes through the atmosphere. In proportional terms, water vapour (H_2O) produces the most attenuation due to absorption (62 % of the total), carbon dioxide (CO_2) is next at 22 %, while other contributors are ozone (O_3 = 7 %), nitrous oxide (N_2O = 4 %), methane

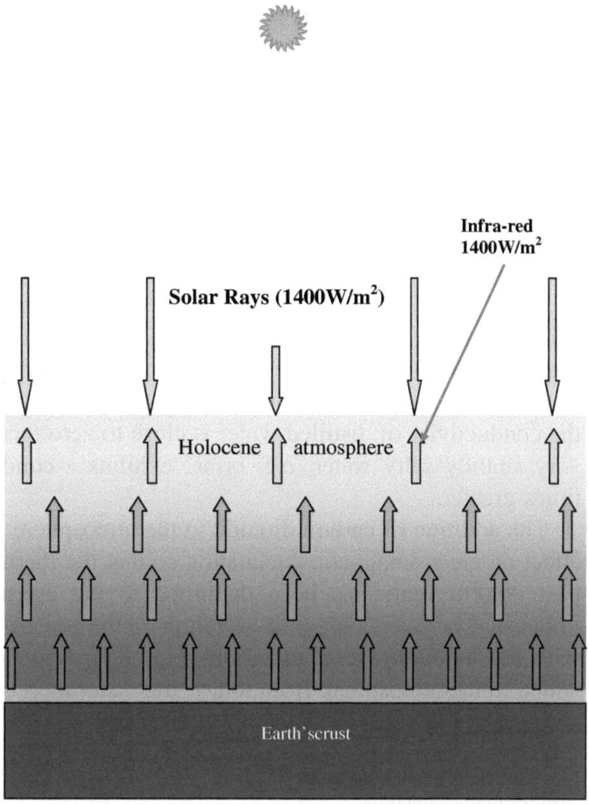

Fig. 1.2 Pre-industrial earth in thermodynamic equilibrium with atmospheric concentration of CO_2 at ~280 ppmv

(CH_4 = 3 %) and others 2 %. The absorbed energy appears as an insignificant rise in temperature of the gases, but more importantly, it means that the re-radiated power reaching space is, in the absence of any correction, much less than the incoming solar power. The law of conservation of energy does not permit this imbalance and requires the earth to heat up until equilibrium is achieved, as suggested in Fig. 1.2, where again the down arrows representing incident solar power density are assumed to penetrate relatively unattenuated to the planetary surface. There, the surface is heated to a mean temperature level where the emitted infrared radiation (closely bunched red up arrows) is strong enough to counteract the atmospheric attenuation as the radiation passes up through the atmosphere (attenuation is represented by increasing arrow separation) so that the power density at the edge of space balances the incoming power density, at a notional level of 1,400 W/m^2. This equilibrium state happily occurred for humanity about 20,000 years ago and has persisted until very recently. The rise in mean global temperature, from Fourier's icy orb, to achieve thermodynamic equilibrium is 33 °C. The mechanism is usually termed the 'greenhouse' effect.

Unfortunately, thermodynamic equilibrium for the 'great earth system' is no longer the current 'state of play', thermally. Since the discovery of coal, and the triggering of the Industrial Revolution, in the eighteenth century, there has been a slow but inexorable rise in the concentration of CO_2 in the atmosphere associated with the combustion of coal, oil and gas. Whereas for the previous 20,000 years of low-technology human existence, the concentration of CO_2 has remained steady at 280 parts per million by volume (ppmv), it has risen steadily to almost 400 ppmv in the last 200 years, according to reliable evidence from the Mauna Loa Observatory in Hawaii [7]. An additional 120 ppmv of CO_2 in the atmosphere hardly seems significant in volume terms, but it has a profound effect electromagnetically. It is not difficult to illustrate this with a quite simple but helpful electrical experiment. If a pair of metal electrodes are placed a small distance apart in a bath of distilled water and fed from a battery via a sensitive ammeter, negligible current is recorded. However, the addition to the water of a trace of salt which occupies a few parts per million of the water causes the current to rise dramatically. This is because while the conductivity of distilled water is close to zero (actually 5.5×10^{-6} s/m), that of very slightly salty water, e.g. brine, exhibits a conductivity of 5 s/m—a million times greater.

The addition of carbon dioxide to the atmosphere has an analogous multiplying effect on electromagnetic attenuation so that the thermodynamic equilibrium of the past 20,000 years has been disturbed to the extent of producing a -4 W/m^2 imbalance between outgoing radiation at the top of the atmosphere as compared with the incoming solar radiation (Fig. 1.3). If no more greenhouse gases were added to the atmosphere from today, the restoration of thermodynamic equilibrium

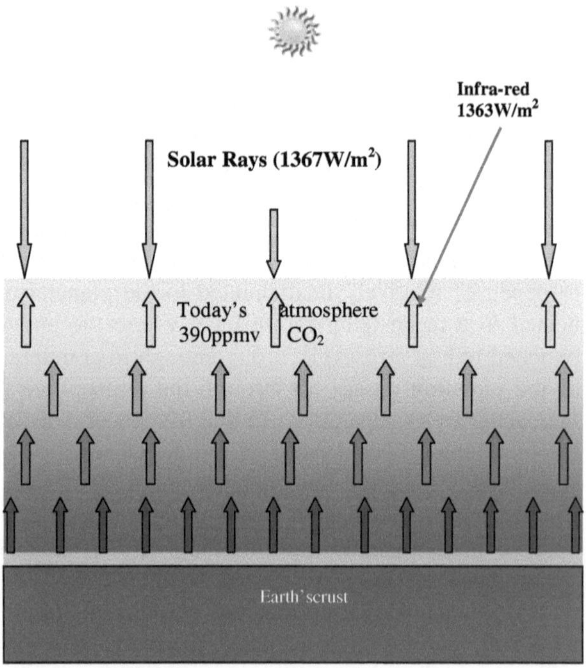

Fig. 1.3 Inter-glacial warmed earth in thermodynamic deficit of -4 W/m^2

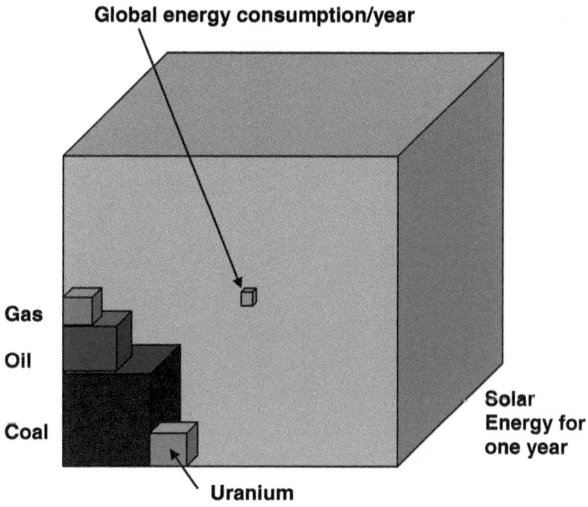

Fig. 1.4 Very approximate volumetric comparison of energy available to mankind and global demand

will require the mean global temperature to rise by about 2 °C relative to pre-industrial levels. This is a risky but possibly tolerable increase if Homo sapiens follows, for the foreseeable future, an urgent strategy of transitioning to renewable energy backed up by population reduction and massive reforestation [8]. This book is aimed at providing a small, but hopefully positive and meaningful, impetus to the implementation of the first of these aims.

It seems inevitable that mankind will be forced to abandon fossil fuels as the primary source of energy underpinning our modern economies. This scenario obviously calls for a massive step by civilisation, but it is by no means impossible, as the graphic reconstructed in Fig. 1.4, and originally attributed to Greenpeace, very clearly shows. When calculable energy resources still buried in the earth, including uranium, are compared volumetrically (i.e. expressing energy content as a volume), with the renewable resource represented by solar power daily impinging on the atmosphere, the latter is much more abundant, and is unlikely to ever be swamped by demand (again see Fig. 1.4).

This text will direct its attention to the physics and engineering of solar power collection by optical means. Of course, it should not be forgotten that wind systems, wave systems and biofuels from vegetation are also mechanisms for gathering energy from the solar power washing over the earth daily.

1.3 Harnessing Radiant Solar Power

To establish the design principles involved in the construction of solar power collectors for incorporation into electrical power distribution systems, it is essential that we possess, a priori, knowledge of how much solar flux there is to collect, not just at the outer reaches of the atmosphere but at lower levels and particularly at

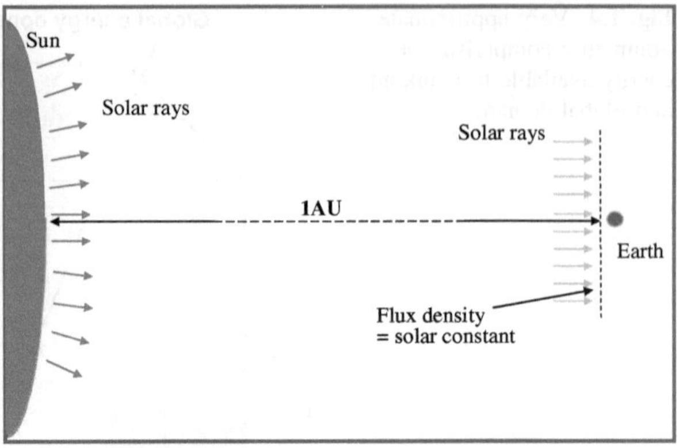

Fig. 1.5 Determination of the solar constant

ground level. And just as importantly we need to know how variable this flux is likely to be. At the edge of space, the power density (irradiance) level has been reasonably accurately known for many years, as indicated in the preceding section. This figure, referred to as the solar constant, is generally deduced from measurements. It is usually presented in the form of an electromagnetic flux density in W/m^2 and is defined as the amount of incoming solar electromagnetic radiation per unit area that would be incident on a plane, perpendicular to the rays, at a distance of one astronomical unit (AU) from the sun (Fig. 1.5). The mean distance from the sun to the earth is approximately one AU.

Today, solar irradiance is measured by scientific instruments mounted on satellites, located at the top of the earth's atmosphere. If the difference between the earth's current distance from the sun, and one AU is known, as it is, then the computation of the magnitude of solar irradiance at one AU and hence the solar constant can be deduced from the measured result, simply by employing the inverse square law, which applies to all fields 'radiating' from a point source. At the distances involved, the sun can be considered to be such a source. The current estimate for the solar constant is 1,367 W/m^2. This flux density exists across the total area of the earth's disc as Fig. 1.5 makes clear.

Actually, the solar 'constant' is misnamed. Scientists have, for many years, suspected that it is influenced by sunspot cycles, but it is only recently, through measurements using satellite-borne instruments, that the extent of the solar variation has been established. Over the last three sunspot cycles, each of which has extended over approximately 11 years, the measurement evidence is that solar irradiation of the upper atmosphere changes by about 1 % from the peak to the trough of the sunspot cycle. For engineers involved in the development of solar power collection systems, this variation is sufficiently small to permit the assumption of solar radiation constancy with little error in their design outcomes.

However, the solar irradiation of the upper atmosphere is only part of the story. The surface area of the earth is four times that of its disc, and we can assume that the mean radiant power to the surface is reduced by a further 50 % both through reflection at the boundary of the earth's atmosphere and space, and by absorption as the solar rays pass through an atmosphere of a notionally average gaseous composition. Consequently, we have to divide the solar constant by eight to get the figure for solar power, sometimes termed irradiance, or insolation, at the earth's surface. On average [9], it exhibits a magnitude of about 170 W/m^2. Nonetheless, in hot equatorial areas such as Arabia, it can be as high as 1,000 W/m^2, in clear conditions. The aim of this book will be to examine established and evolving mechanisms to collect and convert available solar power into an energy source suitable for powering human economies. Several technologies exist for exploiting solar radiation: photovoltaic (PV) methods, solar thermal electric or concentrated solar power (CSP) techniques, solar antennas or passive solar design (PSD), and active solar. Our attention shall be directed mainly towards those technologies which result in the large-scale generation of electricity for human consumption.

The range of technologies which can be identified as underpinning solar radiation collection is introduced in the following sections. The structural and electrical requirements of the available approaches to solar collection, and the fundamental physics upon which each is dependant, are elaborated upon in the relevant chapters. The thrust of these is briefly addressed within the sections below.

1.3.1 Solar Geometry

As has already been intimated in previous sections, the earth's atmosphere has a significant influence on incoming solar radiation at light and near-light frequencies. At wavelengths in the visible range (300–700 nm), electromagnetic waves are absorbed and scattered by air molecules, water vapour particles, aerosols (dust) and of course by the ground itself. The result is that the direct radiation attenuates on its way to the collector, but the incoming light also bends due to scattering and arrives at non-direct angles (see Fig. 1.6). This latter component is termed diffuse radiation.

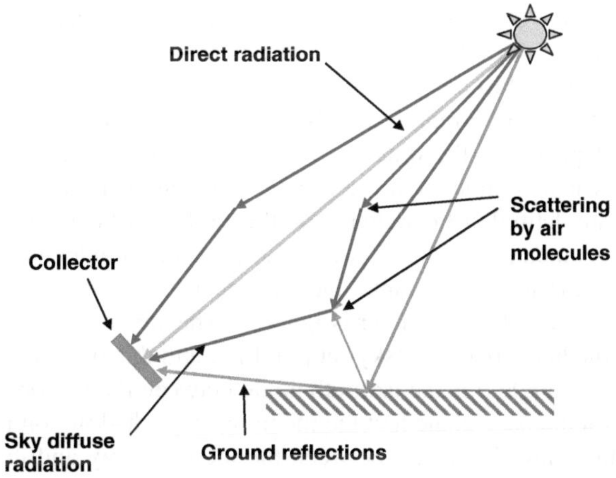

Fig. 1.6 Direct and diffuse sources of solar radiation

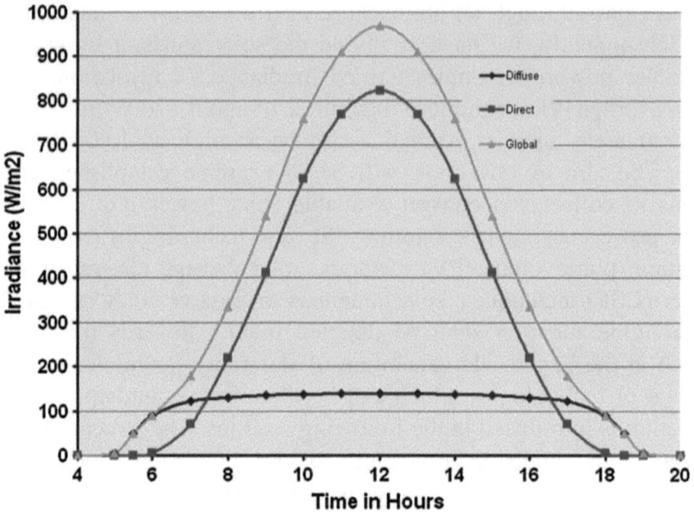

Fig. 1.7 Contributions to solar power density collection from direct and diffuse radiation in a clear sky at latitude $\phi = 36°N$ at spring equinox

On a clear day, it is self-evident that direct radiation is much more powerful than the diffuse component, whereas when it is cloudy, the direct/diffuse proportions are much less obvious. This has not a little relevance to solar power gathering in cloudy Northern and Southern latitudes. A typical solar power density collection profile over the period of one cloudless day, at a 36° latitude with the sun directly overhead at noon, is presented in Fig. 1.7. The direct solar radiation impinging on the collector at noon is almost six times the diffuse contribution, with a maximum direct input of 800 W/m^2.

The basic solar collector problem involves determining and taking full advantage of the local irradiance as a function of time, at a specific location on the earth's surface, as suggested by the irradiance traces in Fig. 1.7. So it is essential that the system designer possesses a reliable means of determining the movement of the sun in the sky above. To do this, it is firstly necessary to establish the geometrical relationship between the collector and the direction of the incident rays. The situation can be represented by the polar diagram shown in Fig. 1.8. The hemisphere depicted in the diagram defines the 'sky' over the specified collecting antenna, which is situated at the centre of the circular base. The collector can be located at any latitude ($\pm\phi$ radians) on the earth's surface, without altering the hemispherical geometry. The longitudinal position for an antenna at fixed latitude on a spinning globe is not relevant, since all receivers at the same latitude are similarly illuminated, if local weather is ignored. The sun's rays are assumed to penetrate through the hemispherical 'sky' at point P_L, and this point is defined by two angles α and γ. The angle α is the azimuth angle between the vector joining the receiver with north on the base plane and the incoming ray 'shadow' on this plane (Fig. 1.8), while γ is the elevation angle between the incoming ray and the base plane.

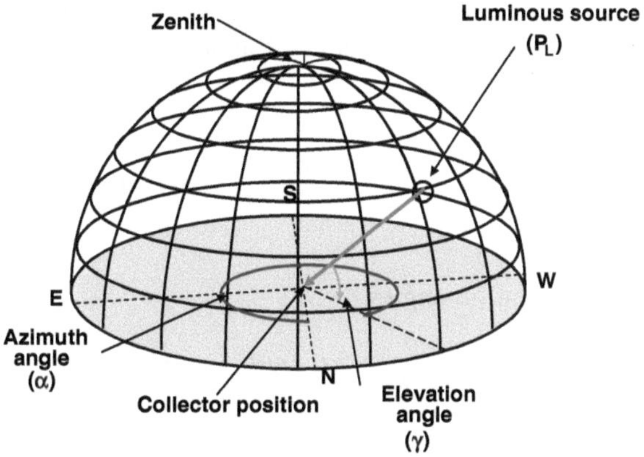

Fig. 1.8 Hemispherical surface from which solar rays can emanate for collector at latitude ϕ on the earth's surface

If we define the direction of the typical incident ray as $\delta = \pi/2 - \gamma$, then the total irradiance (I_{rr}) at the centre of the horizontal plane can be expressed as:

$$I_{rr} = \int_0^{\pi/2} \int_0^{2\pi} p(\delta, \alpha) \cos\delta \sin\delta \, \mathrm{d}\delta \, \mathrm{d}\alpha \tag{1.2}$$

where $p(\delta, \alpha)$ is the power density, in W/m^2, of the incident ray. The angles α and δ are functions of time t. By employing firmly established knowledge from astronomy of the orbital, tilting and wobbling motions of our planet, algorithms for predicting the values of α and γ anywhere on the planet's surface at any time of the day, month, year or century, have become readily available [2]. The following equations represent an efficient distillation of the astronomical data.

$$\alpha = \pi - \cos^{-1}\left[\sin(\gamma)\sin(\phi) - \frac{\sin(\delta_m)}{\cos(\gamma)\cos(\phi)}\right] \text{ radians} \tag{1.3}$$

$$\gamma = \sin^{-1}\left[\cos\left(\frac{15\pi}{180}(12 - t)\right)\cos(\phi)\cos(\delta_m) + \sin\left(\frac{15\pi}{180}(12 - t)\right)\sin(\delta_m)\right] \tag{1.4}$$

$$\delta_m = 0.3948 - 23.2559\cos\left[\frac{2\pi d}{365} + 0.1588\right] - 0.3915\cos\left[\frac{4\pi d}{365} + 0.0942\right] - 0.1764\cos\left[\frac{6\pi d}{365} + 0.4538\right] \tag{1.5}$$

In these equations, d denotes the number of days from the beginning of the year with $d = 1$ at January 1, t denotes local solar time (highest position of sun at 12.00 in 24 h system, ϕ is the latitude of the collector (positive to north of Equator, negative to south) and δ_m is the declination at maximum solar elevation during the day. Further, to take account of atmospheric effects, the following modifications, employing data from the Astronomical Almanac of 1996, have been introduced.

For solar elevation angles $\gamma \geq 15°$

$$\Delta\gamma = \frac{0.00452\, p\pi}{180 T_A \tan\gamma} \tag{1.6}$$

and for elevation angles $\gamma \leq 15°$

$$\Delta\gamma = \frac{(0.1595 + 0.0196\gamma + 0.00002\gamma^2)p\pi}{180(1 + 0.505\gamma + 0.0845\gamma^2)T_A} \tag{1.7}$$

where the following definitions apply: $\Delta\gamma$ denotes the *deviation* in the sun's real elevation angle (γ in radians) due to atmospheric refraction. T_A is the ambient temperature at ground level, while p is the atmospheric air pressure at the collector. Eqs (1.6) and (1.7) are in sufficiently close agreement to ensure a smooth continuation in the angle determinations through the 15° transition value.

It should be noted that in the above solar angle computations, it has been assumed that the collector antenna face, whose direction is conventionally taken as its normal, points vertically upwards towards the zenith in Fig. 1.8. However, in practice, this is unlikely to be the case except at the equator, at the Spring and Autumn equinoxes. Generally, the collector will be directed in such a way as to maximise the time for which the sun's rays strike its face normally or close to normally. In tracking systems, this entails motorising the antenna so that it follows the direction of the sun throughout the daylight hours. If we assume that the antenna face can have an elevation angle γ_M and an azimuth angle α_M, then the solar rays will be incident on the collector face at an angle Φ_{i} given by

$$\Phi_i = \cos^{-1}[\sin\gamma\cos\gamma_M - \cos\gamma\sin\gamma_M(\alpha - \alpha_M)] \tag{1.8}$$

Figure 1.9 has been constructed using the above equations. It shows typical elevation (γ) and azimuth (α) loci, over a 20-h daylight period, for a vertically mounted collector located at a latitude of ϕ = 36°N. As expected, the elevation angle γ is close to zero in the early morning and in the late evening and peaks at noon at a value of ~70°. The azimuth angle reflects the east to west movement of the sun across the sky. When the collector face is elevated to 36° and directed towards the south, the sun's rays are incident at close to normal to the face for a longer period of time as the solid line in Fig. 1.9 demonstrates.

Antenna alignment strategies for optimising power collection will be examined further in Chaps. 7 and 8.

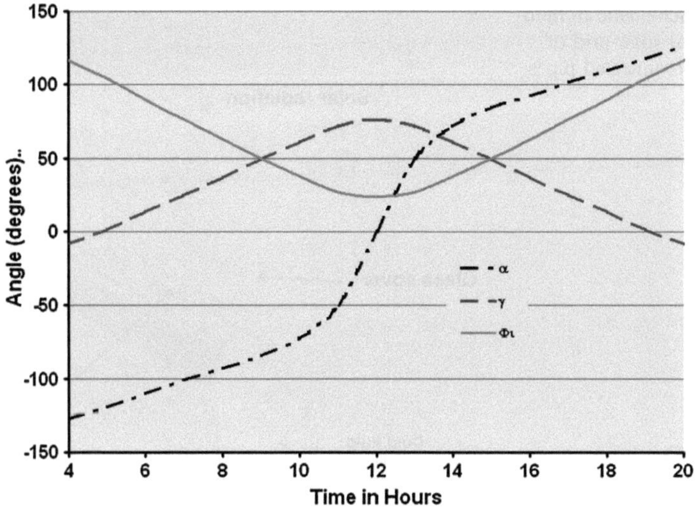

Fig. 1.9 Sun's angular position in the sky as a function of time for a typical northern latitude ($\phi = 36°N$)

1.3.2 Thermal Solar Conversion

Arguably, the technology which pioneered direct and controlled solar power collection was the fluid-filled solar panel. These devices were commonly fitted to suitably directed (i.e. towards the sun) domestic roof surfaces during the 1970s and 1980s, but today solar photovoltaic panels more commonly perform this role. However, the thermal solar panel (Fig. 1.10) provides a useful model for establishing basic geometrical and optical interactions, power relationships, transfer efficiencies and thermal requirements. The essential features of the fluid-based solar collector arrangement shown in Fig. 1.10 are investigated below.

The concept underpinning solar power collection could be described as child's play! Many children, at some stage in their play activity, are likely to have discovered, or been shown, that a magnifying glass creates a bright hot spot on paper, which has sufficient power density to cause the paper to singe and hence to etch a hole. Every scout used to know that this was the only legitimate way to start a fire! Matchsticks were cheating. The magnifying glass if properly shaped concentrates the parallel rays of the sun by bending them, in accordance with Snell's laws (see Chap. 2), through the lens and directing them towards a focus, where the paper should be located. If at the focus there resides a fluid, then the heated fluid can be used to do work, the primary requirement of any energy-gathering technology.

The capacity of a solar power collector employing a fluid as its transfer medium is predominantly determined by the optical properties of the glass cover plate (not a lens in this example) and the tubular panel (Fig. 1.10), and the thermal properties of the fluid. The basic physics is summarised in the relationship:

Fig. 1.10 Schematic of fluid-filled thermal solar grid of parallel interconnected pipes

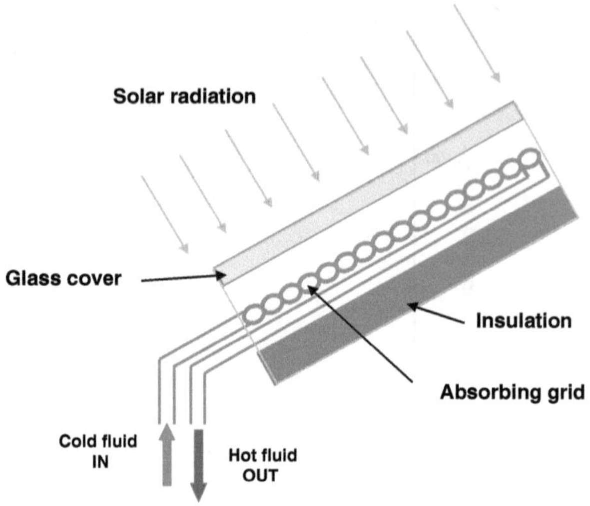

$$\text{Solar power input} = \text{heat losses} + \text{heat collected} \tag{1.9}$$

The key quantity in ascertaining the effectiveness of a flat plate converter of this description is collection efficiency (η), which is essentially the heat collected divided by the radiant power striking the plate. The magnitude of the radiant power gathered by a collector of area A m^2 is quite simply:

$$P_{inc} = p_{si}A \text{ Watts} \tag{1.10}$$

where p_{si} is the solar irradiance in W/m^2 at the gathering surface. However, not all of this radiant power reaches the absorber fluid because of optical deficiencies. The glass cover plate reflects some of the light incident upon it, in accordance with Snell's laws. The proportion passing through to the absorber is usually represented by a transmittance τ. The absorber also scatters light again largely following the dictates of Snell's laws, and in this case, the proportion of the incident light absorbed is represented by an absorptive factor α. Other electromagnetic scattering effects are usually accounted for by a collector efficiency factor F'. So the solar power reaching the absorber surface is:

$$P_{abs} = p_{si}A\tau\alpha F' \text{ Watts} \tag{1.11}$$

The heat absorbed by the fluid flowing through the folded tube structure obviously depends on its temperature rise (ΔT), its specific heat and its rate of flow through it. Temperature rises typically in the range 40–100 °C are possible with well-designed systems. If the fluid passes through the panel at velocity v_f within a folded tube arrangement of cross-sectional area A_t, then the volume of fluid per second passing through the system is:

$$V_{fl} = A_t v_f \ \text{m}^3/\text{s} \tag{1.12}$$

Hence, if the density of the fluid is ρ kg/m^3 and its specific heat is c_f J/kgK, we can state that, for a temperature rise ΔT between the input and output of the panel, the rate at which heat is collected ($\dot{Q}$ in Watts) is given by

$$\dot{Q} = \rho\, c_f V_{fl} \Delta T \ \text{Watts} \tag{1.13}$$

Heat losses, mainly to the atmosphere, can reasonably be assumed to increase linearly with the difference in temperature between the mean value for the panel (T_m) and the atmospheric temperature (T_a), and the area of the panel. So rate of heat loss ($\dot{Q}_{\text{loss}}$ Watts) can be expressed as

$$\dot{Q}_{\text{loss}} = AU(T_m - T_a) \ \text{Watts} \tag{1.14}$$

where U is an empirical factor fitting the equation to experimental evidence.

We define a theoretical efficiency for the flat plate collector as

$$\eta_0 = \frac{P_{abs}}{P_{\text{inc}}} = \tau\alpha F' \tag{1.15}$$

The Eq. (1.9) reduces to

$$P_{abs} = \dot{Q} + \dot{Q}_{\text{loss}} \tag{1.16}$$

or

$$\dot{Q} = P_{abs} - \dot{Q}_{\text{loss}} \tag{1.17}$$

On dividing both sides of the equation by P_{inc} we get

$$\eta = \eta_0 - \frac{U(T_m - T_a)}{p_{si}} \tag{1.18}$$

The equation is widely referred to as the 'Hottel/Whillier/Bliss' equation [10], and in plotted form (Fig. 1.11), as the collector efficiency curve. For the large majority of flat plate collectors of the type suggested by Fig. 1.10, Eq. (1.18) can be treated as a straight line on an η versus $(T_m - T_a)/p_{si}$ plot, with intercept η_O on the η-axis and negative slope U (Fig. 1.11). In practice, U is not a constant but is weakly dependent on temperature and insolation, as is hinted at in the figure, which is typical of a single glazed flat plate collector [10].

This analysis highlights many of the primary features of any device or array of devices aimed at collecting solar radiation and converting the power to a form suitable for human consumption. There is a need to achieve efficient gathering of

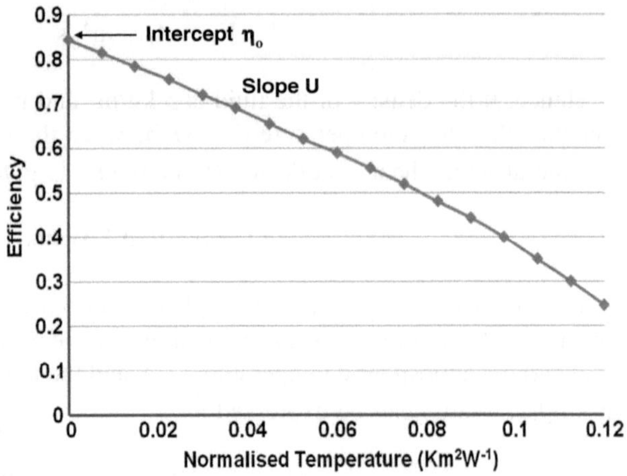

Fig. 1.11 Collector efficiency versus temperature rise, normalised to insolation level (p_{si})

the electromagnetic waves by minimising reflections and other scattering mechanisms. To do this requires a comprehensive understanding of electromagnetism. The relevant topics are addressed in Chaps. 2–4. There is also a need to manage energy conversion, efficiency and thermodynamic issues, and these are dealt with in later segments of the book, such as Chaps. 5–8, which examine issues relating to solar power generation.

1.3.3 Concentrated Solar Power

While flat plate solar/thermal collector systems can provide useful levels of hot water or useful levels of backup space heating in compact buildings, they are generally incapable of generating enough heat to power a steam turbine and hence providing input to an electricity supply system. Temperature rises greater than 150 ° C are required in this case. Not surprisingly, the achievement of high temperatures using solar rays requires the adoption of optical focusing techniques to produce what is termed concentrated solar power (CSP). The obvious way to do this is using lenses, but in very large-scale CSP systems, lenses would be far too expensive and much too cumbersome and heavy to distribute and install over many square miles of desert, so instead, ray concentration is achieved using moulded and electroplated optical reflectors with lightweight materials. In principle, these are not unlike vehicle headlights, but very much larger. If a car headlight reflector were used to collect the rays of the sun on a bright sunny day, a hot spot of light would be formed where the bulb is normally located. In the study of reflector antennas, it is generally assumed that the reflector surface is 'smooth' enough to reflect and transmit the incident electromagnetic waves coherently. The accepted criterion is that any holes or protuberances contributing to surface roughness are in size much less than a wavelength of the incident wave. At optical frequencies, this criterion means 'mirror' quality surfaces, whereas at low radio wave frequencies, the sea surface could be considered smooth. At ultraviolet and higher frequencies, the concept of a reflecting 'surface' breaks down as wavelengths approach the atomic or molecular scale and it becomes necessary to adopt quantum mechanical ideas to explain the relationships between electromagnetic waves and materials. This topic is broached in Chaps. 4 and 10.

In panel sizes appropriate for the forming of large solar arrays, parabolic reflectors are relatively inexpensive, they are not heavy, and, importantly, they can be manoeuvred electronically to track the sun. The concentrated optical beam formed by each reflector of a CSP system must remain focused on the fluid target—usually horizontally or vertically orientated tubes—throughout the day. This means the reflectors are motor driven and controlled to track the sun as it traverses the sky. To do this reliably over extended periods of time, it also means that CSP farms must be located on stable terrain. In addition, they are restricted to land areas where winds are generally light to ensure minimal disturbance to the alignment of the optical reflectors. All of these features of optical collectors and concentrators are elaborated upon in Chaps. 4 and 8.

The technology of CSP farms comprises the following six basic elements: a collector, a receiver, a fluid transporter, an energy convertor, a generator and a transformer. All of these sub-systems can be realised today using well-established and available technology. Needless to say, a range of competing collector topologies are under development, each of which has its advantages and disadvantages. The alternative arrangements are essentially distinguished by the way in which the solar reflectors are organised to concentrate the light onto a receiver containing a working fluid. In parabolic trough systems, the reflectors (curved in one plane only—see Fig. 1.12) are arranged in parallel rows (usually in north–south alignment) directing light onto long straight receiver pipes lying along the focal line of the trough. The changing height of the sun in the sky as the day progresses is accommodated by a tracking system which very slowly rotates the mirrors about a horizontal axis. Fluid flowing, under pressure, through the receiver tube is heated to between 100 and 500 °C and then transported through a well-insulated network of pipes to a boiler, to generate steam.

The other CSP formats which have been proposed envisage mirror arrangements which provide higher optical power density at the focus of the reflectors. In the so-called heliostat system, the individual parabolic reflectors are arranged in rings around a central tower. It is claimed to have two basic advantages over the trough system. Firstly, the sun can be tracked in both elevation and azimuth, and secondly, the fluid passing through the receiver on the central tower is raised to a much higher temperature in the range 800–1,000 °C. This promises greater efficiency, although no full-scale prototypes have been built to establish this. The current state of development of CSP systems of this nature is examined in Chap. 8.

A third system, which circumvents the need for a fluid carrier, is based on solar ray focusing by a circular (typically 40 ft in diameter) dish-shaped parabolic reflector, each of which with its receiver is a stand-alone electricity generator. It is also at the early prototype stage of development. Power station levels of electrical

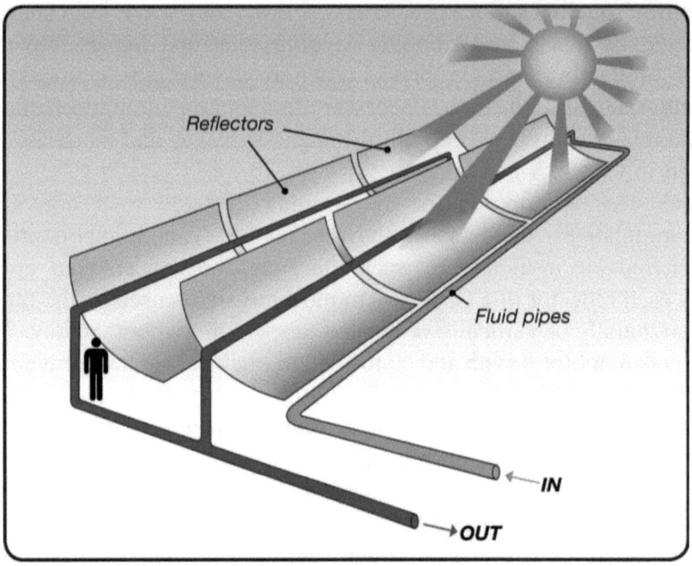

Fig. 1.12 Schematic of parabolic trough CSP system

power are gathered from large numbers of these deployed in an extensive regular grid in a suitable desert scenario. Several prototype installations of limited size have been operating successfully over the past decade. Each dish is like a very large car headlight reflector, and all are automatically controlled to accurately focus the sun's rays onto the receiver. The sun is, again, tracked by tilting the dish in both elevation and azimuth. Despite the additional complexity and manufacturing cost of the large dish-shaped parabolic reflectors and their sophisticated support structures, the arrangement has two distinct advantages. Firstly, the system is modular, in so far as every dish and receiver set is an independent solar power station (rather like a wind generator), and consequently, they can be installed and efficiently operated on hilly terrain, unlike trough and heliostat systems. Secondly, it is possible to replace the conventional fluid mechanism for transporting the heat generated by the focused solar rays, with a device located at the focus of each dish which converts the solar heat directly into electricity. In this way, efficiency improvements can be realised. This device comprises a Stirling engine coupled to an induction generator. The Stirling option becomes feasible with operating temperatures in the range of 700 °C. The technology is fully discussed in Chaps. 4 and 8.

1.3.4 Solar Photovoltaics

Since the 1980s, silicon-based photovoltaic technology for converting direct sunlight into electricity has been becoming an increasingly pervasive influence on modern consumer electronics. Solar cells are now common in watches, radios, calculators, and toys and are to be increasingly found powering street signs, parking meters and traffic lights. Like most technologies, which sustain modern ways of living, the vast majority of users accept it without really ever understanding it. This is, of course, not acceptable for students of electrical science, but even for such students, solar photovoltaics (solar PV) can actually be a difficult subject because it is commonly taught from a quantum mechanical perspective focusing on photon behaviour (see Chap. 6). It is worth noting that at lower than light frequencies, the phenomenon, which could perhaps be termed 'radiovoltaics', is still present [11], but is interpreted as semiconductor diode detection (see Chap. 7). The basic structure of a silicon solar cell is shown in Fig. 1.13. It will be described in detail in Chap. 7.

The growth rate of solar PV technology over the past ten or so years has been little short of phenomenal. It is estimated to currently provide significant levels of electricity in more than a hundred countries. However, the growth has been from a base of virtually zero at the turn of the twenty-first century, so in percentage terms, its share of global energy supply is still rather tiny at ~0.1 %. If mankind awakens to the dangers of fossil fuel combustion, and global warming, this position could rapidly change, with solar PV likely to become a major contributor to the world's electrical power generation systems.

The current technology is predominantly based on silicon although other possibilities exist [12]. Obviously, silicon offers a major advantage over other materials in the fact of its relative abundance in the earth's crust. This in turn implies low cost, which is getting lower, for a large-scale and material-intensive nascent

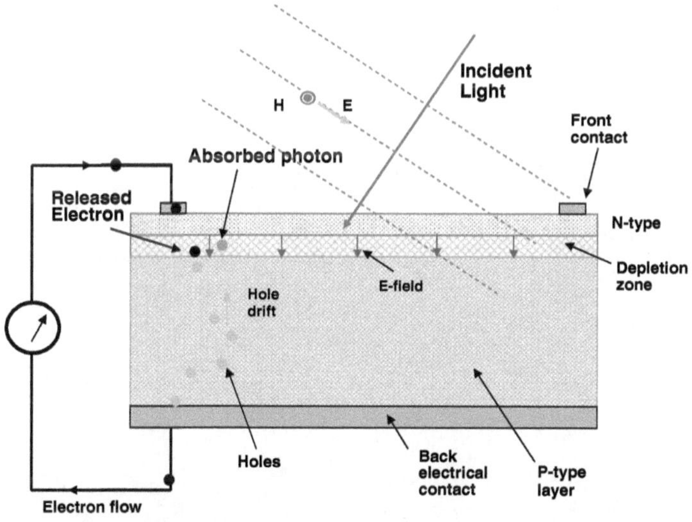

Fig. 1.13 Schematic of photovoltaic crystal showing an electron, released by the EM field, causing holes to drift though the P-layer to the N-layer

technology. After purification, silicon assumes several different forms from mono-crystalline, multi-crystalline, micro-crystalline to an amorphous structure. All of these phases are employed in the production of solar cells. The typical photovoltaic cell fabrication plant is wafer based with about 33 % directed towards mono-crystal silicon, while around 53 % employ techniques based on multi-crystal silicon. The remainder are engaged in non-wafer silicon technologies.

That there is a predominance of the wafer-based methods is largely attributable to the growth of solar PV in parallel with the microelectronics industry which relies heavily on silicon wafer production. While wafer fabrication techniques have advanced rapidly as a result of this synergy, the technology is not optimum for the solar power industry, because typical wafers are thick (200–300 μm) and in the very large-scale production of solar panels the demand for silicon is massive. Silicon wafers are also by no means inexpensive, and the technology offers few avenues for serious cost reductions. Needless to say, new technologies are appearing which avoid using the large amounts of expensively purified silicon as is common in conventional wafer-based methods. For example, two new fabrication techniques have evolved from recent research. These are thin-film hetero-junction silicon solar cells, and solar cells based on polycrystalline thin films (Chap. 7). Devices constructed in these technologies use much less silicon and promise much lower production costs.

The potential of solar PV technology for electricity generation on a large scale is illustrated in Fig. 1.14. It shows the new photovoltaic park opened in Les Mées, in the southern department of Alpes-de-Haute-Provence in 2011. Spread across 36 acres, and built by Belgian firm Enfinity, it joins several other plants located on the vast Puimichel plateau. By the end of 2011, solar panels will cover 200 ha and produce around 100 MW, making it the biggest solar array in France. Enfinity's investment has included work to preserve the landscape with space for grazing by designing a system without a concrete foundation.

Fig. 1.14 Photovoltaic farm in France (*Courtesy* of Alamy)

1.3.5 Orbital Collection

In space, roughly at the boundary between the stratosphere and the mesosphere, it has been noted in Sect. 1.3 that the solar constant has a value of 1,367 W/m^2, yet at the planetary surface, the radiant power density can vary from as low as 200 W/m^2 in cloudy locations to the north and south of the Tropic of Cancer and Tropic of Capricorn, respectively, to 1,000 W/m^2 in suitable desert locations. The situation is shown clearly in the spectrum diagram presented in Fig. 1.15, where the most significant difference between 'top of the atmosphere' radiation (triangular markers) and radiation at sea level (diamond markers) occurs in the visible band, and in some selective absorption bands associated with greenhouse gases.

The approximately tenfold reduction in the visible spectrum is due to electromagnetic scattering and attenuation, associated with the boundaries between atmospheric layers and cloud cover. The process is illustrated crudely in the image in Fig. 1.16 on the left. Needless to say, this power density difference has encouraged engineers to expend considerable effort at devising solutions with the potential to circumvent the atmospheric penalty. Planned developments have mainly revolved around the notion of performing the solar power collection in space and then beaming it to earth on non-optical frequencies (Fig. 1.16, right-hand portion of the image).

A conceptual representation of a solar farm in space orbit is shown in Fig. 1.17. In this example, the intention is for sunlight collection to be primarily secured using arrays of parabolic mirrors (Chap. 8). These will be automatically orientated to direct the solar rays onto photovoltaic collectors (Chap. 7). The DC power formed by the PV array will in most systems which have been proposed be employed to drive electron beam oscillators, such as magnetrons (see Chap. 9), to generate microwave power. This microwave power will be beamed to earth by means of a large planar array antenna, focused on a collector antenna system located on earth, as detailed in Chap. 9.

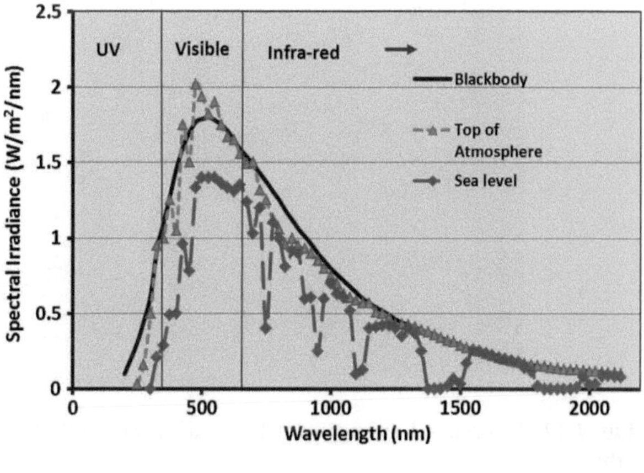

Fig. 1.15 Solar radiation spectrum in space and after passing through the earth's atmosphere

Fig. 1.16 Solar power collection from space to eliminate atmospheric scattering and attenuation

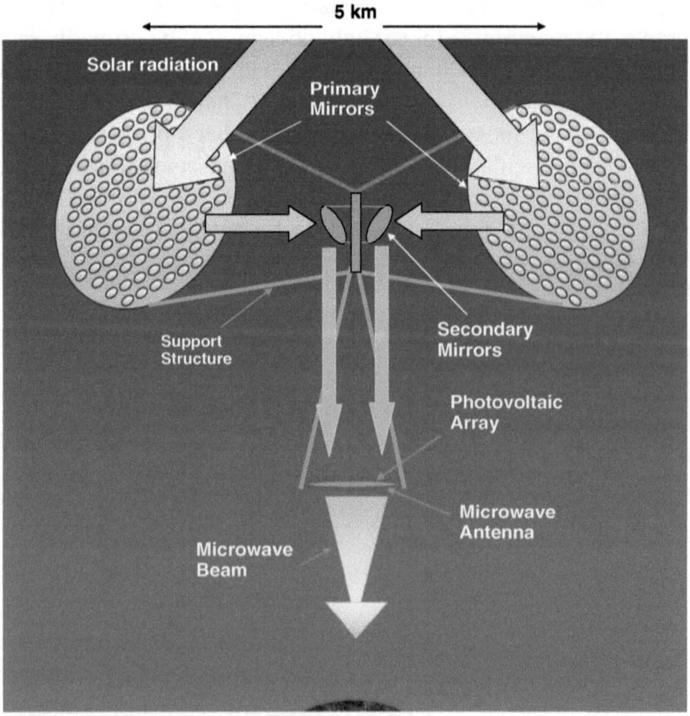

Fig. 1.17 Conceptual arrangement for solar power collection from a platform in geostationary orbit

It is an article of faith with such a system that the very high-power beam focused on the earth cannot stray from its desired path. A beam delivering several megawatts of power to the ground level could be harmful to any humans and animals caught in its full 'glare'. However, proposed collection areas (Chap. 9) are generally sufficiently large to bring power densities down to safe levels ($\sim 10\ mW/cm^2$). Nevertheless, at the distances involved between the space array and the ground, only tiny oscillations at the satellite could be enough to move the beam a long way from its intended path (Chap. 9). Despite the planned low power densities, this could be worrying for nearby communities. Current thinking, as outlined in Chap. 9, is that the adoption of retro-directive technology at the space-based array antenna should provide a level of 'fail-safe' beam guidance which would be reassuring to users.

1.3.6 Nantennas

The nantenna represents a relatively new form of optical reception which relies heavily on antenna techniques commonly used at sub-optical frequencies and in particular microwaves. The name nantenna is a contraction of **nanoantenna**. Given the shortness of the wavelengths at light frequencies, nanotechnology is required to implement the resultant antennas with their extremely small filamentary elements. The nantenna concept is also founded on a particular form of microwave antenna termed the rectenna (rectifying antenna). This device has been developed for use in wireless systems such as electronic tags. It is essentially a specialised radiofrequency antenna which can convert incident radio waves into direct current (see Chap. 10). If this can be done at light frequencies, much more of the solar spectrum can be used for power conversion than is possible with solar PV. The hope is that arrays of nantennas can also be an efficient means of converting sunlight into electric power, producing useable power more efficiently than is possible with conventional solar cells. The idea was first proposed by Robert L. Bailey in 1972 [13, 14].

Any antenna is an electromagnetic collector/receiver which is designed to absorb specific wavelengths that are proportional to the main dimensions of the antenna. A nantenna is no different. A nantenna designed to absorb wavelengths in the range of 3–15 μm has recently (2012) been developed and tested in the USA. These wavelengths correspond to photon energies of 0.08–0.4 eV. Antenna theory (Chap. 3) suggests that a nantenna can absorb any wavelength of light efficiently provided that some dimension of the nantenna is optimised for that specific wavelength. It is likely that nantennas will be used to absorb light at wavelengths between 0.4 and 1.6 μm (see Fig. 1.15) because these wavelengths exhibit relatively high power density and make up about 85 % of the solar radiation spectrum.

Any aerial engineer contemplating the notion of collecting the sun's rays on a conventional antenna structure, with its nominally restricted bandwidth, would be immediately daunted by the prospect of accommodating the vast spectrum of

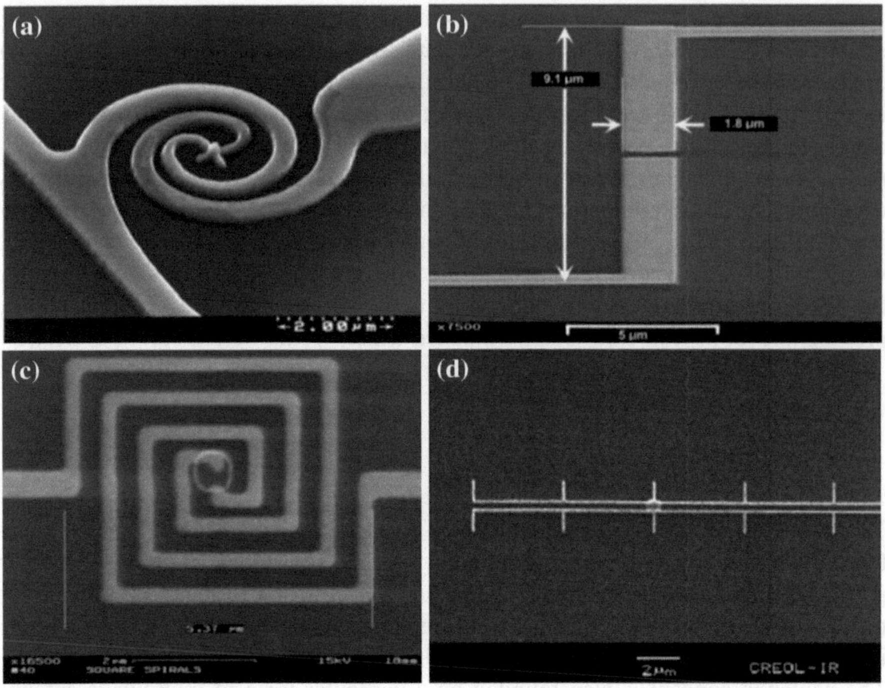

Fig. 1.18 Nantenna options formed from logarithmic spirals (**a**, **c**) and dipoles (**b**, **d**) (*source* reference [15])

frequencies. However, unlike signal reception in communication systems, the optical waves from the sun are incoherent. Phase is irrelevant if only power collection is of interest. The secret of the nantenna, which is evolved from the rectenna (see Chap. 10), is that the currents induced in the antenna are immediately rectified in a semiconducting diode mounted in the antenna terminals. The DC currents are then accumulated in a power building process. Embodiments of some typical nantenna structures based on spiral elements and dipoles are depicted in Fig. 1.18. Note particularly the dimensional scales.

References

1. Morowitz H (1978) Foundations of bioenergetics. Academic Press Inc, London
2. Krauter S (2006) Solar electric power generation. Springer, Berlin
3. Hansen J (2009) Storms of my grandchildren. Bloomsbury Publishing plc, London
4. Sangster AJ (2011) Warming to ecocide. Springer, London
5. Weart SR (2003) The discovery of global warming: new histories of science technology and medicine. Harvard University Press, Massachusetts
6. Muller I (2007) A history of thermodynamics. Springer, Berlin, Heidelberg

7. Keeling CD, Whorf TP (2004) Atmospheric CO_2 concentrations (ppmv) derived from in situ air samples collected at Mauna Loa Observatory, Hawaii, Carbon Dioxide research group scripps institution of oceanography, (SIO). University of California, Oakland
8. Meadows D, Randers J, Meadows D (2004) Limits to growth—the thirty year update (Earthscan)
9. Survey of energy resources—solar energy, World Energy Council, (2007), http://www.worldenergy.org/data.resources
10. Gillet WB, Moon JE (1985) Solar collectors. D. Reidel Publishing Co, Dordrecht
11. Cutler P (1972) Solid-state device theory McGraw-Hill Book Co., New York
12. Reddy PJ (2012) Solar power generation CRC Press/Balkema, Leiden, The Netherlands
13. Bailey RL (1972) A proposed new concept for solar energy converter. J Eng Power **72**
14. Corkish R, Green MA, Puzzer T (2002) Solar energy collection by antennas. Sol Energy 73 (6):395–401
15. Bharadwaj P, Deutsch B, Novotny L (2009) Optical antennas. Adv Optics Photon 1:438–483

Chapter 2
Electromagnetic Waves

The mind of man has perplexed itself with many hard questions. Is space infinite, and in what sense? Is the material world infinite in extent, and are all places within that extent equally full of matter? Do atoms exist or is matter infinitely divisible?
James Clerk Maxwell

Abstract The topic of electromagnetism is extensive and deep. Nevertheless, we have endeavoured to restrict coverage of it to this chapter, largely by focusing only on those aspects which are needed to illuminate later chapters in this text. For example, the Maxwell equations, which are presented in their classical flux and circulation formats in Eqs. (2.1)–(2.4), are expanded into their integral forms in Sect. 2.2.1 and differential forms in Sect. 2.3. It is these differential forms, as we shall see, that are most relevant to the radiation problems encountered repeatedly in ensuing chapters.The process of gathering light from the sun to generate 'green' power generally involves collection structures (see Chap. 8) which exhibit smooth surfaces that are large in wavelength terms. The term 'smooth' is used to define a surface where any imperfections are dimensionally small relative to the wavelength of the incident electromagnetic waves, while 'large' implies a macroscopic dimension which is many hundreds of wavelengths in extent. Under these circumstances, electromagnetic wave scattering reduces to Snell's laws. In this chapter, the laws are developed fully from the Maxwell equations for a 'smooth' interface between two arbitrary non-conducting media. The transverse electromagnetic (TEM) wave equations, which represent interfering waves at such a boundary, are first formulated, and subsequently, the electromagnetic boundary conditions arising from the Maxwell equations are rigorously applied. Complete mathematical representations of the Snell's laws are the result. These are used to investigate surface polarisation effects and the Brewster angle. In the final section, the Snell's laws are employed to examine plane wave reflection at perfectly conducting boundaries. This leads to a set of powerful yet 'simple' equations defining the wave guiding of electromagnetic waves in closed structures.

A.J. Sangster, *Electromagnetic Foundations of Solar Radiation Collection*,
Green Energy and Technology, DOI 10.1007/978-3-319-08512-8_2

2.1 Electromagnetic Spectrum

The study of solar power collection methods is predominantly an exercise in understanding the nature of electromagnetic waves and also in harnessing this widely applicable technology to facilitate the designing, and the optimisation, of optical gathering processes and structures for solar power systems. That light is a form of electromagnetic wave was arguably first established in 1862–1864 by James Clerk Maxwell. The concise set of equations which he developed to explain electromagnetic phenomena (see Sect. 2.2) both predicted the existence of electromagnetic waves and furthermore that these waves would travel with a speed that was very close to the contemporaneously known speed of light. The inference he then made was that visible light and also, by analogy, invisible infrared and ultraviolet rays all represented propagating disturbances (or radiation) occasioned by natural, abrupt changes in electromagnetic fields at some locality in space, such as in the sun. Radio waves, on the other hand, were first detected not from a natural source, but from a wire aerial, into which time-varying currents were deliberately and artificially inserted, from a relatively low-frequency oscillatory circuit. The feat was achieved by the German scientist Heinrich Hertz in 1887.

It is now well established that light (see Fig. 2.1) forms a very small portion of a spectrum of electromagnetic waves which extend from very low-frequency (VLF, MF, VHF at <1 MHz) radio waves, through broadcast waves between 50 and 1,000 MHz, microwaves from 1 to 100 GHz, millimetre waves at about 0.1–1 THz, followed by infrared. The visible spectrum seems narrow when located in the entire electromagnetic spectrum, as presented in Fig. 2.1, but it still encompasses a huge frequency range from 0.43×10^{15} to 0.75×10^{15} Hz (430–750 THz). Beyond the visible section are the ultraviolet, the X-ray and gamma-ray spectra, with a notionally terminal frequency, in an engineering context, for the whole EM spectrum at about 10^{19} Hz which translates to a miniscule wavelength of 0.1 Å = 0.01 nm. Sub-angstrom dimensions are so far outside of normal engineering practice that we need not consider, any further, EM waves at this extremity of the spectrum.

2.2 Electromagnetic Theory and Maxwell's Equations

In traditional electrical engineering science [1, 2], at the macroscopic level where quantum mechanical influences are generally insignificant, all electrical phenomena can be interpreted as being evolved from the forces acting between stationary, or moving, 'point' charges (electrons and protons). In fact, four concise equations, commonly referred to as the Maxwell equations, are sufficient to describe all known macroscopic field interactions in electrical science including behaviours at optical frequencies. These equations are in a minimalist mathematical form:

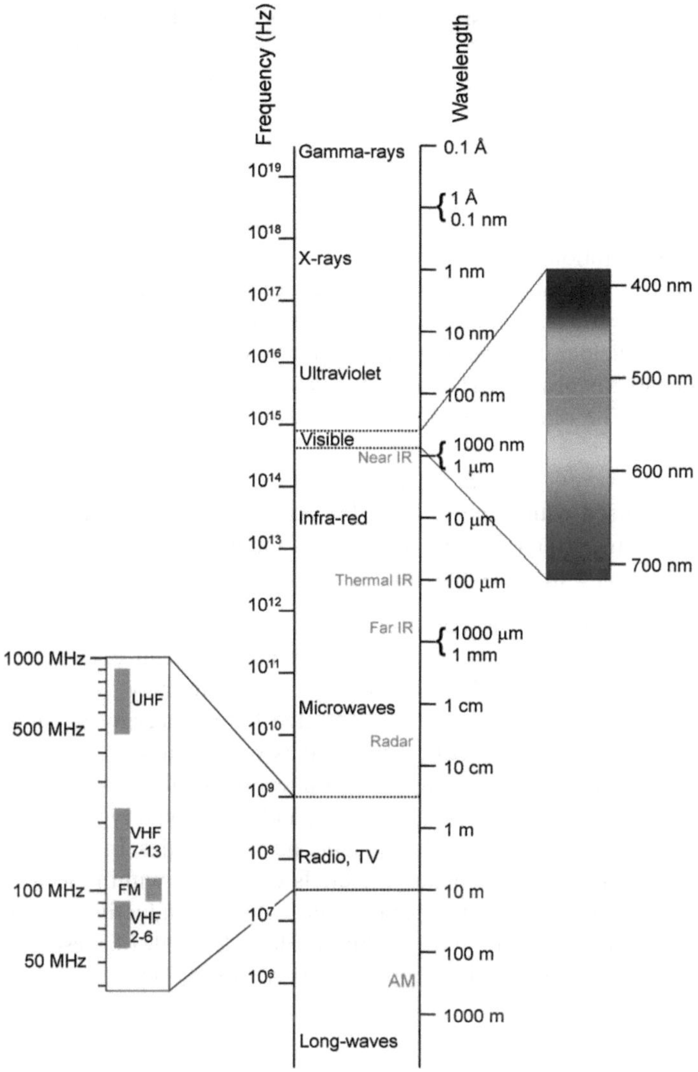

Fig. 2.1 Electromagnetic spectrum (http://en.wikipedia.org/wiki/File:Electromagnetic-Spectrum.png)

$$\text{Flux } \mathbf{D} = \text{charge enclosed} \tag{2.1}$$

$$\text{Flux } \mathbf{B} = 0 \tag{2.2}$$

$$\text{Circ } \mathbf{H} = I + \text{rate of change electric flux} \tag{2.3}$$

$$\text{Circ } \mathbf{E} = -\text{rate of change of magnetic flux} \tag{2.4}$$

Although concisely expressed as they are here, these four equations can, at first sight, still seem rather mystifying, and perhaps a little off-putting, to anyone proposing to study the subject. However, once the symbols and the terminology are established, and the historical development is explained, their obscurity should disappear as their potency is revealed. To this end, the following symbol identifications and further definitions are appropriate.

1. Electric charge (Q), which may be either positive or negative, is conserved in all electrical operations.
2. The electric current through any surface is the rate at which charge passes through the surface, that is, $I = \rho v A$, where ρ is the charge density in coulomb/m^3, v is the velocity of moving charge in m/s, while A is the area in m^2 through which charge is passing. The velocity v and the surface normal are presumed to be aligned. The dimensions of I are coulomb/s, which is an amp in the m.k.s. system.
3. The electric current through any *closed* surface is minus the rate of change of the charge enclosed within the surface ($I = -\mathrm{d}Q/\mathrm{d}t$). This is a general statement of Kirchoff's law, which for circuit engineers mainly appears in the more familiar form $\Sigma I = 0$ at a network junction.

For the purposes of solar engineering, the basic carrier of charge, namely the electron, is considered to be a particle, essentially because the electron wave function in quantum mechanics displays an extremely small wavelength, $\lambda_e = 0.165$ nm. This is too short to be detected by engineering instruments. Consequently, the electron's wavelike behaviour is rarely encountered in engineering applications, even in those encompassing optical interactions.

In Eqs. (2.1)–(2.4), the vector quantities **E** and **B** (the bold type denotes a vector) are the fundamental electric and magnetic field quantities in electromagnetism, while **D** and **H** are auxiliary fields. Materials embedded within electrical systems are defined electrically by three parameters, namely conductivity σ (mhos · m), permittivity ε (Farad/m) and permeability μ (Henry/m). All of these quantities are defined and dimensioned more comprehensively in Ref. [1].

2.2.1 Flux and Circulation

The flux and circulation integrals embedded in Eqs. (2.1)–(2.4) can be defined, rather helpfully, in a relatively non-mathematical form, if averaging (in essence integration) can be considered to be a process which is not unduly remote from 'common sense' (see Refs. [3, 4]). Thus, we have for an arbitrary vector **A**:

$$\begin{aligned}\text{Flux } \mathbf{A} &= \text{average normal component of } \mathbf{A} \text{ over a} \\ &\quad \text{surface area } \mathbf{dS}\ (A_n \text{ say}) \text{ multiplied} \\ &\quad \text{by area } dS \\ &= A_n \times dS\end{aligned} \tag{2.5}$$

The introduction of vector algebra into Eq. (2.5) permits a 'shorthand' representation of the process. For an infinitesimally small area (dS), which can be considered (see Fig. 2.2) to be directionally aligned with a unit vector **n** normal to its surface, then a simple dot product gives

$$\text{Flux } \mathbf{A} = \mathbf{A} \cdot \mathbf{dS} \tag{2.6}$$

For a surface area S of finite size, we then have [4]

$$\text{Flux } \mathbf{A} = \sum_S \mathbf{A} \cdot \mathbf{dS} = \iint_S \mathbf{A} \cdot \mathbf{dS} \tag{2.7}$$

If the surface of interest is not open, as above, but closed like the surface of a balloon, then Eq. (2.7) takes the form:

$$\text{Flux } \mathbf{A} = \oiint_S \mathbf{A} \cdot \mathbf{dS} \tag{2.8}$$

The mathematical form of circulation can be constructed in a similar manner from the basic definition (Fig. 2.3):

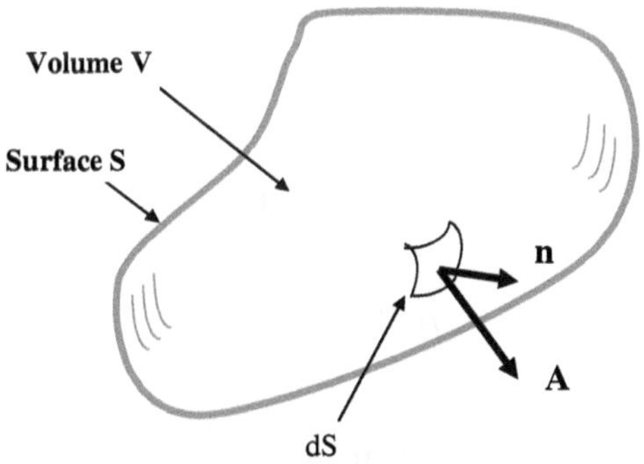

Fig. 2.2 The closed surface S defines the volume V. The direction of the elemental surface dA is defined by the unit vector **n**, and vector **A** represents in magnitude and direction an arbitrary field passing through it

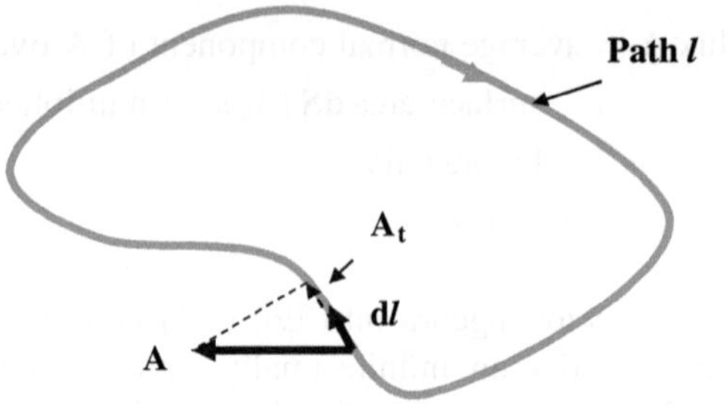

Fig. 2.3 Circulation of **A** around the path ℓ is the line integral of the tangential component A_t

$$\begin{aligned}\text{Circ}\,\mathbf{A} &= \text{average tangential component of } \mathbf{A} \text{ along} \\ &\quad \text{path } \mathbf{d}l\ (A_t \text{ say}) \text{ times the length of path d}l \\ &= A_t \times \mathrm{d}l \end{aligned} \tag{2.9}$$

In vector notation, the 'circulation' (or perhaps it should be 'translation' for an open path) along the elemental path $\mathbf{d}\ell$ is given by

$$\text{Circ}\,\mathbf{A} = \mathbf{A} \cdot \mathbf{d}\ell \tag{2.10}$$

For an arbitrary path of length ℓ, circulation is expressed mathematically in the form:

$$\text{Circ}\,\mathbf{A} = \int_{\ell} \mathbf{A} \cdot \mathbf{d}\ell \tag{2.11}$$

For a closed path or loop, which is much more common in electrical calculations, we get

$$\text{Circ}\,\mathbf{A} = \oint_{\ell} \mathbf{A} \cdot \mathbf{d}\ell \tag{2.12}$$

With the above vector definitions in place, we can now express the Maxwell equations in their vector integral form as follows:

$$\oint\!\!\oint_S \mathbf{D} \cdot \mathbf{dS} = Q_{\text{free}} \tag{2.13}$$

$$\oint\!\!\oint_S \mathbf{B} \cdot \mathbf{dS} = 0 \tag{2.14}$$

$$\oint_C \mathbf{H} \cdot \mathbf{d}\ell = I_{\text{cond}} + \frac{\partial}{\partial t} \iint_A \mathbf{D} \cdot \mathbf{dA} \tag{2.15}$$

$$\oint_C \mathbf{E} \cdot \mathbf{d}\ell = -\frac{\partial}{\partial t} \iint_A \mathbf{B} \cdot \mathbf{dA} \tag{2.16}$$

In Eq. (2.13), Q_{free} denotes the free, unbounded charge within the closed surface S, while in Eq. (2.15), I_{cond} denotes the conducting current, or free charge passing through the open surface A which spans the circuital path C. That is, for positive charge flow,

$$I_{\text{cond}} = \iint_A \rho \mathbf{v} \cdot \mathbf{dA} \tag{2.17}$$

The second term on the right of Eq. (2.15) is Maxwell's displacement current which also threads through the surface A.

Finally, it is important to note that in electromagnetism, a fundamental force equation linking fields and charge also exists as almost a fifth Maxwell equation. It is attributed to Lorentz [5] and is defined in the next section.

2.2.2 Boundary Conditions

At a 'smooth' interface between two different materials (say samples 1 and 2) where smooth implies that surface roughness features are very much less than the free-space wavelength at the frequency of interest, the above four equations reduce to the following boundary conditions:

$$\hat{\mathbf{n}} \cdot \mathbf{D}_1 = \hat{\mathbf{n}} \cdot \mathbf{D}_2 \tag{2.18}$$

$$\hat{\mathbf{n}} \cdot \mathbf{B}_1 = \hat{\mathbf{n}} \cdot \mathbf{B}_2 \tag{2.19}$$

$$\hat{\mathbf{n}} \times \mathbf{E}_1 = \hat{\mathbf{n}} \times \mathbf{E}_2 \tag{2.20}$$

$$\hat{\mathbf{n}} \times \mathbf{B}_1 = \hat{\mathbf{n}} \times \mathbf{B}_2 \tag{2.21}$$

If material 2 is a 'good conductor', the following forms apply:

$$\hat{\mathbf{n}} \cdot \mathbf{D}_1 = \rho_s \tag{2.22}$$

$$\hat{\mathbf{n}} \cdot \mathbf{B}_1 = 0 \tag{2.23}$$

$$\hat{\mathbf{n}} \times \mathbf{E}_1 = 0 \tag{2.24}$$

$$\hat{\mathbf{n}} \times \mathbf{H}_1 = \mathbf{J}_s \tag{2.25}$$

In these equations, $\hat{\mathbf{n}}$ is the unit normal to the surface, ρ_s is the charge density on the surface, and $\mathbf{J}_s$ is the surface current density.

In electromagnetism, the fundamental force equation attributed to Lorentz [3] can be expressed in vectorial form as follows:

$$[\mathbf{F} = Q(\mathbf{E} + \mathbf{u} \times \mathbf{B})] \tag{2.26}$$

In the m.k.s. system, we already know that force is expressed in newtons, Q in coulombs and velocity $\mathbf{u}$ in m/s. In this system, therefore, electric field has the dimension newton/coulomb (N/C), while magnetic flux density $\mathbf{B}$ has the dimension N·s/m·C. Needless to say, we do not use these clumsy forms. In the m.k.s. system, electric field has the basic dimension volt/m, while magnetic flux density gets the dimension tesla (T). The relationship between a volt/m and an N/C and between a tesla and an N·s/m·C can be found in Ref. [1].

2.3 Plane Wave Solution

All materials contain electric charges bound loosely or otherwise within atoms and molecules. If these materials exist in an environment which naturally or artificially causes agitation of the charge, and hence changes in the associated electric and magnetic fields, then electromagnetic waves are unavoidable. These can appear in quite complex trapped, surface, evanescent and radiant embodiments. In these circumstances, the integral forms of Maxwell's equations, developed above, become inappropriate since the finite volumes, surfaces and paths over which integrations have to be performed are no longer identifiable. What is required in this case is a set of equations which represent the field behaviour at a point in space. The conversion from the integral forms to these point forms (differential forms) of Maxwell's equations is developed in most textbooks on the topic (see References) and essentially entails the recruitment of well-known vector-differential theorems such as the divergence theorem and Stokes' law to accomplish the transitions.

Many solar power gathering problems are of the source-free variety, which implies that the source, in this case the sun, is so far distant that the waves of interest here on earth are plane waves. These waves, also termed TEM waves, are described as 'plane' because the radius of curvature of the wave front (see Fig. 1.5) is very large, and thus, the natural rate of curvature of the front can be deemed mathematically insignificant, allowing it to be fully described by means of Cartesian coordinates. In this scenario, the EM problem reduces to a boundary value problem, for which Maxwell's equations, in differential form, become

$$\nabla \cdot \mathbf{D} = 0 \tag{2.27}$$

$$\nabla \cdot \mathbf{B} = 0 \tag{2.28}$$

$$\nabla \times \mathbf{E} = -\frac{\partial \mathbf{B}}{\partial \mathbf{t}} \tag{2.29}$$

$$\nabla \times \mathbf{H} = \frac{\partial \mathbf{D}}{\partial \mathbf{t}} \tag{2.30}$$

where **E** and **H** represent the electric and magnetic field intensities in the region of interest. As before, $\mathbf{D} = \varepsilon\mathbf{E}$ is the electric flux density, while $\mathbf{B} = \mu\mathbf{H}$ is the magnetic flux density. The 'del' operator (∇) expresses directional derivatives in the three space directions. It is a vector, which in the Cartesian system (for example) has the form:

$$\nabla = \hat{\mathbf{a}}_\mathbf{x}\frac{\partial}{\partial \mathbf{x}} + \hat{\mathbf{a}}_\mathbf{y}\frac{\partial}{\partial \mathbf{y}} + \hat{\mathbf{a}}_\mathbf{z}\frac{\partial}{\partial \mathbf{z}} \tag{2.31}$$

where $\hat{\mathbf{a}}_\mathbf{x}, \hat{\mathbf{a}}_\mathbf{y}$ and $\hat{\mathbf{a}}_\mathbf{z}$ are unit vectors directed along x, y, and z, respectively. When the del operator is multiplied by a scalar [ϕ(x, y, z) say], the result is a vector which expresses the gradient or slope of the function ϕ in all three space directions, i.e.

$$\nabla\phi = \hat{\mathbf{a}}_\mathbf{x}\frac{\partial\phi}{\partial \mathbf{x}} + \hat{\mathbf{a}}_\mathbf{y}\frac{\partial\phi}{\partial \mathbf{y}} + \hat{\mathbf{a}}_\mathbf{z}\frac{\partial\phi}{\partial \mathbf{z}} \tag{2.32}$$

Cross multiplication of del with a vector produces the operation of 'curl', while dot multiplication produces the operation of 'divergence' ('div'). Crudely, curl is circulation at a point, while divergence is flux at a point.

2.3.1 Second-Order Differential Equation

To solve the Maxwell equations for *E*-field or *H*-field behaviour in a bounded region, it is first necessary to form an equation either *E* or *H* alone. The standard procedure for achieving this conversion is to perform a curl operation on either the curl equation for *E* or the corresponding equation for *H*. This gives, for example, using Eq. (2.29)

$$\begin{aligned}\nabla \times \nabla \times \mathbf{E} &= -\frac{\partial}{\partial t}\mu\nabla \times \mathbf{H} \\ &= -\frac{\partial}{\partial t}\left[\frac{\partial}{\partial t}\mu\varepsilon\,\mathbf{E}\right] \\ &= -\mu\varepsilon\frac{\partial^2\mathbf{E}}{\partial \mathbf{t}^2}\end{aligned} \tag{2.33}$$

Hence, on using a convenient vector identity, which states that for any vector **A**,

$$\nabla \times \nabla \times \mathbf{A} = \nabla\nabla \cdot \mathbf{A} - \nabla^2 \mathbf{A} \tag{2.34}$$

Equation (2.33) can be re-expressed as follows:

$$\nabla\nabla \cdot \mathbf{E} - \nabla^2 \mathbf{E} = -\mu\varepsilon \frac{\partial^2 \mathbf{E}}{\partial \mathbf{t}^2} \tag{2.35}$$

But, from Eq. (2.27), $\nabla \cdot \mathbf{E} = 0$, for a linear, homogeneous medium for which μ and ε are constants. Therefore,

$$\nabla^2 \mathbf{E} = \mu\varepsilon \frac{\partial^2 \mathbf{E}}{\partial \mathbf{t}^2} \tag{2.36}$$

and by analogy:

$$\nabla^2 \mathbf{H} = \mu\varepsilon \frac{\partial^2 \mathbf{H}}{\partial \mathbf{t}^2} \tag{2.37}$$

Equations (2.36) and (2.37) are wave equations. Equations of this nature, with appropriate variables, appear in most branches of science and engineering, and their solutions have been studied widely. Solutions depend very much on the boundary conditions, namely the conditions imposed on the variables at the periphery or containing surface of the solution region. They can fix the magnitude of the variable (Dirichlet condition) or the rate of change of the variable (Newman condition) or a mixture of both. A unique solution depends on the conditions being neither underspecified or overspecified.

For example, let us consider formulating a solution to Eq. (2.36), and inevitably Eq. (2.37) because of the Maxwell linkages, for a region of free space ($\mu = \mu_0$: $\varepsilon = \varepsilon_0$) which is large enough to presume that all boundaries are effectively at infinity. In this case, we can choose to represent the region mathematically using Cartesian coordinates, and furthermore, since we anticipate that the solution is a waveform, we can arbitrarily determine that the waves travel in the z-direction. This implies that the rates of change of the E-field in x and y are zero, and using (2.27), it follows that $E_z = 0$. The equation to be solved, therefore, is

$$\frac{\partial^2 \mathbf{E}}{\partial \mathbf{z}^2} = \mu_0\varepsilon_0 \frac{\partial^2 \mathbf{E}}{\partial \mathbf{t}^2} \tag{2.38}$$

where, in general, $\mathbf{E} = \hat{\mathbf{a}}_\mathbf{x} E_x + \hat{\mathbf{a}}_\mathbf{y} E_y$. However, if we choose to align the coordinate system so that $\mathbf{E}$ lies along the x-axis (x-polarised solution), then $E_y = 0$ and the wave equation reduces to the scalar form:

$$\frac{\partial^2 E_x}{\partial \mathbf{z}^2} = \frac{1}{c^2} \frac{\partial^2 E_x}{\partial \mathbf{t}^2} \tag{2.39}$$

if $c = {}^{1}\!/\!\sqrt{\mu_0 \varepsilon_0}$.

2.3.2 General Solution

Equation (2.39) has a wave solution of the general form:

$$E_x = Af(z - ct) + Bf(z + ct) \tag{2.40}$$

This is easily demonstrated by substitution back into the equation. The first term represents a wave travelling in the +z-direction, while the second allows for a reflected wave, if such exists. Given that velocity is the rate of change of z with respect to time, it is evident that c represents velocity (actually phase velocity) of the electromagnetic wave in 'free space'. For vacuum, it is equal to 3×10^8 m/s. The application of Maxwell's equations also gives $H_z = 0$ and

$$H_y = \frac{A}{\eta} f(z - ct) + \frac{B}{\eta} f(z + ct) \tag{2.41}$$

Also,

$$\frac{E_x}{H_y} = \pm \sqrt{\frac{\mu_0}{\varepsilon_0}} = \pm \eta \tag{2.42}$$

η is termed the free-space wave impedance which for air or vacuum has the value $120\pi\ \Omega$. The resultant solution is a plane electromagnetic wave, also termed a TEM wave, for which E and H are transverse to the direction of propagation and orthogonal to each other. E and H are also in time phase, as Eq. (2.42) attests (see Fig. 2.4).

Electrical engineers are generally very familiar with the relationship between power (P), voltage (V) and current (I) in the form:

$$P = \frac{1}{2} VI \quad \text{W} \tag{2.43}$$

where V and I are defined in peak, rather than in the more common r.m.s., format. But, voltage is simply integrated electric field E (V/m), and from ampere, current is integrated magnetic field intensity H (A/m), so by analogy, we can suggest that for the plane wave,

$$p = \frac{1}{2} EH = \frac{1}{2} c\varepsilon_0 E^2 \quad \text{W/m}^2 \tag{2.44}$$

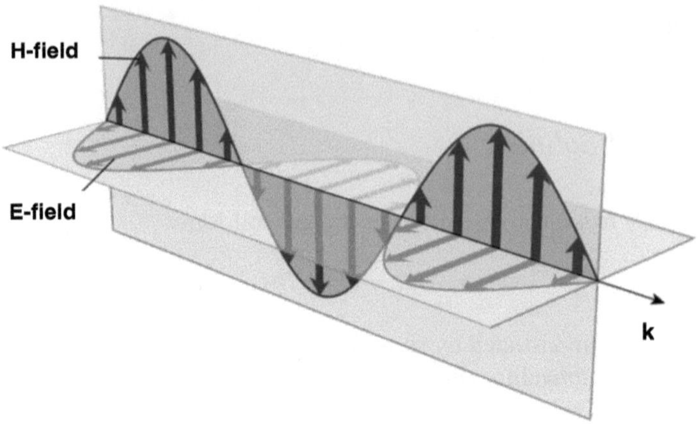

Fig. 2.4 TEM wave field and direction relationships

This means that p is the real power flow density in the TEM wave. In general, complex power flow density in an electromagnetic wave is given by the Poynting vector **S**, where

$$\mathbf{S} = \frac{1}{2}\mathbf{E} \times \mathbf{H} \quad \text{W/m}^2 \tag{2.45}$$

In electrical engineering, it is much more usual to examine wave solutions at a single frequency (ω rad/s), namely sinusoidal solutions. This actually incurs little loss of generality, since any arbitrary time variation carried on a radio wave can be resolved into a spectrum of single-frequency components. The adoption of a single frequency, or a spectral frequency, in carrying through time-varying computations has the distinct advantage that the time variable can be omitted. The calculations are then progressed in phasor notation. In trigonometric form, Eq. (2.41) becomes

$$E_x = A \exp j(\omega t - \beta z) + B \exp j(\omega t + \beta z) \tag{2.46}$$

where A and B are complex constants. The phasor form is

$$E_x = |A| \exp(-j\beta z + \varphi) + |B| \exp(j\beta z + \theta) \tag{2.47}$$

with φ and θ representing the phases, respectively, of A and B.

2.3.3 Snell's Laws

When a plane electromagnetic wave at the frequency of light, or in fact any radio frequency, is incident upon a smooth interface (by 'smooth', it is meant that any surface undulations or protuberances are in size very much less than the wavelength of the impinging waves) between two extended propagating media, part of the wave is reflected back into the incident medium, while part is transmitted or refracted into the second medium, usually with a change of direction.

Analytically, the relationships between the incident and reflected waves can be developed by considering a plane electromagnetic wave, incident at a physically real angle θ_1 to the normal, at the interface between two semi-infinite regions of space, as suggested in Fig. 2.5. Each region is presumed to comprise a linear homogeneous medium with a different index of refraction (n). The index of refraction is defined as follows:

$$n = \frac{c}{v} \tag{2.48}$$

where c is the speed of light in vacuum, or free space, while v is its speed within the specified medium. Also, with reference to Fig. 2.5, the following definitions apply:

$$c = \frac{1}{\sqrt{\mu_0 \varepsilon_0}} \tag{2.49}$$

and

$$v_1 = \frac{1}{\sqrt{\mu_0 \varepsilon_0 \varepsilon_{r1}}} \tag{2.50}$$

$$v_2 = \frac{1}{\sqrt{\mu_0 \varepsilon_0 \varepsilon_{r2}}} \tag{2.51}$$

Here, ε_0 and μ_0 are the free-space permittivity and permeability, respectively, while ε_{r1} is the relative permittivity of medium 1 and ε_{r2} is the relative permittivity of medium 2. Both media are assumed to be lossless and non-magnetic in which case $\mu_1 = \mu_2 = \mu_0$. The indices of refraction for the two media then become

$$n_1 = \sqrt{\varepsilon_{r1}}; \quad n_2 = \sqrt{\varepsilon_{r2}} \tag{2.52}$$

Maxwell's equations in the semi-infinite regions remote from the interface are, as we have seen above, fully satisfied by TEM plane waves. It remains then to satisfy the Maxwell boundary conditions at the interface. If this can be done, the resultant solutions represent complete EM solutions for the specified boundary value problem. For an incident TEM wave, as depicted in Fig. 2.5, the E-field vector and the H-field vector must be mutually orthogonal to each other and to the

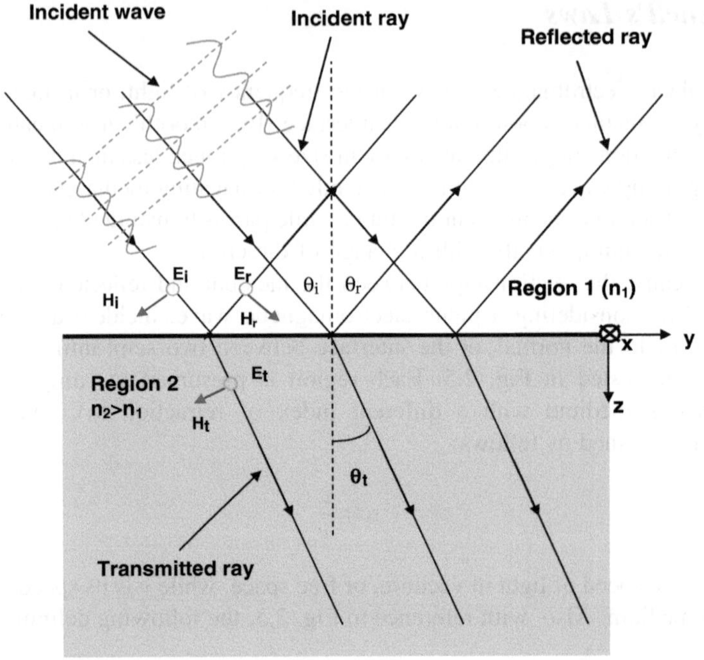

Fig. 2.5 Reflection and refraction at a dielectric interface—perpendicularly polarised case

direction of propagation, usually defined by a unit vector $\hat{\mathbf{k}}$, directed in the direction of the relevant ray.

In this case, we can write

$$\mathbf{H} = \frac{1}{\eta}(\hat{\mathbf{k}} \times \mathbf{E}) \tag{2.53}$$

where η is the wave impedance for the medium containing the wave. Hence, for regions 1 and 2, respectively,

$$\eta_1 = \sqrt{\frac{\mu_0}{\varepsilon_{r1}\varepsilon_0}} \tag{2.54}$$

$$\eta_2 = \sqrt{\frac{\mu_0}{\varepsilon_{r2}\varepsilon_0}} \tag{2.55}$$

However, this condition does not fully establish the polarisation direction, which must also be specified. There are two basic choices from which any other polarisation possibilities can be deduced. We can choose the *E*-field vector of the incident wave to be either normal to the *yz*-plane, or parallel to it. The *yz*-plane in Fig. 2.5 is generally termed the plane of incidence for the incoming wave, being the plane that contains both the direction vector $\hat{\mathbf{k}}$ and the unit normal to the interface ($\hat{\mathbf{n}}$). When the electric field in the incident TEM wave is normal to the plane of incidence, the wave is said to be perpendicularly polarised, and when it is parallel to this plane, it is described as parallel polarised. Note that in relation to the surface of the earth, while parallel polarisation equates to horizontal polarisation, perpendicular polarisation can be termed vertical polarisation only if θ_i approaches 90°. Perpendicular polarisation is often termed transverse electric (TE) propagation, while parallel polarised waves get the complementary description of transverse magnetic (TM) waves.

Now that we know the electromagnetic field forms (TEM waves) remote from the interface between regions 1 and 2 in Fig. 2.5, we can examine the field conditions (boundary conditions) precisely at the interface. For the diffraction set-up depicted in Fig. 2.5 with a perpendicularly polarised TEM wave incident at θ_i, the field directions at a given instant in time can be represented vectorially as shown. Just at the interface, a typical ray of the incident TEM wave is both reflected off the surface and transmitted through it. Also, for a 'smooth' surface, 'common sense' suggests that it is safe to presume that the reflected and transmitted waves retain the polarisation of the incident wave. Furthermore, there will be a single reflected ray and a single transmitted ray. Actually, this latter assumption is not strictly necessary as we will show presently.

When the TEM wave direction (or ray) lies in paths other than along the coordinate axes, it is usual to define the ray direction by the vector $\mathbf{k}$ which is chosen to be equal in magnitude to the wave coefficient k. That is $\mathbf{k} = \hat{\mathbf{k}}k$. Hence, we can express mathematically the wave component in any other direction ($\mathbf{r}$ say). For the case shown in Fig. 2.6, where the electric field is x-directed, the expression has the form:

$$\mathbf{E_i} = \hat{\mathbf{a}}_\mathbf{x} E_i \exp(-j\mathbf{k}\cdot\mathbf{r}) \tag{2.56}$$

Consequently, if $\mathbf{r}$ and $\mathbf{k}$ lie in the yz-plane as suggested in Fig. 2.6, then clearly

$$\mathbf{k} = \hat{\mathbf{a}}_\mathbf{y} k_y + \hat{\mathbf{a}}_\mathbf{z} k_z \tag{2.57}$$

$$\mathbf{r} = \hat{\mathbf{a}}_\mathbf{y} y + \hat{\mathbf{a}}_\mathbf{z} z \tag{2.58}$$

Also,

$$k^2 = k_y^2 + k_z^2 \tag{2.59}$$

so we can conveniently write

$$k_y = k\sin\theta \tag{2.60}$$

and

$$k_z = k\cos\theta \tag{2.61}$$

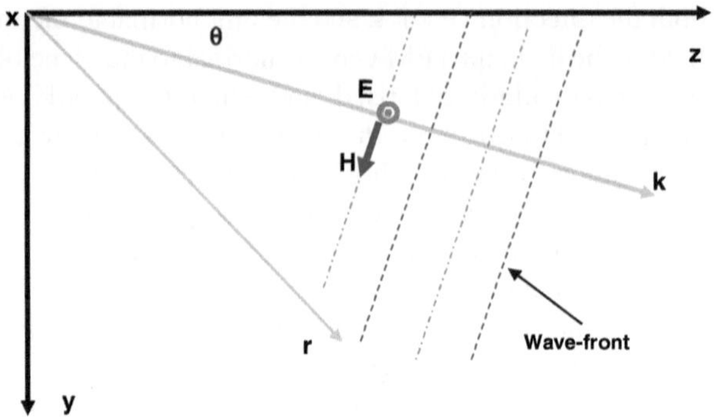

Fig. 2.6 Representation of TEM wave with **E**, **H** and **k** in mutually orthogonal directions

Hence, employing these relationships, Eq. (2.56) can be expanded into the non-vectorial form:

$$E_{xi} = -E_i \exp(j(\omega t - k_1 z \cos\theta_i - k_1 y \sin\theta_i) \tag{2.62}$$

where

$$k_1 = \frac{\omega}{v_1} = \frac{\omega}{c} n_1 \tag{2.63}$$

For a TEM wave, the electric and magnetic fields are related through Eq. (2.53). Hence, on combining Eqs. (2.62) and (2.53), and observing the field directions in Fig. 2.5, we obtain for magnetic fields:

$$H_{yi} = -H_i \cos\theta_i \exp[j(\omega t - k_1 z \cos\theta_i - k_1 y \sin\theta_i)] \tag{2.64}$$

$$H_{zi} = H_i \sin\theta_i \exp[j(\omega t - k_1 z \cos\theta_i - k_1 y \sin\theta_i)] \tag{2.65}$$

Also, we note that if these field components represent a TEM wave, then we must have

$$\frac{E_i}{H_i} = \eta_1 = \frac{\eta_0}{n_1} \tag{2.66}$$

Similar constructions lead to the following equations for the reflected and transmitted field components:

$$E_{xr} = -E_r \exp(j(\omega t + k_1 z \cos\theta_i - k_1 y \sin\theta_i) \tag{2.67}$$

$$H_{yr} = H_r \cos\theta_i \exp[j(\omega t + k_1 z \cos\theta_i - k_1 y \sin\theta_i)] \tag{2.68}$$

$$H_{zr} = H_r \sin\theta_i \exp[j(\omega t + k_1 z \cos\theta_i - k_1 y \sin\theta_i)] \tag{2.69}$$

with

$$\frac{E_r}{H_r} = \eta_1 \tag{2.70}$$

and

$$E_{xt} = -E_t \exp(j(\omega t - k_2 z \cos\theta_t - k_2 y \sin\theta_t) \tag{2.71}$$

$$H_{yt} = -H_t \cos\theta_t \exp[j(\omega t - k_2 z \cos\theta_t - k_2 y \sin\theta_t)] \tag{2.72}$$

$$H_{zt} = H_t \sin\theta_t \exp[j(\omega t - k_2 z \cos\theta_t - k_2 y \sin\theta_t)] \tag{2.73}$$

where

$$\frac{E_t}{H_t} = \eta_2 \tag{2.74}$$

and

$$k_2 = \frac{\omega}{c} n_2 \tag{2.75}$$

The above field expressions for the incident and reflected waves in region 1 and the transmitted waves in region 2 each separately satisfy Maxwell's equations in these regions. A solution that satisfies Maxwell's equations for the entire volume including the interface is achieved by enforcing the electromagnetic field boundary conditions, given in Eqs. (2.18)–(2.22), at the interface. That is, at $z = 0$, we require that across the divide between regions 1 and 2:

$$E_x \text{ is continuous} \tag{2.76}$$

$$H_y \text{ is continuous} \tag{2.77}$$

$$B_z \text{ is continuous} \tag{2.78}$$

On combining Eq. (2.76) with the field expressions (2.62), (2.67) and (2.71), we obtain with little difficulty:

$$E_{xi} + E_{xr} = E_{xt} \tag{2.79}$$

on the $z = 0$ plane. The implication is that

$$E_i \exp(-jk_1 y \sin\theta_i) + E_r \exp(-jk_1 y \sin\theta_r) = E_t \exp(-jk_2 y \sin\theta_t) \tag{2.80}$$

This equation must remain true over the entire $z = 0$ boundary, from $-\infty \leq y \leq +\infty$. This is only possible if

$$k_1 \sin\theta_i = k_1 \sin\theta_r = k_2 \sin\theta_t \tag{2.81}$$

It is pertinent to note here that if at the commencement of this derivation, we had, without pre-knowledge of refraction rules, chosen to presume that several reflected waves at angles θ_{r1}, θ_{r2}, θ_{r3} …, and several transmitted waves at angles θ_{t1}, θ_{t2}, θ_{t3} …, were possible, then the equivalent form of Eq. (2.80) would lead to

$$k_1 \sin\theta_{r1} = k_1 \sin\theta_{r2} = k_1 \sin\theta_{r3} = \ldots\ldots\ldots \tag{2.82}$$

and

$$k_2 \sin\theta_{t1} = k_2 \sin\theta_{t2} = k_2 \sin\theta_{t3} = \ldots\ldots\ldots \tag{2.83}$$

These equations clearly dictate that $\theta_{r1} = \theta_{r2} = \theta_{r3} = \ldots$, and $\theta_{t1} = \theta_{t2} = \theta_{t3} = \ldots$ In other words, an 'optically smooth' surface produces only one reflected wave and one transmitted wave.

Equation (2.81) is the source of Snell's laws which state that at an optically smooth interface between two lossless media,

$$\theta_r = \theta_i \tag{2.84}$$

$$\frac{\sin\theta_t}{\sin\theta_i} = \frac{k_1}{k_2} = \frac{n_1}{n_2} \tag{2.85}$$

However, these laws govern only the reflection and refraction angles. We also need to have knowledge of the relative magnitudes of the reflected and transmitted waves, and how these are influenced by material properties.

When Eqs. (2.76)–(2.78) are applied to the TEM field components at the boundary, while also applying Snell's laws, the following relations are generated:

$$E_i + E_r = E_t \tag{2.86}$$

$$(H_i - H_r)\cos\theta_i = H_t \cos\theta_t \tag{2.87}$$

$$(B_i - B_r)\sin\theta_i = B_t \sin\theta_t \tag{2.88}$$

Equation (2.86) can be converted to magnetic field form by employing Eqs. (2.70) and 2.74) leading to

$$\eta_1(H_i + H_r) = \eta_2 H_t \tag{2.89}$$

Consequently, if we choose to define reflection coefficient for this perpendicularly polarised example (TE case) as

$$\rho_{\mathrm{TE}} = \frac{H_r}{H_i} \tag{2.90}$$

then making use of Eqs. (2.87) and (2.89), the following useful relationship is deduced:

$$\rho_{\mathrm{TE}} = \frac{\eta_2 \cos\theta_i - \eta_1 \cos\theta_t}{\eta_2 \cos\theta_i + \eta_1 \cos\theta_t} \tag{2.91}$$

This can also be expressed in a slightly more familiar form, which explicitly incorporates the indices of refraction, namely

$$\rho_{\mathrm{TE}} = \frac{n_1 \cos\theta_i - n_2 \cos\theta_t}{n_1 \cos\theta_i + n_2 \cos\theta_t} \tag{2.92}$$

Similarly, if we choose to define the transmission coefficient as

$$\tau_{\mathrm{TE}} = \frac{H_t}{H_i} \tag{2.93}$$

then

$$\tau_{\mathrm{TE}} = \frac{2n_2 \cos\theta_i}{n_1 \cos\theta_i + n_2 \cos\theta_t} \tag{2.94}$$

It is not difficult to demonstrate that

$$\rho_{\mathrm{TE}} = \frac{H_r}{H_i} = \frac{E_r}{E_i} \tag{2.95}$$

and

$$\frac{E_t}{E_i} = \frac{n_1}{n_2} \tau_{\mathrm{TE}} \tag{2.96}$$

An analogous derivation can also be followed through for the parallel polarisation case (TM case). If this is done, we obtain

$$|\rho_{\mathrm{TM}}| = \frac{E_r}{E_i} = \frac{H_r}{H_i} = \left| \frac{n_2 \cos\theta_i - n_1 \cos\theta_t}{n_2 \cos\theta_i + n_1 \cos\theta_t} \right| \tag{2.97}$$

and

$$\tau_{\mathrm{TM}} = \frac{2n_1 \cos\theta_i}{n_2 \cos\theta_i + n_1 \cos\theta_t} \tag{2.98}$$

The reflection coefficient, as a function of incident angle for both TE and TM cases, is plotted in Fig. 2.7. Clearly, for an interface between lossless dielectrics of differing refractive indices, the reflection behaviours are distinct. While for the TE

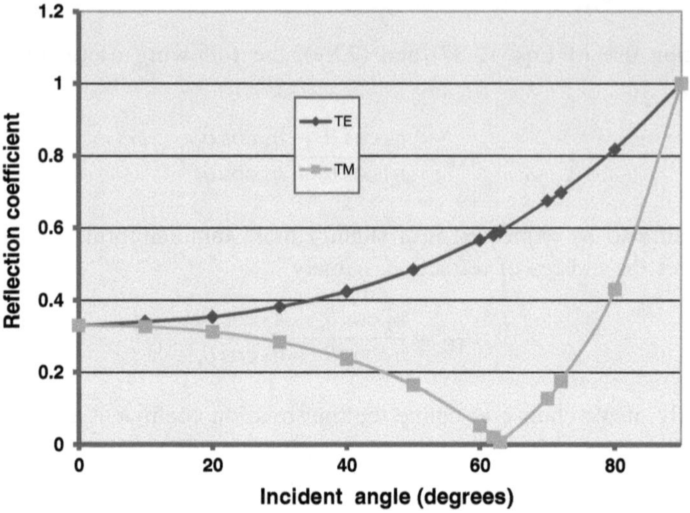

Fig. 2.7 Perpendicular (*diamonds*) and parallel (*squares*) polarised reflection coefficients as a function of incident angle ($n_2 > n_1$)

case, it increases monotonically from a magnitude of 0.33 ($n_1 = 1$ *and* $n_2 = 2$) at $\theta_i = 0$, to unity at $\theta_i = 90°$, it drops to zero close to 60° in the TM case. At the zero reflection angle, the two surfaces are said to be 'matched' for surface-normal wave components. It is termed the Brewster angle, a physical property which underpins the design of light polarisers.

2.3.4 Wave Guiding

Snell's laws can also be used to explore the processes behind electromagnetic wave trapping or guidance, concepts which are needed in later chapters. While it is well understood that at low frequencies, TEM waves can be guided by a pair of conductors, such as in power lines, in parallel wire telephone lines or in coaxial lines, high-frequency wave trapping in hollow conducting pipes is not so easy to comprehend. Such waveguides are increasingly being used in many of the antenna configurations employed in solar power applications. This method of guidance is very efficient and is especially applicable to high power transmission [2, 7–10]. It relies on the nature of plane wave interference patterns and can, perhaps, best be explained by consideration of Fig. 2.8.

Figure 2.8 depicts (in two dimensions for simplicity) a pair of plane electromagnetic waves (TEM waves) of equal magnitude travelling in different directions A and B. The waves are represented by their wave fronts, with the wave peaks in each case denoted by solid transverse lines (planes in 3D) and wave troughs by dashed lines. The distance between a wave peak and wave trough is, of course, half

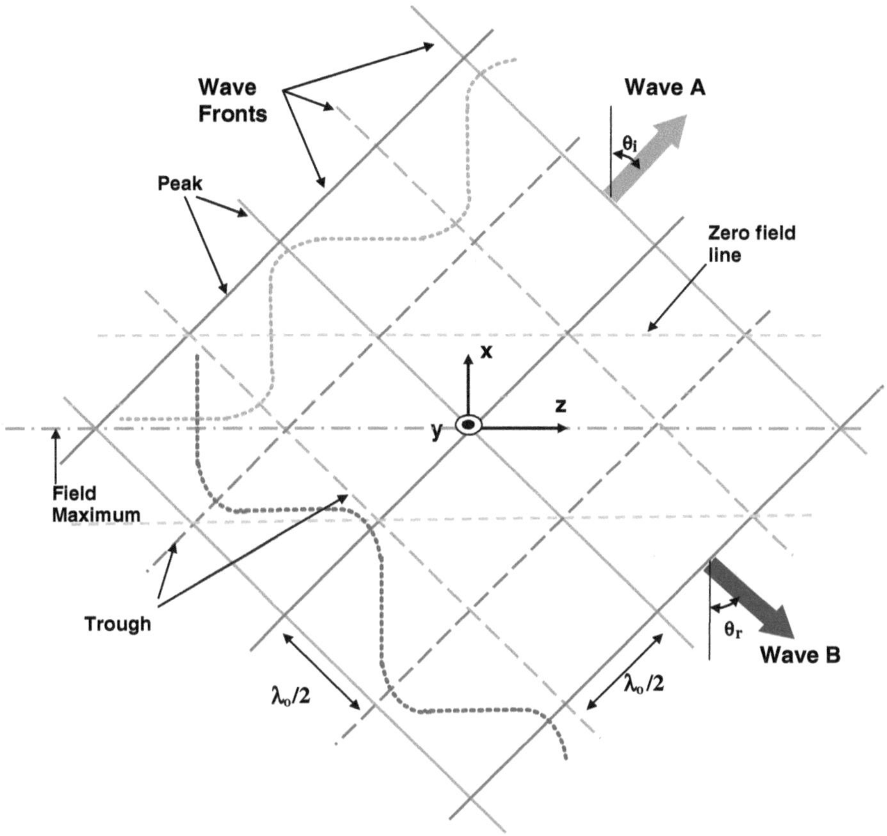

Fig. 2.8 Plane wave interference pattern

of the free-space wavelength ($\lambda_o/2$). The waves are travelling at the velocity of light (c) in the directions of the large arrows. On examination of this wave pattern, it is not too difficult to observe that along the horizontal chain-dotted line (or in three dimensions—the *yz*-plane), peaks of wave *A* coincide with those of wave *B*, and troughs coincide with troughs—and this is independent of the movement of the waves. This line (or plane) represents a stationary (in the *x*-direction) field maximum 'independent of time', while the waves continue to exist.

In contrast, along the green z-directed dashed line, peaks of wave *A* coincide with troughs of wave *B*, and vice versa, resulting in a stationary field null at these positions. Consequently, if a perfectly conducting sheet of infinite extent, orientated normal to the *x*-axis, is located at the stationary null position, the field pattern remains unchanged. For a sheet at the upper dashed line, the red direction arrow (wave A) then represents an incident wave and the blue arrow (wave B) a reflected wave, which, according to Snell's laws at a perfect mirror, reflects with a magnitude equal to the incident wave and at an angle such that $\theta_r = \theta_i$, as is required to retain the pattern. For perpendicularly polarised plane waves with the *E*-field confined to the y-direction, the E-field pattern forms a cosine distribution between the null planes. This pattern can be trapped or guided by introducing a second conducting sheet at the lower null locus in Fig. 2.8. The trapped pattern travels in the z-direction with a phase velocity:

$$v_p = c/\sin\theta \tag{2.99}$$

and a wavelength

$$\lambda_p = \lambda_o/\sin\theta \tag{2.100}$$

where c is the speed of light, λ_o is the TEM wavelength in free space and $\theta = \theta_i = \theta_r$. The magnetic field distribution can easily be deduced by applying trigonometrical rules, and the total E/H pattern is termed a TE guided wave. A dual TM guided wave can be formed by commencing with parallel polarised TEM components.

A TE wave between parallel conducting planes separated by a distance a is illustrated in Fig. 2.9c. The sinusoidal field variations in x are clearly shown. The relationship between plane separation a and wavelength λ_o can again be deduced from trigonometry and yields a

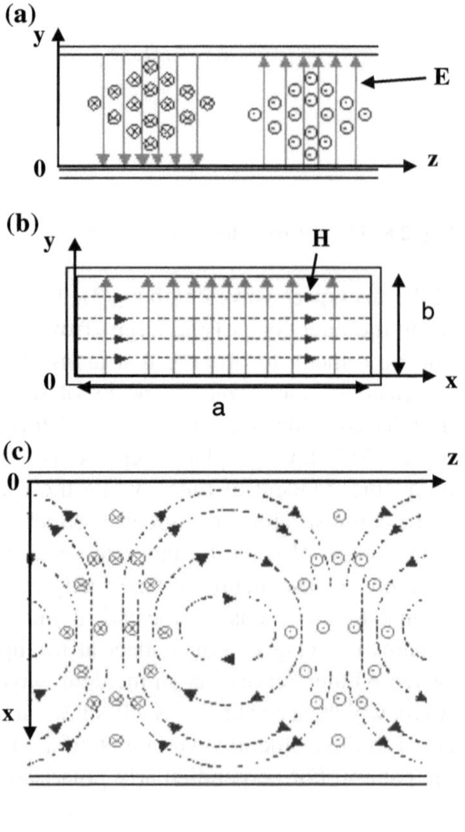

Fig. 2.9 Dominant TE_{10} mode in rectangular waveguide (*solid green vectors* = E-field; *dashed blue* = H-field). **a** Side view (TE_{10}). **b** End view (TE_{10}). **c** Top view (TE_{10})

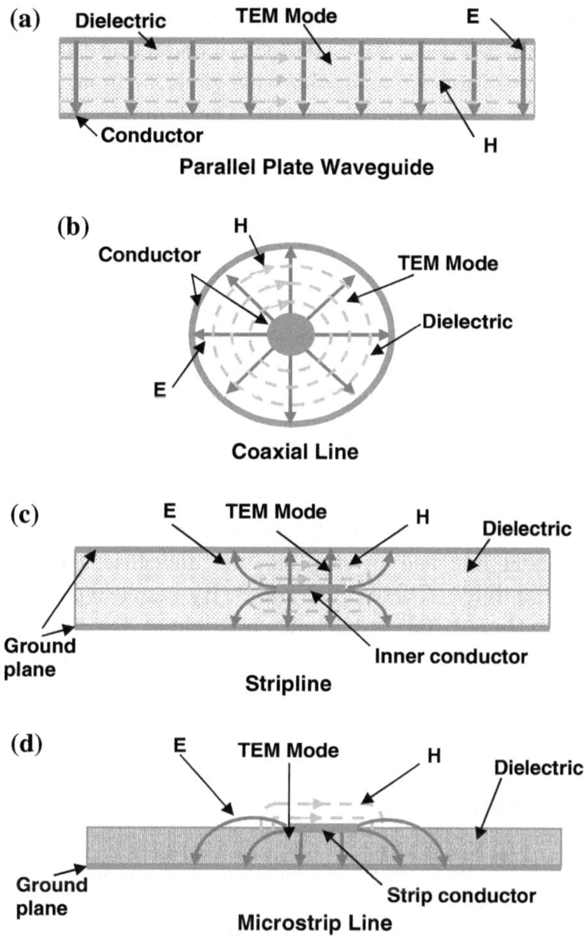

Fig. 2.10 TEM-mode transmission lines

$$\cos\theta = \frac{m\lambda_o}{2a} \tag{2.101}$$

m is the number of half-sinusoids of field pattern between the null planes. In Fig. 2.8, it is not necessary to choose the nearest null planes to create a trapped pattern. Equation (2.101) only has meaning for $m\lambda_o < 2a$, so that for $m = 1$, the case depicted in Fig. 2.9c, the free-space wavelength must be less than $2a$ for propagation to occur. The corollary is that the frequency of the wave f ($=1/\lambda_o$) must be greater than a certain critical value or cut-off value corresponding to the cut-off wavelength $\lambda_c = 2a$. Furthermore, if $a < \lambda_o < 2a$ the $m = 2, 3, 4$—solutions all yield the impossible requirement that $\cos\theta > 1$. This means that in the prescribed frequency range, only the $m = 1$ solution is possible. The solution is termed the dominant mode for the parallel plane waveguide of separation a and is defined as the TE_{10} mode—with one *E*-field variation in *x* and zero variation in *y*. This mode is shown in Figs. 2.9a–c. Perfectly conducting 'lids' can be introduced at $y = 0$ and $y = b$ to form a rectangular waveguide, without altering the pattern, because the *E*-field is normal to these walls. The *b*-dimension is usually chosen to be approximately half the *a*-dimension to maximise bandwidth. (For further elucidation, see Sect. 6.5.)

It is relevant to emphasise here that guided electromagnetic waves can also be procured by simply trapping the TEM wave [6] between conductors that lie normal to the electric field vectors, thus satisfying the boundary conditions. The most common alternatives are shown in Fig. 2.10. Parallel plate waveguide (Fig. 2.10a) provides only limited guidance in the direction normal to the plates, but clearly shows how the insertion of smooth conducting planes has negligible effect on the propagation conditions for the TEM mode. Full trapping is provided by coaxial line (Fig. 2.10b) but at the expense of phase velocity reduction and the potential for power loss in the dielectric which is necessary to separate the inner from the outer conductor. Stripline (Fig. 2.10c) is essentially 'flattened' coaxial line and has the advantage of ease of fabrication using printed circuit board (PCB) techniques. In coaxial line and in stripline, the dielectric separator usually displays a relative permittivity of between 2 and 3. By increasing this to between 6 and 10 in microstrip (Fig. 2.10d), it becomes possible to dispense with the upper ground plane and create an open structure into which microwave components can relatively easily be inserted.

References

1. Ferrari R (1975) An introduction to electromagnetic fields. Van Nostrand Reinhold Co., Ltd., New York
2. Hammond P (1971) Applied electromagnetism. Pergamon Press Ltd., Oxford
3. Feynman RP, Leighton RB, Sands M (1972) Lectures in physics, vol. II. Addison-Wesley Publishing Co., London
4. Morse PM, Feshbach H (1953) Methods of theoretical physics. McGraw-Hill Book Co., New York
5. Johnk CTA (1988) Engineering electromagnetic fields and waves. Wiley, New York
6. Kraus JD (1984) Electromagnetics. McGraw-Hill Book Co., London
7. Lorrain P, Corson D (1962) Electromagnetic field and waves. W.H. Freeman & Co., San Francisco
8. Baden-Fuller AJ (1993) Engineering electromagnetism. Wiley, New York
9. Bevensee RM (1964) Slow-wave structures. Wiley, New York
10. Stratton JA (2007) Electromagnetic theory. Wiley, New Jersey

Chapter 3
Classical Radiation Theory

The tendency of modern physics is to resolve the whole material universe into waves, and nothing but waves. These waves are of two kinds: bottled-up waves, which we call matter, and unbottled waves, which we call radiation or light.

Sir James Jeans

Abstract Electromagnetic boundary value problems that incorporate a source of, or a sink for, electromagnetic waves are examined in Chap. 3 by adopting the 'source-present' Maxwell equations. It is shown that the resultant inhomogeneous second-order vector differential equations can be solved by employing analogies with statics, and by borrowing the mathematical forms derived there, particularly those which relate static electric and magnetic fields to charge and current. This leads to the formulation of the retarded scalar electric, and vector magnetic, potentials, which are demonstrated to be solutions of the source-present Maxwell equations.

The retarded potentials are applied to the classical 'short current element' or Hertzian dipole, which is a useful building block in the mathematical modelling of a range of antennas of the 'wire' or reflector type (see Chap. 4). It is shown that the fields surrounding the current element resolve into near-field and far-field forms. In the far-field $k_o r > 1$, the solution is predominantly in the form of TEM waves radiating maximally in the plane bisecting the element and minimally in the axial directions. The pattern is 'dough-ring' shaped. The radiated energy density obeys the inverse square law, and in circuit terms, the power radiated can be represented by a radiation resistance. In the near field, components of E and H other than the radiated fields exist. These fields diminish very rapidly with distance from the source and are negligible for $k_o r > 1$. They represent capacitive (electric dipole) and inductive stored energy, which means that in circuit terms, an antenna can present reactive impedance to the feed line.

The significance of radiation pattern and radiation resistance to the collection of electromagnetic waves, when the antenna becomes a sink rather than a source, is also elaborated upon in the chapter. This, of course, is the essential problem represented by solar power collection, which is examined in Chaps. 7, 8, 9 and 10.

A.J. Sangster, *Electromagnetic Foundations of Solar Radiation Collection*,
Green Energy and Technology, DOI 10.1007/978-3-319-08512-8_3

3.1 Radiation Mechanism

The fundamental ‘brick’ in the edifice that is classical electrical science, is the notion, or idea, that charge can be either positive or negative and that the measureable forces (attractive or repulsive) between charges can be represented by fields. For an isolated charge accumulation, the relevant field is the electric field, or if it is in motion, it is enhanced by a magnetic field. These fields are fully accommodated by, respectively, the Maxwell first equation (Eq. 2.1) termed the Gauss law and third Eq. (2.3) termed Ampere’s law. It should be noted that this simple concept is significantly challenged by quantum mechanics, which views a field as merely a convenient representation of the fact that an exchange of particles (photons) has occurred. Fortunately, however, electrical science can largely be formulated, as we shall see, without the quantum mechanical interpretation, which really only intervenes where it overlaps with material science, such as in the physics of semiconductors. An engineering evaluation of the quantum mechanical viewpoint will be provided in Chap. 6.

The essential mechanism underpinning radiation can actually be demonstrated pictorially, and quite succinctly, by considering the fields of an isolated charge in vacuum, which is initially stationary and is then moved rapidly (see Fig. 3.1) from position A to position B and back again in time τ. When it returns to A, it remains stationary for all time $t > \tau$; evidently, during the motion, it accelerates and decelerates. At time $t = 0$, while the charge is stationary, the Gauss law dictates that for a spherical charge, the E-field must be spherically symmetric and radial (E_r). This spreading field must diminish in intensity with radial distance (r) in accordance with the inverse square law.

Now, let us consider what happens at time t $(0 < t < \tau/2)$ with the charge on the move, at a moment when it is positioned between A and B, as suggested in Fig. 3.1a. At this instant, only the electric field very close to the charge obeys the Gauss law and forms a quasi-static spherically symmetric field pattern there. Some distant away from the charge, the field remains distributed as if the field source was still positioned at A. Since the ‘information’ that the source is no longer at A can only be carried electromagnetically on a wave travelling at velocity c (see Chap. 2), the transition obviously occurs at radius ct beyond which the field will be in its original distribution. To represent this situation graphically, the continuous field lines must be ‘kinked’, resulting in a transverse electric field component E_t that carries the ‘information’ about the change in source location as indicated in Fig. 3.1a.

However, the Ampere law, which is expressed in its full form by Maxwell’s equation (2.29), dictates that the time changing electric field E_t cannot exist on its own. It generates a magnetic field H_t that is orthogonal to E_t and to the direction of propagation. Not surprisingly, as already suggested in Chap. 2, the change information is transmitted out into space by a TEM wave. The process is not too unlike the way in which a sharp physical displacement of the end of a taught wire can be made to propagate along the wire. The elasticity of the wire that underpins the ‘whip action’ is

Fig. 3.1 Field distortion due to rapid lateral shift of free positive charge in a vacuum

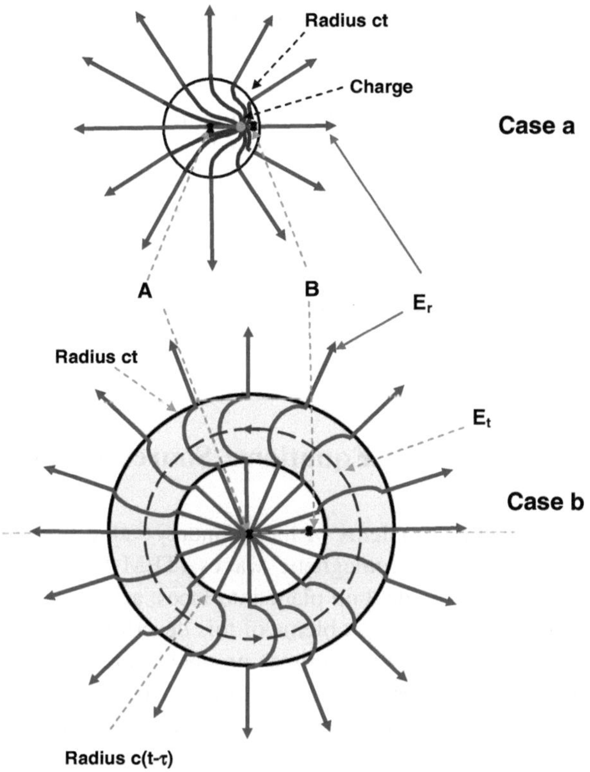

replicated in the EM case by the 'elasticity' of the E/H interaction embodied in the Maxwell equations.

During the period $\tau/2 < t < \tau$ as the charge returns to the origin at A, the field 'kink' and the changing E-field E_t will be in the opposite direction (see Fig. 3.1b), so that when the charge again becomes stationary at A, the field distribution shown in Fig. 3.1b will prevail. For the stationary charge at A, the field returns to a radial distribution, as required by the Gauss law, but only out to a radius of $c(t-\tau)$. Also, as with Fig. 3.1a, the field is radial beyond $r = ct$, where the effect of the original charge movement has still not reached. In the transition region (shaded), a spherical shell between $r = c(t-\tau)$ and $r = ct$, a time-delayed 'information pulse' exists, travelling at velocity c in the form of a *radiated* TEM wave. The time delay is an essential feature of the radiation mechanism, as we shall see.

Needless to say, in order to gain a thorough understanding and conception of the radiation process in more complex radiation situations, a much more rigorous treatment is required (see Sect. 3.1). Missing from this qualitative description is the very important aspect of the radiation mechanism which dictates that the transverse fields forming the radiated pulse or wave diminish with distance more slowly than the radially directed fields. While the transverse fields are inversely proportional to r, the radial fields obey the inverse square law and are inversely proportional to r^2.

This means that at sufficiently large distances from the radiation source, only the transverse (TEM) fields exist. Given that the power density in the outgoing wave is proportional to E_t^2 (see Eq. 2.43), it follows that the power density in the wave is proportional to $1/r^2$. But the volume of the spherical shell containing the wave increases in proportion to r^2, which implies that in vacuum, or in a lossless medium, the *total* power in the wave is unchanging with radial distance. This power will generally be equivalent to the power delivered to the antenna or radiator from a transmitting device connected to its input terminals. Without this feature of radiation, Earth would not be in the 'Goldilocks' zone of the sun and life on our planet would not exist. But more prosaically, it is also critical to the transmission of information across space by electromagnetic waves.

3.2 Maxwell's Equations: Source Form

Field theory problems of the boundary value type that were explored in Chap. 2, comprising electromagnetic waves (TEM waves) in free space, or in the atmosphere, and incident on plane interfaces, are generally termed source-free problems. This is because, in problems of this description, the region of interest is considered to be remote from the sources producing the fields, and the problem reduces, as we have seen, to a problem of electromagnetic waves intercepted by bounding surfaces. For such problems, Maxwell's equations, in differential form, are (see Chap. 2):

$$\nabla \times \mathbf{E} = -j\omega\mu\mathbf{H} \tag{3.1}$$

$$\nabla \times \mathbf{H} = j\omega\varepsilon\mathbf{E} \tag{3.2}$$

$$\nabla \cdot \mathbf{D} = 0 \tag{3.3}$$

$$\nabla \cdot \mathbf{B} = 0 \tag{3.4}$$

Here, **E** and **H** represent the electric and magnetic field intensities in the region of interest. $\mathbf{D} = \varepsilon\mathbf{E}$ is the electric flux density while $\mathbf{B} = \mu\mathbf{H}$ is the magnetic flux density. Note that in vacuum, or air, $\varepsilon = \varepsilon_0$ and $\mu = \mu_0$.

In radiation problems, on the other hand, the source becomes an integral part of the field problem; for example, in such problems, as the dipole antenna and the reflector antenna. It should, perhaps, be noted that the normal convention, in formulating the mathematical equations associated with radiation, is to presume that the waves emanate from a 'source'. However, the transmit/receive process is entirely reciprocal as demonstrated in Chap. 5, so we could equally presume that the focus of the waves is a 'sink' (see Fig. 3.2), and the mathematics would not change. In the collection of solar waves, of course, the sun is the source and the optical receiving antenna is the sink.

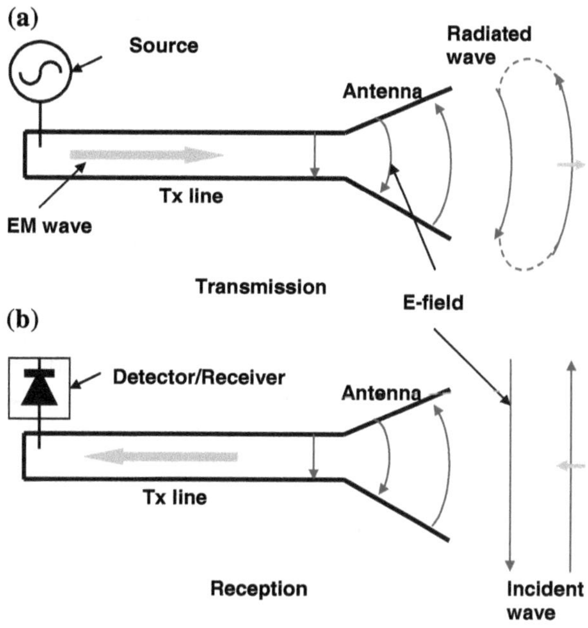

Fig. 3.2 **a** Radiation source. **b** Radiation sink

In these source/sink present electromagnetic boundary value problems, the most general forms of Maxwell's equations are required. These are:

$$\nabla \times \mathbf{E} = -j\omega\mu\mathbf{H} \tag{3.5}$$

$$\nabla \times \mathbf{H} = j\omega\varepsilon\mathbf{E} + \mathbf{J} \tag{3.6}$$

$$\nabla \cdot \mathbf{D} = \rho \tag{3.7}$$

$$\nabla \cdot \mathbf{B} = 0 \tag{3.8}$$

where **J** and ρ are the *impressed* 'electric' current density and 'electric' charge density, respectively, which form the source terms for the differential equations.

In some radiation problems, the basic source of the radiation is a 'magnetic' current (electric field) rather than conventional 'electric' current (magnetic field), in which case it is more appropriate to use the following forms for Maxwell's equations:

$$\nabla \times \mathbf{E} = -j\omega\mu\mathbf{H} - \mathbf{J}_m \tag{3.9}$$

$$\nabla \times \mathbf{H} = j\omega\varepsilon\mathbf{E} \tag{3.10}$$

$$\nabla \cdot \mathbf{D} = 0 \tag{3.11}$$

$$\nabla \cdot \mathbf{B} = \rho_m \tag{3.12}$$

where $\mathbf{J}_m$ and ρ_m are the impressed 'magnetic' current density and 'magnetic' charge density, respectively.

It is important, particularly in the 'electric' case to distinguish between the impressed currents and charges, and currents and charges arise because of the presence of the fields in a medium having finite conductivity. Conduction current density $\mathbf{J}$ is proportional to the electric field and is given by $\sigma\mathbf{E}$, where σ is the conductivity of the material. This current component can readily be accounted for, in the Maxwell equations, by replacing the permittivity ε by the complex form $\varepsilon(1-j\tan\delta) = \varepsilon(1-j\sigma/\omega\varepsilon)$. The density of free charge, apart from that associated with impressed currents, may be presumed to be zero. Thus, in subsequent discussion, the charge density ρ and the current density $\mathbf{J}$ appearing in the field equations can be taken as comprising purely impressed charges and currents. Any other currents, if they arise, will be accounted for by the complex electric permittivity, which will be written simply as ε for convenience.

The impressed currents and charges, expressed explicitly in the source-present form of Maxwell's equations, must be related to each other through the equation of continuity for current and charge. That is

$$\nabla \cdot \mathbf{J} = -\frac{\partial \rho}{\partial t} \tag{3.13}$$

The dual relationship for 'magnetic current' and 'magnetic charge' is

$$\nabla \cdot \mathbf{J}_m = -\frac{\partial \rho_m}{\partial t} \tag{3.14}$$

To solve the Maxwell equations for E-field or H-field behaviour in a bounded region, it is first necessary to form an equation in either E or H alone. The standard procedure for achieving this conversion is to perform a curl operation on either the curl equation for E or the corresponding equation for H. This gives, for example, using Eq. (3.5)

$$\begin{aligned} \nabla \times \nabla \times \mathbf{E} &= -j\omega\mu\nabla \times \mathbf{E} \\ &= -j\omega\mu[j\omega\varepsilon\,\mathbf{E} + \mathbf{J}] \\ &= k^2\mathbf{E} - j\omega\varepsilon\mu\mathbf{J} \end{aligned} \tag{3.15}$$

where $k^2 = \omega^2\mu\varepsilon$. Hence, on using a convenient vector identity, which states that for any vector $\mathbf{S}$

$$\nabla \times \nabla \times \mathbf{S} = \nabla\nabla \cdot \mathbf{S} - \nabla^2 \mathbf{S} \tag{3.16}$$

Equation (3.15) can be re-expressed as follows:

$$\nabla\nabla \cdot \mathbf{E} - \nabla^2 \mathbf{E} = k^2 \mathbf{E} - j\omega\mu \mathbf{J} \tag{3.17}$$

But, from Eq. (3.7), $\nabla \cdot \mathbf{E} = \rho/\varepsilon$; therefore

$$\begin{aligned} &\frac{\nabla\rho}{\varepsilon} - \nabla^2 \mathbf{E} = k^2 \mathbf{E} - j\omega\mu \mathbf{J} \\ \text{or} \quad &\nabla^2 \mathbf{E} + k^2 \mathbf{E} = \frac{\nabla\rho}{\varepsilon} + j\omega\mu \mathbf{J} \end{aligned} \tag{3.18}$$

This is an *inhomogeneous* vector wave equation in which the term on the right-hand side of the equals sign forms a *source* function. By employing the equation of continuity for charge and current, the source function can be expressed in terms of the impressed current alone. That is

$$\nabla^2 \mathbf{E} + k^2 \mathbf{E} = j\omega\mu \mathbf{J} - \frac{\nabla\nabla \cdot \mathbf{J}}{j\omega\varepsilon} \tag{3.19}$$

If we commence with Eq. (3.6), rather than Eq. (3.5), and follow an entirely analogous procedure, a corresponding inhomogeneous vector wave equation for *H* can be derived, which has the form:

$$\nabla^2 \mathbf{H} + k^2 \mathbf{H} = -\nabla \times \mathbf{J} \tag{3.20}$$

In deriving the above equations, it has been presumed that μ and ε are invariant with time. In free space or vacuum, the predominant media for solar radiation, $\mu = \mu_0$ and $\varepsilon = \varepsilon_0$.

In a source-free region, for which the impressed current J is non-existent, Eqs. (3.19) and (3.20) reduce, as they should, to the well-known homogeneous vector Helmholtz equations, namely:

$$\begin{aligned} &\nabla^2 \mathbf{E} + k^2 \mathbf{E} = 0 \quad \text{and} \\ &\nabla^2 \mathbf{H} + k^2 \mathbf{H} = 0 \end{aligned} \tag{3.21}$$

These are the phasor forms of Eqs. (2.36) and (2.37).

When a radiating source of the magnetic current type exists within the region of interest, a similar derivation when applied to Eqs. (3.9) and (3.10) leads to the following pair of inhomogeneous vector differential equations

$$\nabla^2\mathbf{E} + k^2\mathbf{E} = \nabla \times \mathbf{J}_m \tag{3.22}$$

and

$$\nabla^2\mathbf{H} + k^2\mathbf{H} = j\omega\varepsilon\mathbf{J_m} - \frac{\nabla\nabla \cdot \mathbf{J_m}}{j\omega\mu} \tag{3.23}$$

It is clear on examining their forms that the impressed current density vectors $\mathbf{J}$ and $\mathbf{J}_m$ enter into the inhomogeneous differential Eqs. (3.19), (3.20), (3.22) and (3.23) in a particularly complicated manner, making their direct solution extremely difficult, although with modern computational solvers employing finite difference or finite element techniques—not impossible. Nevertheless, this difficulty is normally circumvented by resorting to the use of *auxiliary potential functions,* which generate more easily solvable wave equation forms. These auxiliary potential functions may, or may not, represent clearly definable physical entities (especially is this so in the absence of sources), and so it is usual to adopt the viewpoint that these potentials are merely useful functions from which the electromagnetic fields can be derived. The two most common potential functions that arise in radiation problems are discussed in the next section.

3.3 Auxiliary Potential Functions

In formulating auxiliary potential functions, with the aim of facilitating the solution of the source-present second-order differential equations, as represented by (3.19) and (3.20), it is usually advisable to choose forms that satisfy at least one or more of the basic the Maxwell equations. Such a choice is:

$$\mathbf{B} = \nabla \times \mathbf{A} \tag{3.24}$$

which immediately satisfies Eq. (3.8), since by definition $\nabla \cdot \nabla \times \mathbf{S} = 0$ for any vector $\mathbf{S}$. In addition the choice:

$$\mathbf{E} = -\frac{\partial \mathbf{A}}{\partial \mathbf{t}} + \nabla\phi \tag{3.25}$$

satisfies the Maxwell Eq. (3.5) since $\nabla \times \nabla\phi = 0$ for any scalar function. The auxiliary vector $\mathbf{A}$ is usually termed the magnetic vector potential, which has relevance in magnetostatics, while auxiliary scalar ϕ is termed the electric potential (sometimes denoted by *V*) and has its origins in electrostatics. Consequently, if we can develop solvable source-present second-order differential equations for $\mathbf{A}$ and ϕ, then the associated electric and magnetic field forms can be deduced subsequently from the above two equations.

To develop differential forms in **A** and ϕ, we again enrol the Maxwell equations as expressed in Eqs. (3.5)–(3.8). On substituting **D** (= ε_0**E** for vacuum) in Eq. (3.7) with the expression for **E** given in Eq. (3.25), we obtain:

$$-\nabla^2\phi - \frac{\partial}{\partial t}(\nabla \cdot \mathbf{A}) = \frac{\rho}{\varepsilon_0} \tag{3.26}$$

which hardly seems an advance on Eqs. (3.19) and (3.20). However, Eqs. (3.24) and (3.25) only partially define **A** and ϕ. Equation (3.19) delineates only a rotational component of **A** ($\mathbf{A}_r$ say) while Eq. (3.20) defines a part of ϕ, which has finite gradient in the space domain. Therefore, we can add to **A** any component that has a zero curl (termed a lamellar component) without altering Eq. (3.24), and we can add to ϕ any function whose gradient is zero, leaving Eq. (3.25) unchanged. Provided the Maxwell relationships are not disrupted, these arbitrary introductions to **A** and ϕ can be chosen in a manner, which simplifies Eq. (3.26). Such a choice is the Lorentz condition whereby:

$$\nabla \cdot \mathbf{A} = -\frac{1}{c^2}\frac{\partial\phi}{\partial t} \tag{3.27}$$

Actually, this choice is not totally arbitrary. The Lorentz condition can be identified with the Kirchoff's law, which relates conduction current to rate of change of charge (see Chap. 2). It also brings the resultant radiation equations into line with the requirements of special relativity [1].

On substituting Eq. (3.27) into Eq. (3.26), a second-order differential in ϕ alone is formed, namely:

$$\nabla^2\phi - \frac{1}{c^2}\frac{\partial^2\phi}{\partial t^2} = -\frac{\rho}{\varepsilon_0} \tag{3.28}$$

or in phasor form:

$$\nabla^2\phi + k_0^2\phi = -\frac{\rho}{\varepsilon_0} \tag{3.29}$$

where

$$k_0 = \frac{\omega}{c} = \omega\sqrt{\mu_0\varepsilon_0} \tag{3.30}$$

Recalling that $\mathbf{H} = \mathbf{B}/\mu_0$ and $\varepsilon = \varepsilon_0$ in free space then substituting Eqs. (3.24) and (3.25), together with the Lorentz condition, into Eq. (3.10), yields an equivalent second-order differential equation for **A**, again in phasor form, namely:

$$\nabla^2 \mathbf{A} + k_0^2 \mathbf{A} = -\mu_0 \mathbf{J} \tag{3.31}$$

Equations (3.29) and (3.31) are clearly wave equations, but with source functions on the right of the equals sign, which are much less convoluted than those in Eqs. (3.19) and (3.20). These source functions have now been simplified to the impressed charge density (divided by a constant (ε_o)) in the region of interest in Eq. (3.29), and impressed current density (multiplied by a constant (μ_o)) in Eq. (3.31). If ρ and **J** are known in space and time, on an antenna located in a volume of space (say), then in principle, the above equations can be solved directly for ϕ or **A**, respectively. In practice, only one of the two wave equations need be solved since ϕ and **A** are related through the Lorentz condition. However, rather than demonstrate the formal solution, it is actually more instructive to construct a solution by employing electrostatic and magnetostatic analogies.

3.3.1 Electrostatics Analogy

If Eq. (2.13) is applied to a small spherical volume of stationary and isolated free charge (q), located in a magnetic field-free region, then symmetry and the circulation law for E (Eq. 2.16) dictate that the electric vector **D** will be everywhere radial and unvarying over an enveloping spherical surface (S) of radius r (Fig. 3.3). Under these circumstances:

$$\oiint_S \mathbf{D} \cdot \mathbf{dA} = 4\pi r^2 D_r \tag{3.32}$$

Clearly, Maxwell's first equation reduces to the inverse square law, or Gauss' law, which is usually written:

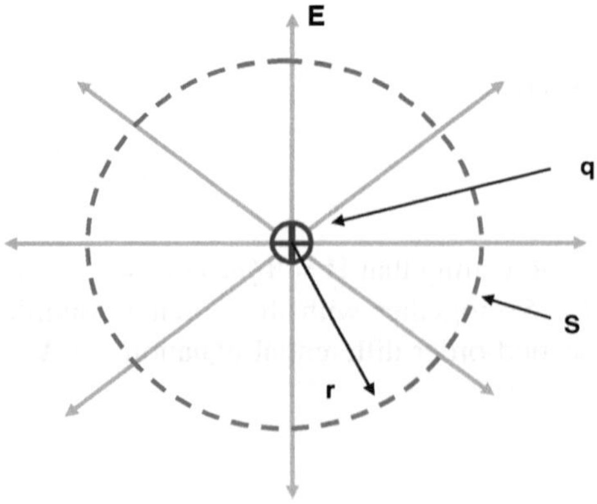

Fig. 3.3 Electric field flux through a spherical surface containing a point charge q

$$E_r = \frac{q}{4\pi r^2 \varepsilon_0} \tag{3.33}$$

provided that, for air, or vacuum:

$$\mathbf{D} = \varepsilon_0 \mathbf{E} \tag{3.34}$$

In the m.k.s. system, **D** has the dimension C/m^2, while the constant of proportionality $\boldsymbol{\varepsilon}_0$ is termed the absolute permittivity of free space. Measurement reveals that it has the value 8.84 × 10^{-12} F/m, where in the m.k.s. system, a Farad in a Coulomb/volt. More generally, $\mathbf{D} = \boldsymbol{\varepsilon}_r \varepsilon_0 \mathbf{E}$, where $\boldsymbol{\varepsilon}_r$ is termed the relative permittivity of the material concerned. It need not be a constant and could be a complex tensor to represent a general non-linear, anisotropic, inhomogeneous medium.

In conjunction with Eq. (2.26), Eq. (3.33) yields the Coulomb force law that can be expressed as:

$$F = \frac{qQ}{4\pi r^2 \varepsilon_0} \text{newtons} \tag{3.35}$$

for isolated point charges q and Q (in coulombs) distance r m apart in a vacuum.

Equation (3.33) really only applies if q is concentrated at a point, and so the situation in Fig. 3.4 where a distributed charge of density ρ is assumed to surround the origin of the Cartesian coordinate system (shaded region). To determine **E** in this case, we must first dice the shaded volume up into elemental volumes such a $\mathrm{d}V'$ $(= \mathrm{d}x'\mathrm{d}y'\mathrm{d}z')$ at r'. For such a volume, the charge enclosed is $q = \rho(x, y, z)\mathrm{d}V'$ coulombs and if it can be considered to reside at a point in space, then given that potential $\phi = -E\mathrm{d}r$, where $\mathrm{d}r = |r - r'|$, the potential at P due to this element is:

$$\mathrm{d}\phi = \frac{\rho(x, y, z)dV'}{4\pi\varepsilon_0|r - r'|} \tag{3.36}$$

The potential at P due to the entire charge volume (V) at O is the summation of all contributions $\rho(x, y, z)\mathrm{d}V'$ at all possible positions O' throughout the charge volume. In the limit, this summation becomes an integral, giving:

$$\phi = \frac{1}{4\pi\varepsilon_0} \iiint_V \frac{\rho \mathrm{d}V'}{|r - r'|} \tag{3.37}$$

While this equation exhibits the correct spatial form for electric potential at distance from an arbitrary charge distribution, it is applicable only to time invariant charge concentrations. The discussion in relation to Fig. 3.1 suggests that if the charge alters with time then Eq. (3.37) could justifiably be applied to fields in close proximity to the charge, but at more distant locations, a time retardation mechanism is required. So can we build time delay into Eq. (3.37)? It turns out that we can.

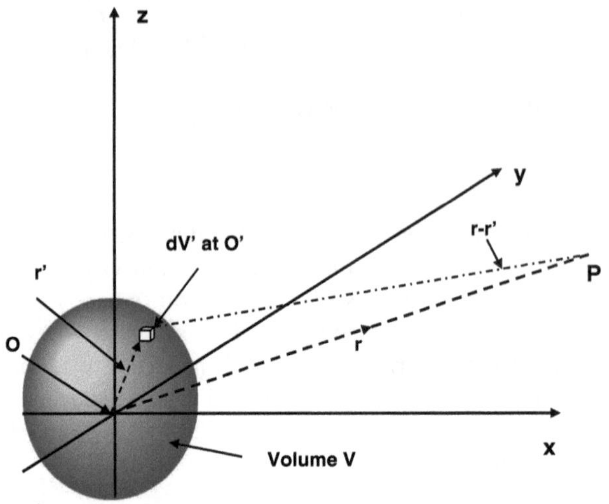

Fig. 3.4 Geometry for construction of retarded potentials

If the charge distribution at O' in Fig. 3.4 alters with time, the information that this change has occurred can be felt at point P only after a period of time $|r - r'|/c$, which is the time it takes for a TEM wave to travel distance $|r - r'|$. Hence, it is not implausible to suggest that at time t and at position **r,** the contribution to ϕ (**r**,t) at P can be calculated using Eq. (3.37) if we insert the charge distribution at time $t - |r - r'|/c$. This is true for all volume elements $\mathrm{d}V'$ within V. Consequently, the desired time varying solution to Eq. (3.37) can be formulated by simply incorporating the time retardation into Eq. (3.37), to give:

$$\phi = \frac{1}{4\pi\varepsilon_0} \iiint_V \frac{\rho(\mathbf{r}', t - |\mathbf{r} - \mathbf{r}'|/c)\mathrm{d}V'}{|\mathbf{r} - \mathbf{r}'|} \tag{3.38}$$

Eq. (3.38) is a solution to Eq. (3.28) that can be established by substitution. The solution to the phasor form Eq. (3.29) simplifies to:

$$\phi = \frac{1}{4\pi\varepsilon_0} \iiint_V \frac{\rho(\mathbf{r}')\mathrm{d}V'}{|\mathbf{r} - \mathbf{r}'|} \tag{3.39}$$

with time expressed through the common factor exp($j\omega t$) which can be suppressed.

3.3.2 Magnetostatics Analogy

The fundamental measurement-based laws of magnetostatics, as deduced by the early pioneers can be constructed in similar manner from Eqs. (2.14) and (2.15). Equation (2.14), which is usually referred to as the flux law for magnetic field, simply states that, however, generated—by a permanent magnet, by the Earth's core, by current on a wire, by a current-carrying coil—magnet flux density forms closed loops. At the macroscopic level, there are no magnetic field sources performing the same role as charge in electrostatics. Hence, if we apply Eq. (2.15) to a steady (DC) current (I) on a long straight wire, we can confidently say that the magnetic field which the law predicts must form closed circular loops centred on the wire—assuming the wire has a cylindrical cross section. We therefore obtain:

$$\oint_C \mathbf{H} \cdot \mathbf{d}\ell = 2\pi r H_t = I$$

$$H_t = \frac{I}{2\pi r} \tag{3.40}$$

$$B_t = \frac{\mu_0 I}{2\pi r}$$

Equation (3.40) is Ampere's law, and since I is expressed in amps (A) in m.k.s. system, the dimension of magnetic field strength H is A/m. With H in A/m and B in Tesla. Consistency of the magnetic field laws requires that for vacuum:

$$\mathbf{B} = \mu_0 \mathbf{H} \tag{3.41}$$

where the constant μ_0 is termed the absolute permeability of free space and has the magnitude $4\pi \times 10^{-7}$ Henries/m (Henry = T m^2/A). As with electrostatics, we can say that more generally, $\mathbf{B} = \mu_r\mu\mathbf{H}$, where μ_r is termed the relative permeability of the material concerned. It need not be a constant, being a function of **H** in a ferrite material, and it could be a complex tensor to represent a general, anisotropic, inhomogeneous medium. Application of the circuital law for H to a current element $Id\ell$ rather than to a current on a long straight wire leads to the Biot and Savart law (see Ref. [3]). Referring to Fig. 3.5 where a current element Idl' is located at O′, a distance $r'$from the origin at O. In determining the magnetic flux density at P distance r from O, we can assume from Ampere's law that the flux forms circular loops centred on the current I. The contribution **dB** at P must be proportional to I in magnitude and inversely proportional to the square of the distance between O′and P, namely $|\mathbf{r} - \mathbf{r}'|$. Directionally, the vector **dB** must be orthogonal to both $Idl'$and the vector $(\mathbf{r} - \mathbf{r}')$. So, mathematical reasoning suggests that we can express **dB** in the following form:

$$\mathbf{dB} = \frac{\mu_0 I}{4\pi} \frac{\mathbf{dl}' \times (\mathbf{r} - \mathbf{r}')}{|\mathbf{r} - \mathbf{r}'|^3} \tag{3.42}$$

This the Biot–Savart law is most commonly expressed in the form:

$$dB = \frac{\mu_0 I d\ell}{4\pi r^2} \tag{3.43}$$

Fig. 3.5 Magnetic field at P due to current element Idl'

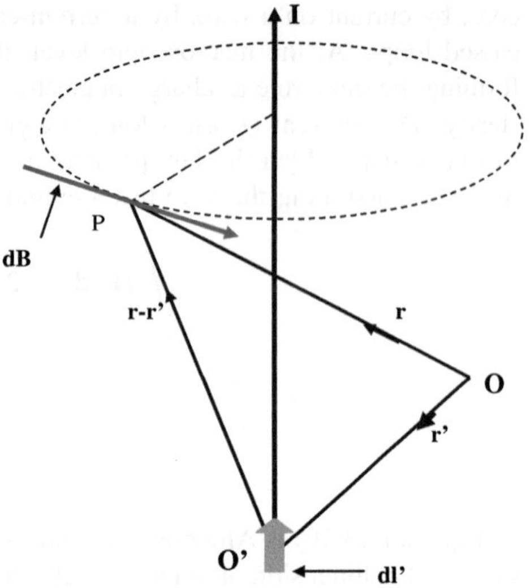

for a current element at the centre of the magnetic field loop. If the straight portion of wire carrying I is actually part of a complete electrical circuit s, then on summing all the Idl'contributions around s leads, in the limit of infinitesimally small dl', to an integration giving:

$$\mathbf{B}(\mathbf{r}) = \frac{\mu_0 I}{4\pi} \oint_s \frac{\mathbf{dl}' \times (\mathbf{r} - \mathbf{r}')}{|\mathbf{r} - \mathbf{r}'|^3} \tag{3.44}$$

Remembering that we have chosen to define the vector magnetic potential through Eq. (3.24) and noting that [1]:

$$\nabla \times \frac{\mathbf{dl}'}{|\mathbf{r} - \mathbf{r}'|} = -\frac{(\mathbf{r} - \mathbf{r}')}{|\mathbf{r} - \mathbf{r}'|^3} \times \mathbf{dl}'$$

thence, combining Eqs. (3.24) and (3.44) yields the following:

$$\mathbf{A}(\mathbf{r}) = \frac{\mu_0 I}{4\pi} \oint_s \frac{\mathbf{dl}'}{|\mathbf{r} - \mathbf{r}'|} \tag{3.45}$$

The time retarded form for $\mathbf{A}$ can be deduced in much the same way as we have already done for ϕ with the additional observation that current and current density are related through:

$$I\mathbf{dl}' = \mathbf{J}\mathrm{d}V' \tag{3.46}$$

Hence, we obtain:

$$\mathbf{A}(\mathbf{r}, t) = \frac{\mu_0}{4\pi} \iiint_V \frac{\mathbf{J}(\mathbf{r}', t - |\mathbf{r} - \mathbf{r}'|/c)}{|\mathbf{r} - \mathbf{r}'|} \mathrm{d}V' \tag{3.47}$$

or in phasor form:

$$\mathbf{A}(\mathbf{r},) = \frac{\mu_0}{4\pi} \iiint_V \frac{\mathbf{J}(\mathbf{r}')}{|\mathbf{r} - \mathbf{r}'|} \mathrm{d}V' \tag{3.48}$$

The retarded potentials **A** and ϕ represent powerful tools in the search for solutions to complex electromagnetic radiation and reception problems and are widely used in various simulation packages by providing the 'building blocks', such as the 'short current element', to construct field patterns for distributed current sources, which are known in space and time. This is particularly true of those packages employing moment methods. The short current element, or Hertzian dipole, is examined in the next section.

Needless to say, equivalent forms to **A** and ϕ also exist, which provide expedient routes to the solutions to Eqs. (3.22) and (3.23). These are usually defined by the parameters $\mathbf{A}_m$—an auxiliary potential function associated with 'magnetic current' $\mathbf{J}_m$, and ϕ_m—an auxiliary potential linked to 'magnetic charge' ρ_m. These forms will be considered in more detail in Chap. 4.

3.4 Hertzian Dipole

The Hertzian dipole and the 'short current element' are essentially synonymous. The concept is presented schematically in Fig. 3.6. The short current element is viewed as a filament of current Idl, which exhibits constant magnitude along its length dl. As such, it is physically impossible to realise, but remains a helpful mathematical construct. It functions as an exceptionally useful building block in antenna modelling that enables the field patterns of practical antenna structures to be determined by integration. Consequently, it is instructive to examine the field behaviour of this current filament, and this is implemented by applying the appropriate potential function to the example shown in Fig. 3.6.

The current element is assumed to be located at the origin of a Cartesian coordinate system and that it takes the form a constant z-directed current varying in time at frequency ω rad/s—i.e. the current $I = I_0 \cos \omega t$. If time retardation is incorporated into the definition of current, then this becomes:

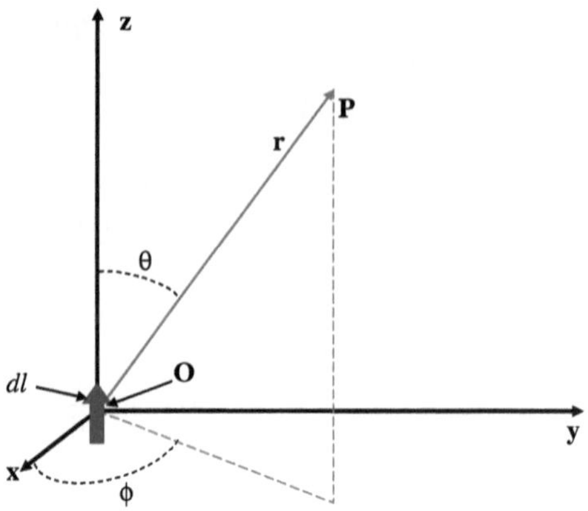

Fig. 3.6 Short current element at origin of Cartesian coordinate system

$$I = I_0 \cos\left(\omega\left(t - \frac{r}{c}\right)\right) = I_0 \cos(\omega t - k_0 r)$$

In complex notation, this can be written:

$$I = \mathrm{Re}[I_0 \exp(j(\omega t - k_0 r))]$$

In these two equations, the phase coefficient $k_0 = {}^{\omega}/_{c} = \omega\sqrt{\mu_0 \varepsilon_0}$ for free space. So, for this one-dimensional element of length dl in the z-direction Eq. (3.45) becomes:

$$\mathbf{A} = \frac{\mu_0 I_0 \mathbf{dl}}{4\pi} \exp(j(\omega t - k_0 r)) \tag{3.49}$$

If we adopt the unit vector notation, for the Cartesian system, namely $\left(\hat{\mathbf{a}}_\mathbf{x}, \hat{\mathbf{a}}_\mathbf{y}, \hat{\mathbf{a}}_\mathbf{z}\right)$ then:

$$\mathbf{A} = A_z \hat{\mathbf{a}}_\mathbf{z} = \frac{\mu_0 I_0 dl \hat{\mathbf{a}}_\mathbf{z}}{4\pi} \exp(j(\omega t - k_0 r)) \tag{3.50}$$

It is evident from Eq. (3.50) that at distant point P (Fig. 3.6), the magnetic vector potential is z-directed. However, we know from our earlier discussion on the radiation mechanism that the important field components are orthogonal to the direction vector $\mathbf{r}$ and are therefore best represented by spherical components. Consequently, we let:

$$\mathbf{A} = \hat{\mathbf{a}}_\mathbf{r} A_r + \hat{\mathbf{a}}_\theta A_\theta + \hat{\mathbf{a}}_\phi A_\phi \tag{3.51}$$

where $A_r = A_z \cos\theta$, $A_\theta = -A_z \sin\theta$ and $A_\varphi = 0$. Hence, the magnetic field components (see Ref. [2]) of the radiated wave at P can be derived using $\mathbf{B} = \mu_0 \mathbf{H} = \nabla \times \mathbf{A}$ resulting in:

$$H_\phi = \frac{I_0 \mathrm{d}l}{4\pi}\left[\frac{jk_0}{r} + \frac{1}{r^2}\right]\sin\theta \exp(-jk_0 r) \tag{3.52}$$

$$H_\theta = H_r = 0 \tag{3.53}$$

The electric field components of the same radiated wave can be constructed by applying the Maxwell Eq. (3.2) to generate:

$$E_\theta = \sqrt{\frac{\mu_0}{\varepsilon_0}}\frac{I_0 \mathrm{d}l}{4\pi}\left[\frac{jk_0}{r} + \frac{1}{r^2} - \frac{j}{k_0 r^3}\right]\sin\theta \exp(-jk_0 r) \tag{3.54}$$

$$E_r = \sqrt{\frac{\mu_0}{\varepsilon_0}}\frac{I_0 \mathrm{d}l}{2\pi}\left[\frac{1}{r^2} - \frac{j}{k_0 r^3}\right]\cos\theta \exp(-jk_0 r) \tag{3.55}$$

$$E_\phi = 0 \tag{3.56}$$

From the discussion in Sect. 3.1, the fields radiating away from the short current element are expected to be normal to the direction vector r, and to each other. They are also predicted to diminish in magnitude at a rate which is inversely proportion to distance r from the source. The only field components that meet this requirement are the first term of H_ϕ in Eq. (3.52) and the first term of E_θ (Eq. 3.54). So, in the far field, generally defined as occupying the region given by $k_0 r > 1$:

$$H_\phi = \frac{jk_0 I_0 \mathrm{d}l}{4\pi r}\sin\theta \exp(-jk_0 r) \tag{3.57}$$

$$E_\theta = \sqrt{\frac{\mu_0}{\varepsilon_0}}H_\phi = \eta_0 H_\phi \tag{3.58}$$

A pictorial representation of the E_θ field distribution is shown in Fig. 3.7. Note that the pattern is circularly symmetric in the 'horizontal' plane (x–y plane in Fig. 3.6), which slices through the current element, but it is not 'isotropic'—radiating equally in all directions. The 'solid' pattern is 'dough-ring' shaped with electric (and magnetic) field nulls along the z-axis ($\theta = 0^\circ, 180^\circ$) of the source current. The H_ϕ field is everywhere normal to the E-field, and normal to the plane of the page containing Fig. 3.7.

So, what is the significance of the other field components, which diminish with distance as $1/r^2$ and $1/r^3$ in Eqs. (3.54) and (3.55)? The $1/r^3$ term is reminiscent of the electrostatic field behaviour for an electric dipole [2]. It arises because the current I must, in accordance with Kirchoff's law, be associated with positive- and negative-charge build-up ($I = -\frac{\mathrm{d}q}{\mathrm{d}t}$; see Chap. 2) at its extremities, so forming, at any given instant in time, an electric dipole. The field exists only in the 'near'-field region ($k_o r < 1$) of the source, being negligible in the far field. It supports electrostatic or capacitive energy storage.

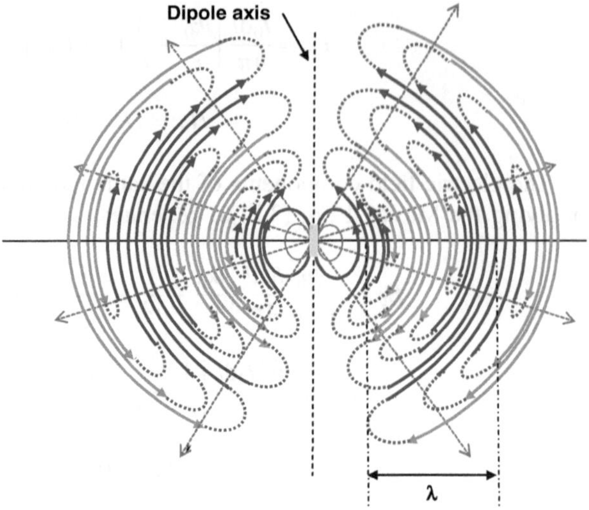

Fig. 3.7 Electric field contours for radiation from a short current element at a specific instant in time (*red* = peaks; *blue* = troughs in the wave)

The $1/r^2$ that exists in both the electric and magnetic field formulations [Eqs. (3.52), (3.54) and (3.55)] exhibits the characteristics of the stored inductive magnetic field in the vicinity of a current-carrying wire. In magnetostatics, it is derived using the Biot–Savart law. The electric field components arise because the antenna current, and hence this inductive magnetic field, is time varying and are derivable using the Faraday's law (Eq. 3.5). Again these fields diminish rapidly with distance from the source and are essentially zero in the far field.

The nature of these field components can be further illuminated by examining complex power flow in the near field of the source. Complex power flow density in W/m^2 is generally expressed in the form:

$$\mathbf{p} = \frac{1}{2}\mathbf{E} \times \mathbf{H}^* \ \mathrm{W/m^2} \tag{3.59}$$

where $\mathbf{p}$ is a vector directed normal to $\mathbf{E}$ and $\mathbf{H}$. Following convention, $\mathbf{H}$ is conjugated to suppress the exponential term, in much the same way as in electrical circuit theory, complex power is expressed as $P = \frac{1}{2}VI^*$. Hence, expressing $\mathbf{E}$ and $\mathbf{H}$ in their r, θ and ϕ components, we obtain:

$$\mathbf{p} = \frac{1}{2}(\hat{\mathbf{a}}_\phi E_r H_\theta^* - \hat{\mathbf{a}}_\theta E_r H_\phi^* - \hat{\mathbf{a}}_\phi E_\theta H_r^* + \hat{\mathbf{a}}_\mathbf{r} E_\theta H_\phi^* + \hat{\mathbf{a}}_\theta E_\phi H_r^* - \hat{\mathbf{a}}_\mathbf{r} E_\phi H_\theta^*) \tag{3.60}$$

On introducing Eq. (3.52) to Eq. (3.56) into Eq. (3.60), the following components to the complex power density p for the short current element are derived:

$$\mathbf{p} = \hat{\mathbf{a}}_\theta \frac{-jk_0 I_0^2 dl^2}{16\pi^2}\sqrt{\frac{\mu_0}{\varepsilon_0}}\left[\frac{1}{r^3} + \frac{1}{k_0^2 r^5}\right]\sin\theta\cos\theta$$
$$+ \hat{\mathbf{a}}_\mathbf{r} \frac{k_0 I_0^2 dl^2}{32\pi^2}\sqrt{\frac{\mu_0}{\varepsilon_0}}\left[\frac{k_0}{r^2} - \frac{j}{k_0^2 r^5}\right]\sin^2\theta \tag{3.61}$$

The real power flow density radiating in the r-direction away from the source is given by $p_r = \mathrm{Re}\mathbf{p} = \frac{1}{2}\mathrm{Re}(\mathbf{E} \times \mathbf{H}^*)$ and has the form:

$$p_r = \frac{k_0^2 I_0^2 dl^2}{32\pi^2}\sqrt{\frac{\mu_0}{\varepsilon_0}}\frac{\sin^2\theta}{r^2} \tag{3.62}$$

The remaining components of Eq. (3.61) are imaginary in character and represent stored energy in the near field on the source. These components can be extracted by expressing $p_i = \mathrm{Im}\mathbf{p}$. This leads to an r-directed component:

$$p_i|_r = \frac{-jI_0^2 dl^2}{32k_0\pi^2}\sqrt{\frac{\mu_0}{\varepsilon_0}}\frac{\sin^2\theta}{r^5} \tag{3.63}$$

and a θ-directed component:

$$p_i|_\theta = \frac{-jk_0 I_0^2 dl^2}{16\pi^2}\sqrt{\frac{\mu_0}{\varepsilon_0}}\left[\frac{1}{r^3} + \frac{1}{k_0^2 r^5}\right]\sin\theta\cos\theta \tag{3.64}$$

The real power flow density (Eq. (3.62)) diminishes, as expected, in proportion to the inverse radius squared (inverse square law), while the sine squared dependence of the equation denotes that the radiating lobe maximises in the x–y plane (at $\theta = 90^\circ$) for all values of ϕ. The stored energy density term $p_i|_r$ (Eq. (3.63)), which decays in inverse proportion to radius to the power of five, also exists mainly in the x−y plane. This electric dipole energy essentially oscillates at the frequency of the source, between electric and magnetic stored forms, with the instantaneous power flow density moving in a radial direction. On the other hand, the stored energy density represented by Eq. (3.64), which displays θ-directed instantaneous power flow, is concentrate in ‘energy lobes’ located at $\theta = \pm 45^0$, for all values of ϕ see Fig. 3.8). Both electric dipole ($1/r^3$) and inductive ($1/r^5$) energy forms are represented.

To complete the picture of the radiation behaviour of the short current element, it is instructive to examine to total power delivered by the element for a given current. To determine the total power, it is necessary to integrate the power density over a spherical surface centred on the element and enclosing it. That is:

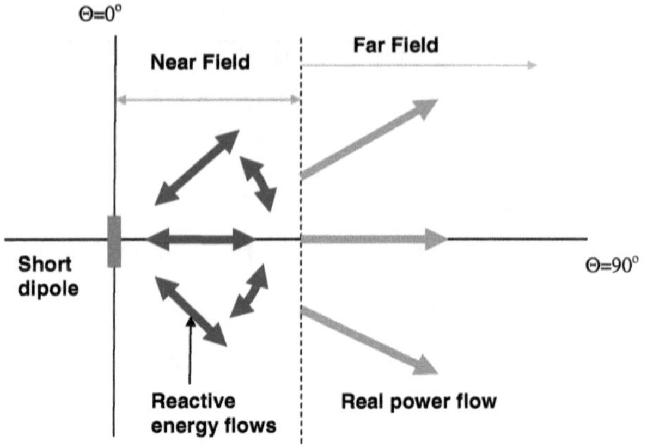

Fig. 3.8 Power and energy flows in near- and far-field regions for short current element

$$P_{rad} = \iint_S p_r \hat{\mathbf{a}}_\mathbf{r} \cdot dS = \int_0^{2\pi}\int_0^{\pi} \frac{k_0^2 I_0^2 dl^2}{32\pi^2}\sqrt{\frac{\mu_0}{\varepsilon_0}}\frac{\sin^2\theta}{r^2} r^2 \sin\theta \, d\theta \, d\phi$$
$$= \frac{k_0^2 I_0^2 dl^2}{32\pi^2}\sqrt{\frac{\mu_0}{\varepsilon_0}} 2\pi \int_0^{\pi} \sin^3\theta \, d\theta \tag{3.65}$$

But $\int_0^{\pi} \sin^3\theta \, d\theta = 4/3$ and $k_0 = \frac{2\pi}{\lambda_0}$, where λ_o is the free-space wavelength. Hence:

$$P_{\text{rad}} = \frac{\pi\eta_0 I_0^2}{3}\left(\frac{dl}{\lambda_0}\right)^2 \tag{3.66}$$

This can be further reduced to a now traditional form:

$$P_{\text{rad}} = 40\pi^2\left(\frac{dl}{\lambda_0}\right)^2 I_0^2 \text{ Watts} \tag{3.67}$$

by using the relation $\eta_O = 120\pi$. Not unexpectedly, the power radiated is proportional to the square of the current applied, but more importantly, it is proportional to the square of the element length normalised to the free-space wavelength. This ratio occurs repeatedly in antenna theory and underpins the unmistakeable trend that antennas largely diminish in size as frequency rises.

In electrical circuits, it is useful to express power in terms of a resistance, and so it is convenient to make the relationship:

$$P_{\text{rad}} = I_{rms}^2 R_{\text{rad}} \tag{3.68}$$

where I_{rms} is the root mean squared current in the source, to yield:

$$P_{\rm rad} = \frac{1}{2} I_0^2 R_{\rm rad} \tag{3.69}$$

$R_{\rm rad}$ is termed the radiation resistance of the source. On combining Eqs. (3.69) and (3.67), the following classical equation is obtained:

$$R_{\rm rad} = 80\pi^2 \left(\frac{dl}{\lambda_0}\right)^2 \text{ Ohms} \tag{3.70}$$

If an antenna is viewed as a device that provides a transitioning mechanism from a guided wave in a feed line supplied from a transmitter to a free-space (TEM) wave, then radiation resistance represents the coupling between the antenna and free space through its radiation pattern. It should be understood that $R_{\rm rad}$ has nothing to do with ohmic loss in the antenna structure. For optimum coupling to free space, the output impedance of the transmitter, the characteristic impedance of the feed line and the radiation resistance should be matched. For the short current element, however, this is not possible. The assumptions incorporated into the development of the above equations require $^{dl}/_{\lambda_0} < 0.1$, which gives $R_{\rm rad} < 8\ \Omega$. An antenna feed line typically exhibits a characteristic impedance between 50 and 100 Ω obviously very different from $R_{\rm rad}$. Clearly, techniques are required to embed and match practical antennas into real transmission systems, and these will be explored in Chaps. 6, 7 and 8.

The antenna format that is more applicable to solar power collection scenarios is that of reception, where the transmission of power is now from an antenna to a receiver. This operating mode raises not dissimilar issues. To take the simplest case of an isotropic antenna that in transmission mode radiates power equally in all directions—the antenna pattern forms a perfect sphere around the source with the power density at radius r given by:

$$p_{\rm r} = \frac{P_{\rm T}}{4\pi r^2} \text{ W/m}^2 \tag{3.71}$$

where P_T is the power in watts from the transmitter. It is assumed that the 'theoretical' set-up is properly matched from transmitter to radiator.

In reception, total collection of the incoming electromagnetic waves requires two conditions to be met. Firstly, the incoming plane waves must 'match' the radiation pattern by arriving, in this case, in a spherically symmetric manner as suggested in Fig. 3.9. Secondly, the 'isotropic' antenna radiation resistance must be impedance matched to the feed line and to the input impedance of the receiver; otherwise power is re-radiated back into space.

For the short current element, the same conditions would apply with the requirement that the incoming waves must arrive from the directions dictated by the pattern in Fig. 3.7 and with the designated relative power density levels.

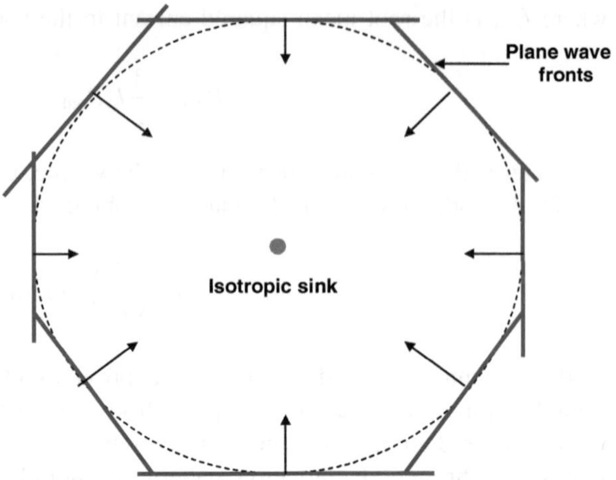

Fig. 3.9 Spherically distributed plane waves incident on an isotropic sink

The significance for solar power collection is that a gathering antenna, in order to be truly effective, must have a radiation pattern that focuses on the sun only, with ideally a single beam pattern that precisely radiates into the solid angle subtended by the sun's disc at the earth's surface. This is, of course, not physically possible, but as we shall see in Chap. 7, modern high-gain antennas can go a long way to meet this condition. Impedance matching is well developed and this aspect of the collection process will be fully examined in ensuing chapters.

References

1. Grant IS, Phillips WR (1975) Electromagnetism. Wiley, Chichester
2. Kraus JD (1984) Electromagnetics. McGraw-Hill Book Co., Singapore
3. Balanis CA (1982) Antenna theory. Wiley, New York
4. Collin RE, Zucker FJ (1969) Antenna theory. McGraw-Hill Book Co., New York
5. Read FH (1980) Electromagnetic radiation. Wiley, Chichester
6. Ulaby FT (1997) Fundamentals of applied electromagnetics. Prentice Hall Inc., New Jersey

Chapter 4
Aperture Antennas for Solar Systems

> *Engineering refers to the practice of organizing the design and construction [and, I would add operation] of any artifice which transforms the physical world around us to meet some recognized need.*
>
> G.F.C. Rogers

Abstract In antenna terms, solar power collectors arguably reside largely in the 'aperture antenna' camp. Consequently, Chap. 4 endeavours to provide the reader with a fundamental appreciation of the radiation characteristics of aperture antennas by deriving the key integral equation for the magnetic vector potential $\mathbf{A}_m$ from the Maxwell equations. The $\mathbf{A}_m$ equation, as we shall see in Chap. 8, is the route into geometrical optics, an analytical technique which represents the 'bedrock' of reflector antenna design. With a knowledge of $\mathbf{A}_m$ that has been secured, the radiation properties of some regularly shaped apertures are then examined, and this leads to the derivation of spectrum functions which relate the far radiation field of the antenna, to the near or aperture field, through a Fourier transform pair. The relationships that exist between antenna directivity, gain, radiation efficiency and sidelobes and how these basic properties are influenced by the aperture field distribution of the given antenna are studied in detail, for both two-dimensional and three-dimensional radiation pattern embodiments.

4.1 Introduction

The auxiliary potential functions are introduced in Chap. 3 and help to facilitate the analysis and modelling of wire, or current carrying, antennas, including simple reflectors as used in solar systems. Their successful adoption is dependent on knowing the surface current density everywhere on the metallic structure. This is very often not possible as with lens antennas and dielectric rod antennas for which there is no identifiable conduction current flow, while it is very difficult to arrange, for antennas such as horns. These antenna types (see Fig. 4.1) are generally referred

A.J. Sangster, *Electromagnetic Foundations of Solar Radiation Collection*,
Green Energy and Technology, DOI 10.1007/978-3-319-08512-8_4

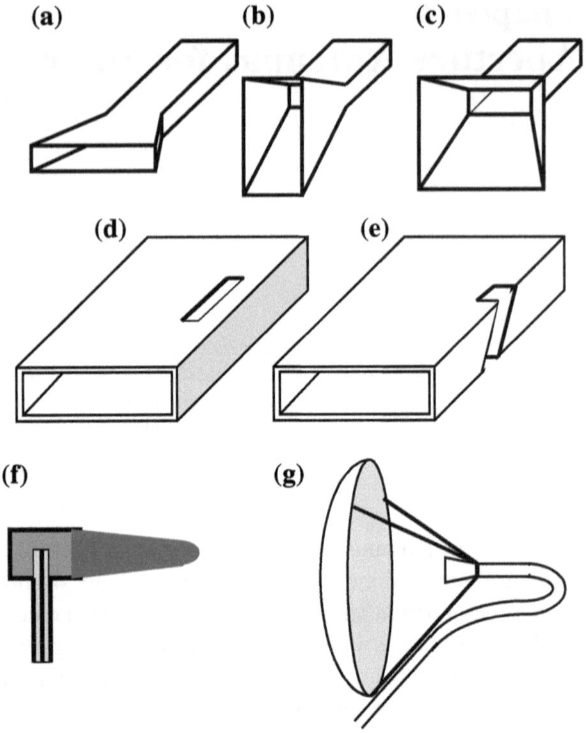

Fig. 4.1 Examples of aperture antennas: **a** H-plane horn, **b** E-plane horn, **c** sectoral horn, **d** shunt-slot in waveguide, **e** edge-slot in waveguide, **f** dielectric rod, **g** horn-fed parabolic antenna

to as aperture antennas. Those represented by sketches a, b and c in Fig. 4.1 are horn types in which an empty rectangular waveguide opens out into a rectangular or square aperture depending on the application. In sketches d and e, the slots cut into the broad and side walls of the waveguide form the aperture radiators, while f represents a leaky-wave dielectric rod radiator, where the aperture is the surface of the dielectric rod. The classification also includes reflector antennas (g), although these are also amenable to the techniques described in Chap. 3.

To determine the fields radiated by an aperture antenna, it is usually necessary to employ a 'field equivalence principle' [1–3]. By adopting such a procedure, the antenna structure of interest, which would represent a largely intractable mathematical problem in its original manifestation, is typically reconfigured to a much simpler more amenable geometrical form, while remaining in essence electromagnetically equivalent to the original. While this modified form may no longer lead to an exact solution on analysis, it can be modelled more readily and, if constructed insightfully and intelligently, generally yields more than acceptable results. The basic concept is illustrated in Fig. 4.2a. In the left-hand-side diagram, input current densities $\mathbf{J}$ and $\mathbf{J}_m$ (e.g. on and inside a horn antenna) set up a radiation field $\mathbf{E}$, $\mathbf{H}$ in a large area of space (V_2). Within this area, an arbitrary surface S of volume V_1 (say) is inserted, which is invisible to the field distribution $\mathbf{E}$, $\mathbf{H}$. The equivalence theorem states that if on the *surface* S surface currents $\mathbf{J}_s$ and

$\mathbf{J_{ms}}$ are imposed such that the external field **E**, **H** is unaltered, then the introduced currents are equivalent sources for the region V_2. Therefore, on S:

$$\mathbf{J_s} = \hat{\mathbf{n}} \times (\mathbf{H} - \mathbf{H_i}) \text{ and } \mathbf{J_{ms}} = -\hat{\mathbf{n}} \times (\mathbf{E} - \mathbf{E_i})$$

where the interior of S is presumed to contain a field distribution $\mathbf{E_i}$, $\mathbf{H_i}$ which is specified only on S by means of the above boundary relations. Since the interior volume V_1 is not of interest in setting up the equivalent sources, the interior fields $\mathbf{E_i}$, $\mathbf{H_i}$ can be arbitrarily chosen. The most common choice is $\mathbf{E_i} = \mathbf{H_i} = 0$ which is termed Love's equivalence principle [3]. In this case on S, $\mathbf{J_s} = \hat{\mathbf{n}} \times \mathbf{H}$ *and* $\mathbf{J_{ms}} = -\hat{\mathbf{n}} \times \mathbf{E}$. Furthermore, this means that the medium within S can also be chosen arbitrarily. For example, a conductor can be judiciously placed there with no electromagnetic effect on the arrangement.

In Fig. 4.2a it should be noted that both electric current density $\mathbf{J_s}$ and magnetic current density $\mathbf{J_{ms}}$ have been inserted onto the surface S. However, according to the uniqueness theorem [4], we need only specify $\mathbf{J_s}$ or $\mathbf{J_{ms}}$ on S, not both, to ensure that the resultant field solution is unique. For example, consider the conducting body S with an aperture A as shown in Fig. 4.2b for which the field in the aperture ($\mathbf{E}_a \times \hat{\mathbf{n}}$) is known. The radiated field in the exterior region is unaffected if the aperture in the conducting body is closed by a conducting wall as depicted in Fig. 4.2b (right-hand-side diagram) and replaced by the magnetic current $\mathbf{J_{ms}} = \mathbf{E}_a \times \hat{\mathbf{n}}$ placed closely adjacent to the position of the original aperture—essentially on the surface of the aperture wall.

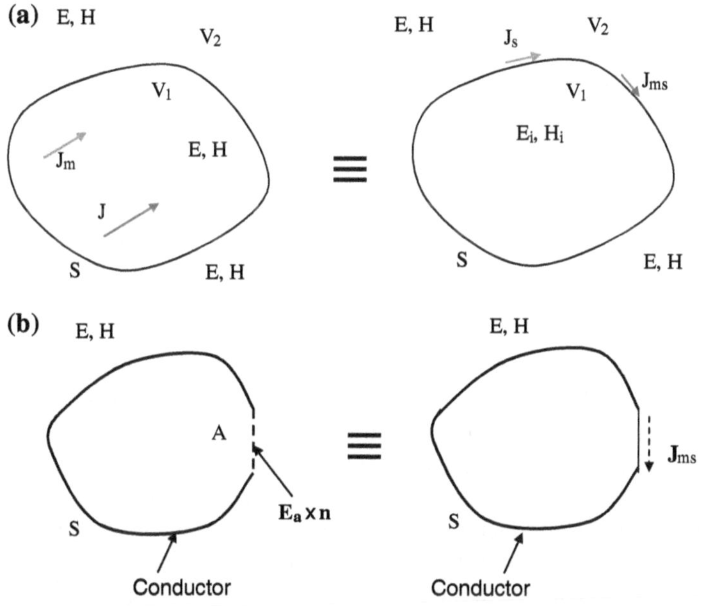

Fig. 4.2 Equivalence principles

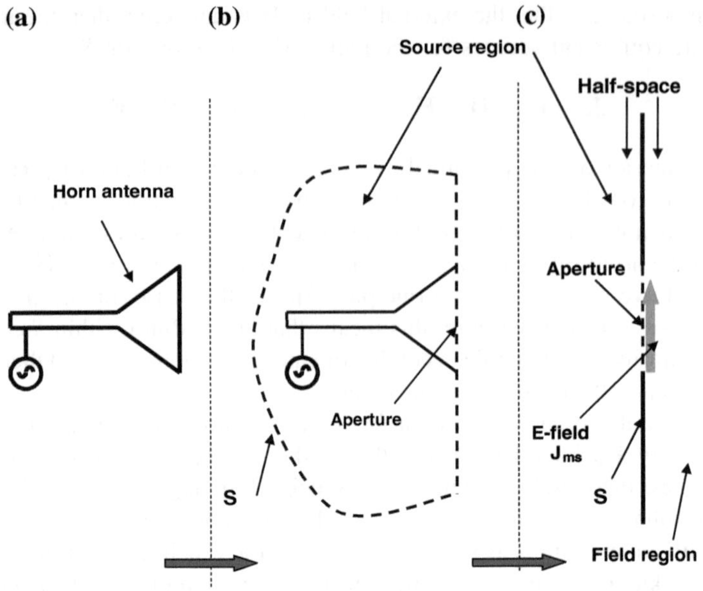

Fig. 4.3 Application of equivalence principle **a** original horn structure **b** enclosing surface *S* **c** equivalent aperture in infinite screen

Let us consider applying this equivalence technique to the horn antenna. Firstly, the horn—the electromagnetic wave source—is enclosed (Fig. 4.3b) by an arbitrary surface which both accommodate the 'radiating aperture' of the source and completely encase it. It is commonly referred to as the Huygen's surface. A field-free region is introduced within the surface *S* if surface currents are introduced to form fields external to *S* equivalent to those generated by the original source. This entails placing on *S* a distribution of electric and magnetic surface current densities $\mathbf{J_s}$ and $\mathbf{J_{ms}}$. If selected appropriately, these current replacements generate the original external field. Given that on a defined surface, the electromagnetic boundary conditions dictate that $\mathbf{J_s} = \hat{\mathbf{n}} \times \mathbf{H}$ and $\mathbf{J_{ms}} = -\hat{\mathbf{n}} \times \mathbf{E}$, where $\hat{\mathbf{n}}$ denotes the unit normal to the surface, the application of the equivalence principle requires precise knowledge of both **E** and **H** over the surface *S*. These field definitions must obviously be dictated by the original source geometry. Actually, because of the relationship between **E** and **H,** built into the Maxwell equations, it is not necessary to define both field quantities over *S*. It is usually enough to define either **E** or **H** alone over the entire surface, or more commonly **E** alone over part of *S* and **H** alone over the rest. The resultant external field solution is unique, provided that on no part of *S* is surface current density unspecified. This means that *S* can be viewed as a perfect conductor, with $\mathbf{J_{ms}}$ specified over an area adjacent to the position of the horn aperture where E_a is finite (Fig. 4.3b). In principle, this is a simpler problem to solve using auxiliary potential functions, particularly as we are relatively free in the way we choose the shape and form of S [3]. The E-field is imaged in the conducting surface to satisfy the boundary conditions so that

$$\mathbf{J_{ms}} = -2\hat{\mathbf{n}} \times E_a$$

For example, if the portion of the surface which intercepts the aperture is presumed to be a plane perfectly conducting surface of infinite extent, then the equivalent model now comprises two half-spaces separated by an infinitesimally thin conducting plane, with the horn aperture field replaced by a magnetic current sheet adjacent to the conducting plane as suggested in Fig. 4.3c. This is clearly a much simpler radiation problem to solve than the raw original. To solve the problem, all we now need is a knowledge of the E-field, and hence $\mathbf{J_{ms}}$, in the horn aperture, which is generally available from waveguide theory. Hence, using an appropriate auxiliary function, the radiated field in the right-hand-side half-space can be determined (see Sect. 4.2). The disadvantage of the technique is that only field patterns in the forward half-space of the antenna are generated—back radiation is ignored. However, this is seldom a serious limitation since it is the forward pattern that is of primary concern in the majority of antenna applications whether operated in transmission or reception mode.

4.2 Auxiliary Potentials

In very much the same way as was expounded in Chap. 3, if we resort to an auxiliary potential function to simplify the electromagnetic boundary value problem created by a radiation source, the choice should satisfy at least one of the source-free Maxwell equations in an electric charge-free space. Here, we are concerned with magnetic current (electric field) sources and consequently an apt choice is to set (see [3]):

$$\mathbf{E} = -\frac{1}{\varepsilon_0}\nabla \times \mathbf{A_m} \tag{4.1}$$

where $\mathbf{A_m}$ is the auxiliary potential function. Unlike the magnetic potential function $\mathbf{A}$ in Chap. 3, $\mathbf{A_m}$ has little electrical significance, other than as a useful mathematical generator of electric field, and hence magnetic field.

The Maxwell equations which are applicable to the problem of radiation from a magnetic current source of density $\mathbf{J}_m$ and magnetic charge density ρ_m are:

$$\nabla \times \mathbf{E} = -j\omega\mu\mathbf{H} - \mathbf{J}_m \tag{4.2}$$

$$\nabla \times \mathbf{H} = j\omega\varepsilon\mathbf{E} \tag{4.3}$$

$$\nabla \cdot \mathbf{D} = 0 \tag{4.4}$$

$$\nabla \cdot \mathbf{B} = \rho_m \tag{4.5}$$

The magnetic current density $\mathbf{J}_m$ and magnetic charge density ρ_m are linked through the equation:

$$\nabla \cdot \mathbf{J}_m = -j\omega\rho_m \tag{4.6}$$

which is suggestive of the Kirchoff relationship between electric current and charge (see Chap. 3).

Following a derivation paralleling Chap. 3, these lead to two second-order vector-differential equations for electric and magnetic fields, namely

$$\nabla^2\mathbf{E} + k^2\mathbf{E} = \nabla \times \mathbf{J}_m \tag{4.7}$$

and

$$\nabla^2\mathbf{H} + k^2\mathbf{H} = j\omega\varepsilon\mathbf{J}_m - \frac{\nabla\nabla \cdot \mathbf{J}_m}{j\omega\mu} \tag{4.8}$$

Clearly, the right-hand-side source functions in these equations are unappealingly complex. However, the auxiliary potential function choice delineated above, when substituted into the derivation process as outline in Chap. 3, generates a much more amenable second-order vector-differential equation, as follows:

$$\nabla^2\mathbf{A}_m + k_0^2\mathbf{A}_m = -\varepsilon_0\mathbf{J}_m \tag{4.9}$$

where $k = k_0 = \omega\sqrt{\mu_0\varepsilon_0}$ for radiation into free space or vacuum. By making an obvious parallel with the treatment of the analogous equation in Chap. 3 (Eq. 3.31), it is not difficult to assert that the solution to Eq. (4.9) must have the phasor form

$$\mathbf{A}_m(\mathbf{r}, t) = \frac{\varepsilon_0}{4\pi}\iiint_V \frac{\mathbf{J}_m(\mathbf{r}')}{|\mathbf{r} - \mathbf{r}'|}\mathrm{d}V' \tag{4.10}$$

That Eq. (4.10) is a solution to Eq. (4.9) can be established by substitution.

The evident mathematical duality that exists between radiation solutions emanating from surface conduction current densities ($\mathbf{J} \equiv \mathbf{J_s}$) on 'wire' structures, and magnetic current densities ($\mathbf{J_m} \equiv \mathbf{J_{ms}}$) representing field sources in apertures, is highlighted in Table 4.1.

Table 4.1 Duality of equations associated with electric and magnetic current sources

Electric sources ($J \neq 0$, $J_m = 0$)	Magnetic sources ($J_m \neq 0$, $J = 0$)
$\nabla \times \mathbf{E} = -j\omega\mu_0\mathbf{H}$	$\nabla \times \mathbf{H} = -j\omega\varepsilon_0\mathbf{E}$
$\nabla \times \mathbf{H} = \mathbf{J} + j\omega\varepsilon_0\mathbf{E}$	$-\nabla \times \mathbf{E} = \mathbf{J_m} + j\omega\mu_0\mathbf{H}$
$\mathbf{H} = \frac{1}{\mu_0}\nabla \times \mathbf{A}$	$\mathbf{E} = -\frac{1}{\varepsilon_0}\nabla \times \mathbf{A_m}$
$\mathbf{E} = -j\omega\mathbf{A} - j\frac{1}{\omega\mu_0\varepsilon_0}\nabla(\nabla \cdot \mathbf{A})$	$\mathbf{H} = -j\omega\mathbf{A}_m - j\frac{1}{\omega\mu_0\varepsilon_0}\nabla(\nabla \cdot \mathbf{A}_m)$
$\nabla^2\mathbf{A} + k_0^2\mathbf{A} = -\mu_0\mathbf{J}$	$\nabla^2\mathbf{A}_m + k_0^2\mathbf{A}_m = -\varepsilon_0\mathbf{J}_m$
$\mathbf{A}(\mathbf{r}, t) = \frac{\mu_0}{4\pi}\iiint_V \frac{\mathbf{J}(\mathbf{r}')}{\lvert\mathbf{r}-\mathbf{r}'\rvert}\mathrm{d}V'$	$\mathbf{A}_m(\mathbf{r}, t) = \frac{\varepsilon_0}{4\pi}\iiint_V \frac{\mathbf{J}_m(\mathbf{r}')}{\lvert\mathbf{r}-\mathbf{r}'\rvert}\mathrm{d}V'$

4.3 Radiation From a Linear Aperture

Radiation mechanisms associated with antennas which fall into the category of 'aperture antennas' are attributable to diffraction from the electric field distribution formed across the aperture at any given instant in time. This aperture field distribution can be modelled in many ways, but the mechanism can be illuminated most clearly by examining the electromagnetic penetration of a narrow one-dimensional slit as depicted in Fig. 4.4. The arrangement comprises a perfectly flat conducting screen of infinite extent in the x–y plane, containing a slot of width a in the x-direction and extending to infinity at $\pm y$. A plane electromagnetic wave in the left-hand-side half-space is incident on the screen and hence the slit and sets up an x-directed electric field E_0 in the aperture, which we can presume to be constant across the narrow aperture.

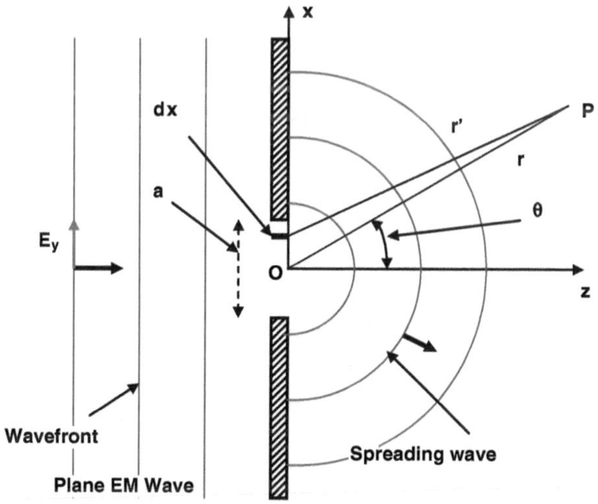

Fig. 4.4 Electromagnetic wave diffraction by a parallel sided slit in a perfectly conducting membrane (Huygen's mechanism)

4.3.1 Huygen's Principle and Equivalent Sources

By Huygen's principle, every point on the E-field in the aperture is a source of spreading waves. In two dimensions, these waves spread into a cylindrical space with $1/\sqrt{r}$ dependence on distance from source, while in three dimensions, they spread out spherically with $1/r$ dependence for the fields. That this is so is evident from the nature of Eq. (4.10). So let us consider an element of the slit of width dx at x in the x-direction (Fig. 4.4). We can assume that the fields everywhere are invariant in the y-direction for a slit of constant width for all y. In accordance with Huygen's principle, the field within the element can be considered to be a source of cylindrical waves in the right-hand-side medium ($z > 0$). The strength of this 'secondary' source will obviously be proportional to E_0 and also to the width dx of the element. Hence, the field at the remote point P in Fig. 4.4, which is located at distance r' from the source at (x, 0, 0) and distance r from the origin at O, can logically be represent by a function of the following form:

$$\mathrm{d}E_P = \frac{AE_0\mathrm{d}x}{\sqrt{r'}}\exp(-jk_0r') \tag{4.11}$$

where A is an amplitude constant, and k_0 is the propagation coefficient for free-space electromagnetic waves in the right-hand-side region. If P is very remote from the aperture such that $r \gg$a, then we can reasonably presume that $r' \approx r$ in the denominator of Eq. (4.11). However, for the much more sensitive-phase shift, we use $r' = r - x\sin\theta$ and so Eq. (4.11) becomes:

$$\mathrm{d}E_P = \frac{AE_0\mathrm{d}x}{\sqrt{r}}\exp(-jk_0(r - x\sin\theta)) \tag{4.12}$$

where θ is the angle between OP and the z-axis in Fig. 4.4.

The total diffracted field at P due to the field E_0 in the aperture can now be determined by summing all of the dx contributions at P for all values of x between $-a/2$ and $+a/2$. This results in the following integral equation:

$$E_P(r,\theta) = \frac{AE_0}{\sqrt{r}}\exp(-jk_0r)\int_{-a/2}^{+a/2}\exp(jk_0\,x\sin\theta)\,\mathrm{d}x \tag{4.13}$$

which, on performing the integration, reduces to:

$$E_P(r,\theta) = \frac{AE_0a}{\sqrt{r}}\exp(-jk_0r)\frac{\sin\left(\pi\frac{a}{\lambda_0}\sin\theta\right)}{\pi\frac{a}{\lambda_0}\sin\theta} \tag{4.14}$$

Significantly, the form of this equation reveals that the diffracted wave at P from the superposition of cylindrical waves, emanating from all parts of the aperture containing the field E_0, combines to form an 'interference pattern' which in this case has a characteristic sin*X*/*X* distribution. Distributions of this nature are ubiquitous in aperture antenna theory, and they will be examined in more detail in the following section.

4.3.2 Plane Wave Spectrum

In several branches of electromagnetism, the wave solutions to the electromagnetic boundary value problem are too complex to recruit directly into device development. Fortunately however, such solutions can usually be resolved into constituent plane waves. A good example [2] is rectangular, or circular cylindrical, waveguide where the behaviours of the TE or TM modes can best be implemented by expressing them in plane wave terms (see Chap. 2). This approach can also be used to advantage in certain radiation problems, particularly those of the aperture type. Mathematically, as we shall see, the plane wave spectrum technique, as it is termed, is not difficult to implement for antennas with known aperture field shapes and can lead to quick and reliable predictions of an antenna's radiation characteristics.

The plane wave spectrum of a radiating aperture is a fundamentally more rigorous means of estimating the diffraction pattern than the Huygen's method. For example, a point on the cylindrical wave front depicted in Fig. 4.4 can locally be viewed as comprising a 'plane' wave within an elemental sector $d\theta$, particularly for large r values. This idea is illustrated in Fig. 4.5. The combination of plane waves representing all sectors $d\theta$ between $\theta = -\pi/2$ and $\theta = +\pi/2$ mathematically reconstructs the original cylindrical wave front. This combination of plane waves is usually termed the angular spectrum, or spectrum function ($F(\theta)$), and for the case of a very narrow slit, it has a constant magnitude. Actually, this function is conventionally expressed as $F(\sin\theta)$ rather than $F(\theta)$.

On the other hand for a radiating slit of finite width, the spectrum function is by no means constant. In this more general case, it can be evaluated be reference to Fig. 4.5, where the slit has a width a in the x-direction. As in the Huygens example, the slit is irradiated from the left by a single plane wave of E-field amplitude E_0. The figure shows a typical plane wave component of the radiated energy in the right-hand-side half-space. It propagates in the direction $\mathbf{k}_0$ which is at an angle θ to the z-axis. If the right-hand-side region is free space, the magnitude of $\mathbf{k}_0$ is given by $\mathbf{k}_0 = \omega\sqrt{\mu_0\varepsilon_0}$, and the direction is given that of the constituent plane wave. The point $P(x, z)$ (Fig. 4.6) is expressed in the cylindrical coordinate frame by the vector $\mathbf{r}$. The plane wave propagating in the direction θ subtends angle $d\theta$ at the origin, and logically the wave magnitude will be proportional to both $F(\sin\theta)$ and $d\sin\theta$, on using $\sin\theta$ rather than θ to denote angle. Hence, the E-field at P due to this representative plane wave can be expressed as:

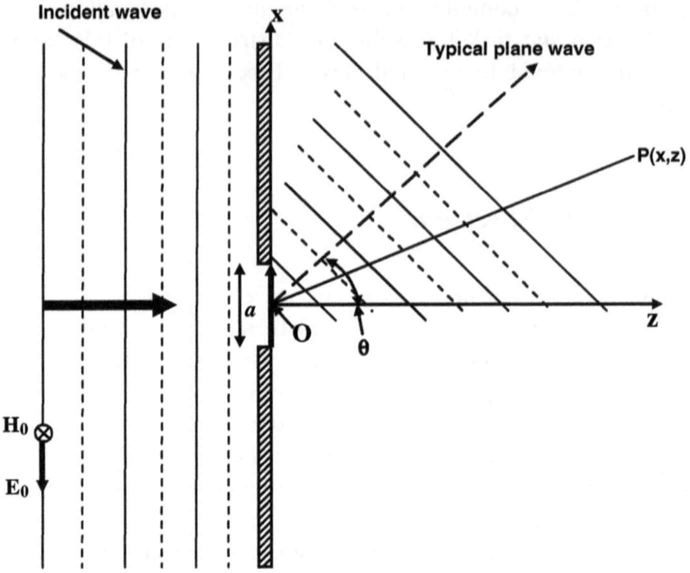

Fig. 4.5 Plane wave diffraction at a narrow slit in an infinite perfectly conducting screen—illustrating the diffraction pattern is a spectrum of plane waves

Fig. 4.6 Cylindrical coordinates (r, ψ) of a field point P

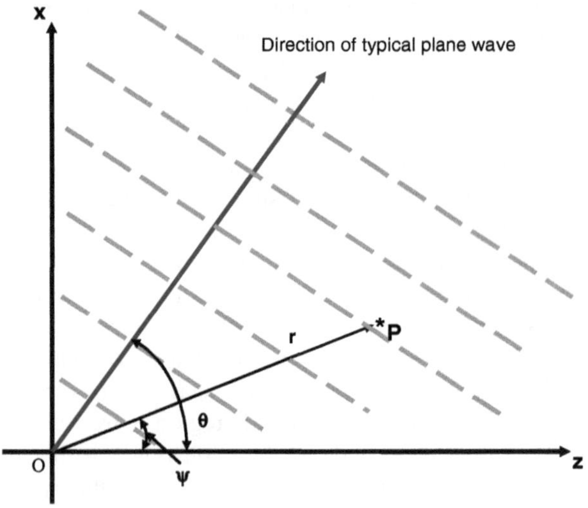

$$\begin{aligned} dE_P(x,z) &= F(\sin\theta)d(\sin\theta)\exp(-j\mathbf{k_0}\cdot\mathbf{r}) \\ &= F(\sin\theta)d(\sin\theta)\exp(-jk_0(x\sin\theta + z\cos\theta)) \end{aligned} \tag{4.15}$$

and on collecting and adding all possible contributions from all typical plane waves, the following integral function is created:

$$E_P(x,z) = \int_{-\infty}^{\infty} F(\sin\theta)\exp(-jk_0(x\sin\theta + z\cos\theta))d(\sin\theta) \tag{4.16}$$

The integration limits are taken from $-\infty$ to $+\infty$ to ensure that all possible contributions to the diffracted field are included—both real and imaginary.

If the field in the aperture is expressed functionally as:

$$f(x) = E_P(x, 0) \tag{4.17}$$

then Eq. (4.16) becomes

$$f(x) = \int_{-\infty}^{\infty} F(\sin\theta)\exp(-jk_0 x\sin\theta)d(\sin\theta) \tag{4.18}$$

In electrical circuit theory, Fourier transforms are regularly used to relate frequency to time and vice versa through the equations:

$$f(t) = \frac{1}{2\pi}\int_{-\infty}^{\infty} F(\omega)\exp(j\omega t)d\omega \tag{4.19}$$

$$F(\omega) = \int_{-\infty}^{\infty} f(t)\exp(-j\omega t)dt \tag{4.20}$$

So by analogy with this time/frequency transform pair, it is reasonable to suggest that Eq. (4.18) has the transform partner:

$$F(\sin\theta) = \frac{1}{\lambda_0}\int_{-\infty}^{\infty} f(x)\exp(jk_0 x\sin\theta)\,dx \tag{4.21}$$

This transform relationship between the angular spectrum and aperture field, admittedly expressed in a limited two-dimensional form here, in Eqs. (4.18) and (4.21), is a very important one in antenna theory and will be considered in more detail below.

In the elementary diffraction example presented in Fig. 4.5, the electric field, and hence $f(x)$, in the range of x from $-\infty$ to $+\infty$, is nonzero only in the slit; therefore,

$$f(x) = \begin{cases} E_0 & \text{for } |x| \leq a/2 \\ 0 & \text{for } |x| > a/2 \end{cases} \tag{4.22}$$

On substituting this form for $f(x)$ into Eq. (4.21), we obtain

$$F(\sin\theta) = \frac{E_0}{\lambda_0} \int_{-\infty}^{\infty} \exp(jk_0 x \sin\theta)\, dx$$

This integral is well known and exhibits the following solution:

$$F(\sin\theta) = \frac{E_0 a}{\lambda_0} \frac{\sin\left(\frac{\pi a \sin\theta}{\lambda_0}\right)}{\frac{\pi a \sin\theta}{\lambda_0}} \tag{4.23}$$

It has the same angular dependence in the far field as the radiation pattern represented by Eq. (4.14) which was deduced from the Huygen's principle. That the spectrum function bears a direct relationship with the far-field pattern of an aperture antenna is more fully developed in many texts on antenna theory [3]. This powerful relationship remains true for all aperture antennas, not just the slit example examined here.

4.4 Spectrum Functions

The Fourier transform pair [Eqs. (4.18) and (4.21)] which expresses the relationship between spectrum function (far field) and an aperture field function (antenna source field) provides an excellent base from which to examine the fundamental radiation characteristics of aperture antennas. They are of course limited to furnishing pattern information for the principal plane of the antenna only—that is the plane containing the major axis of the aperture and the normal to the aperture (the x–z plane in Fig. 4.5). Notwithstanding this limitation, all the essential features of the antenna pattern can be demonstrated. To assess antenna behaviour in planes other than the principal plane, the three-dimensional forms of Eqs. (4.18) and (4.21) must be used. These are developed in Sect. 4.4.

In Fig. 4.7, spectrum functions for four commonly used antenna aperture field distributions are displayed in two-dimensional rectangular rather than polar form. The functions themselves are listed in Table 4.2. The rectangular form is more traditional, but with the growing availability of sophisticated drafting tools within EM solvers, the polar representation is becoming increasingly common place.

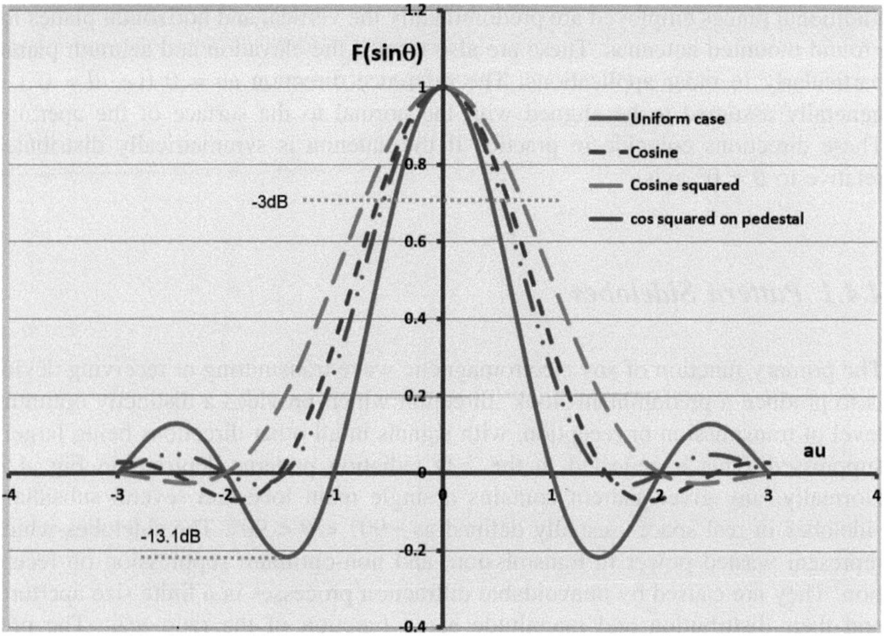

Fig. 4.7 Far field patterns for *uniform*, *cosine*, *cosine squared*, and *cosine squared on a pedestal*, shaped aperture distributions

Table 4.2 Commonly used spectrum functions

	Aperture distribution normalised to λ_0/a $f(x)$	Spectrum function $F(\sin\theta)$
1	Uniform $f(x) = \lambda_0/a$	$\frac{\sin(\pi au)}{\pi au}\left(\mathrm{u} = \frac{\sin\theta}{\lambda_0}\right)$
2	Cosine $f(x) = \frac{\lambda_0}{a}\cos\left(\frac{\pi x}{a}\right)$	$\left(\frac{\pi}{2}\right)^2 \frac{\cos(\pi au)}{\left(\pi/2\right)^2 - (\pi au)^2}$
3	Cosine squared $f(x) = \frac{\lambda_0}{a}\cos^2\left(\frac{\pi x}{a}\right)$	$\frac{\sin(\pi au)}{\pi au}\frac{1}{1-(au)^2}$
4	Parabolic $f(x) = \frac{\lambda_0}{a}\left[1 - \left(\frac{2x}{a}\right)^2\right]$	$\frac{3}{(\pi au)^2}\left[\frac{\sin(\pi au)}{\pi au} - \cos(\pi au)\right]$
5	Parabolic squared $f(x) = \frac{\lambda_0}{a}\left[1 - \left(\frac{2x}{a}\right)^2\right]^2$	$\frac{15}{(\pi au)^2}\left[\frac{3}{(\pi au)^2}\left[\frac{\sin(\pi au)}{\pi au} - \cos(\pi au)\right] - \frac{\sin(\pi au)}{\pi au}\right]$
6	Cosine squared on 10-dB pedestal $f(x) = \frac{\lambda_0}{a}\left[0.3162 + 0.6838\cos^2\left(\frac{\pi x}{a}\right)\right]$	$\frac{\sin(\pi au)}{\pi au}\left[\frac{1-0.481(au)^2}{1-(au)^2}\right]$

Of course, for 'real' antennas, radiation patterns are three-dimensional, and modern drafting software can generate 3-D images. However, 3-D representations on a 2-D page can lose subtle details of a given pattern, and so the 2-D depictions continue to be used to ensure that the pattern of interest is fully recorded. Normally, this means that at least two planar patterns are required representing the major planes (or 'cuts') of the antenna. These are usually chosen to be orthogonal to each other. The traditional planes employed are predominantly the vertical and horizontal planes for ground mounted antennas. These are also termed the elevation and azimuth planes particularly in radar applications. The reference direction au = 0 (i.e. $\theta = 0°$) is generally assumed to be aligned with the normal to the surface of the aperture. These directions coincide in practice if the antenna is symmetrically distributed relative to $\theta = 0°$ axis.

4.4.1 Pattern Sidelobes

The primary function of any electromagnetic wave transmitting or receiving device is to produce a predominant 'look' direction which provides a distinctly optimum level of transmission or reception, with signals in all other directions being largely suppressed. This is reflected in the 2-D radiation patterns depicted in Fig. 4.7. Normally, any given pattern contains a single main lobe and several subsidiary sidelobes in real space—usually defined as $-90° < \theta < 90°$. The sidelobes which represent wasted power in transmission, and non-optimum suppression on reception. They are caused by unavoidable diffraction processes in a finite size aperture, and their distribution and magnitude are a function of the ratio a/λ_0. The predominant requirement of the antenna design engineer is to ensure that the sidelobes are kept as low as possible with respect to the mainlobe, and avenues for achieving this are suggested in the above figures. These show that illumination functions ($f(x)$) which are shaped to produce low-edge fields in the aperture produce low sidelobes. However, at a fixed a/λ_0, this is achieved at the expense of widening the angle subtended by the mainlobe. On the other hand, the main beamwidth, which determines the directivity of the antenna, can be reduced by increasing a/λ_0. This stratagem introduces more nulls into the pattern and hence moves the first null (nearest the mainlobe peak) towards au = 0, that is $\theta = 0$, thus 'squeezing down' the mainlobe width.

For example, if $a = 0.3$ m and $\lambda_0 = 3$ cm, the first null occurs at 5.7° for a uniform aperture illumination (Fig. 4.7). If a is now increased to 0.6 m, the first null shifts to 2.85°, thus narrowing the primary lobe. The magnitude of the first sidelobe relative to the peak of the mainlobe, as can be seen from Fig. 4.7, has the magnitude $0.22 = -13.1$ dB. For most applications, this would normally be viewed as too high and aperture field shaping would be introduced to lower it.

4.4.2 Mainlobe Beamwidth

The angular width of the mainlobe of a directional antenna is arguably the parameter of primary concern in relation to achieving a design which meets a specific electromagnetic wave transmission or reception requirement. In two dimensions, the definition of this lobe-width or beamwidth could simply be the angular distance between first nulls as the spectrum patterns suggest. In three dimensions, as we shall see in Sect. 4.4, beamwidths in at least two orthogonal planes are usually required to define the lobe.

For the example considered above, of a 0.3m wide uniformly illuminated slit radiator, the distance between first nulls can be deduced from Fig. 4.7 (solid red curve) and gives a null beamwidth of 11.4°. While in theory this is a perfectly acceptable definition, in practice this is not so for three very sound reasons:

1. There may not be a null—a feature of some limited pattern types.
2. If the null exists, its precise location may be difficult to determine accurately.
3. The minimum detectable level of an EM signal in any practical measurement is well above zero field, as a result of noise and other signal perturbations.

As a consequence of these difficulties, the convention, which has sensibly evolved, defines beamwidth as the angular separation in a given plane, between the points on either side of the mainlobe, which are 3 dB in power density below the density at the beam *maximum*. This is termed the half-power beamwidth (HPBW) and is identified on the spectrum pattern in Fig. 4.7. Note that the pattern is normalised to give a peak value of unity. Since the spectrum actually represents a field pattern, the 3-dB positions are located where the field is $1/\sqrt{2}$ below the peak value. The HPBW is determined by equating the spectrum function ($f(x)$) to 0.707. In the case of uniform illumination, this gives:

$$\sin \pi a u = 0.707 \pi a u$$

with $\theta = \theta_{3\mathrm{dB}}$. For example, for a 0.3m wide uniformly illuminated slit at 10 GHz (λ_0 = 3.0 cm) the beamwidth from Fig. 4.7 is approximately 5.1°, whereas for a cosine illumination it is 6.8°. Thus, the improved sidelobe performance associated with the cosine illumination is obtained at the expense of some deterioration in beamwidth. It is assumed that a/λ_0 is the same in both cases.

4.4.3 Pattern Gain

A property of a directional antenna that is arguably of primary interest to users, whether in reception or transmission roles, is its gain. In any physically realisable antenna, there will generally be one direction in which the radiated or collected power density is greatest. To give a quantitative value to this optimum operating condition, the power density at this pattern location is compared with some generally accepted reference level. This level is provided by estimating the power density in the same distance and direction produced by an isotropic source or sink delivering (or receiving) the same total power as the original antenna. As we have already noted, an isotropic antenna notionally produces an omni-directional or spherical radiation pattern. So, for a lossless antenna, or for a low loss antenna for which resistive and other losses can be ignored, the ratio of the peak power density delivered or received by the test antenna in the direction of the mainbeam to that produced by an equivalent isotropic antenna is termed the directive gain or directivity (D). On the other hand, when the antenna losses are not ignored, the resultant gain (G) is inevitably lower, and it is usually termed the power gain. Directivity and gain are related through the relation:

$$G = \eta D \tag{4.24}$$

where η (the efficiency) is a measure of the total power transmitted by the given antenna divided by the total power delivered to it, or in reception terms, the total power delivered to the receiver divided by the total power received by the antenna. Directivity and gain will be examined further in Sect. 4.5.

4.5 Rectangular Aperture

While the examination of linear apertures has the advantage of simplifying the basic analysis process associated with the evolution of radiation patterns and with identifying the nature and properties of these patterns, we still need tools to investigate practical aperture antennas. These real antennas are of course two-dimensional, and hence, they generate three-dimensional radiation patterns. Here, it will be demonstrated that two-dimensional apertures can be modelled by revisiting Eq. (4.10) and extending its applicability to two dimensions.

In the computational modelling of two-dimensional aperture problems, the main difficulty lies in performing the integrations embedded in Eq. (4.10). However, if interest is limited to the far field, which is the normal scenario, the complexity of the integrations can be significantly reduced. It is helpful at this stage to make some simplifying definitions. Firstly in the far field, the following approximations can be made with little impact on accuracy:

$$R \approx |\mathbf{r}| = r \text{ (for magnitude changes)} \tag{4.25}$$

$$R = r + r' \cos \xi \text{(for phase changes)} \tag{4.26}$$

here, ξ denotes the angle between the vectors $\mathbf{r}$ and $\mathbf{r'}$ which are depicted in Fig. 4.8.

For far-field computations, $\xi \approx 0$ in which case $\mathbf{R}$ and $\mathbf{r}$ are essentially parallel. Hence, applying Eq. (4.10) to a radiating aperture which can be represented by a surface S, we obtain

$$\mathbf{A}_m = \frac{\varepsilon_0}{4\pi} \iint\limits_S \mathbf{J_m} \frac{\exp(-jk_0 R)}{R} \mathrm{d}S' \tag{4.27}$$

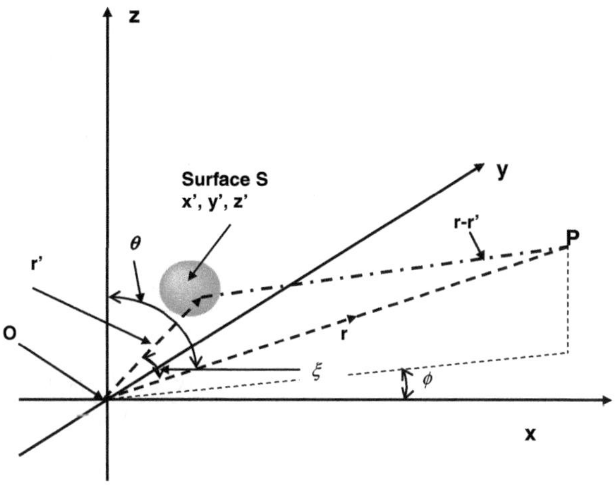

Fig. 4.8 Geometry for two-dimensional aperture antenna analysis

If we now incorporate Eqs. (4.25) and (4.26) into Eq. (4.27), $\mathbf{A_m}$ can be written as:

$$\mathbf{A_m} \approx \frac{\varepsilon_0 \exp(-jk_0 r)}{4\pi r} \iint_S \mathbf{J_m} \exp(+jk_0 r' \cos \xi) \mathrm{d}S' \tag{4.28}$$

In the polar (r, θ, ϕ) coordinate system, the vector $\mathbf{A_m}$ has two components in the far field, namely

$$\mathbf{A_m} = \mathbf{a}_\theta A_{m\theta} + \mathbf{a}_\phi A_{m\phi} \tag{4.29}$$

where $\mathbf{a}_\theta$ and $\mathbf{a}_\phi$ are unit vectors. There is no A_{mr} since in the far field, the radiation is transverse electromagnetic in character. At the position of the aperture surface at r' (x', y', z'), the vector magnetic current can be expressed as:

$$\mathbf{J_m} = \mathbf{a_x} J_{mx} + \mathbf{a_y} J_{my} + \mathbf{a_z} J_{mz} \tag{4.30}$$

Hence, combining Eqs. (4.28), (4.29), and (4.30) yields the following accessible forms:

$$A_{m\theta} = \frac{\varepsilon_0 \exp(-jk_0 r)}{4\pi} \iint_S \left[J_{mx} \cos\theta \cos\phi + J_{my} \cos\theta \sin\phi - J_{mz} \sin\theta\right] \exp(+jk_0 r' \cos \xi) \mathrm{d}S' \tag{4.31}$$

$$A_{m\phi} = \frac{\varepsilon_0 \exp(-jk_0 r)}{4\pi} \iint_S -\left[J_{mx} \sin\phi + J_{my} \cos\phi\right] \exp(+jk_0 r' \cos \xi) \mathrm{d}S' \tag{4.32}$$

The electric and magnetic fields in the far field can be generated using the appropriate equations in Table 4.1—noting that the divergence of $\mathbf{A_m}$ is zero in the far field.

Hence, we obtain, finally, the far-field solution for the general two-dimensional aperture antenna:

$$E_r \approx 0 \tag{4.33}$$

$$E_\theta \approx -\frac{jk_0 \exp(-jk_0 r)}{4\pi r} A_{m\phi} \tag{4.34}$$

$$E_\phi \approx +\frac{jk_0 \exp(-jk_0 r)}{4\pi r} A_{m\theta} \tag{4.35}$$

$$H_r \approx 0 \tag{4.36}$$

$$H_\theta \approx \frac{jk_0 \exp(-jk_0 r)}{4\pi r} \left[-\frac{A_{m\theta}}{\eta_0}\right] \tag{4.37}$$

$$H_\phi \approx -\frac{jk_0 \exp(-jk_0 r)}{4\pi r} \left[\frac{A_{m\phi}}{\eta_0}\right] \tag{4.38}$$

4.5.1 Uniformly Illuminated Rectangular Aperture

It is useful to establish the effectiveness of these equations by examining a standard example which can be found in Balanis, Chap. 12. It comprises a rectangular aperture of width a and height b located within an infinite perfectly conducting ground plane (see Fig. 4.9). The electric field in the aperture is assumed to be constant in both the x- and y-directions and to be y-polarised. This is not a practical field distribution, but it serves the purpose of the demonstration with minimal complexity. This field distribution can be expressed as:

$$\mathbf{E_a} = \mathbf{a_y} E_0 \tag{4.39}$$

existing in the range $-a/2 \le x' \le a/2$ and $-b/2 \le y' \le b/2$. E_0 is a constant in this case.

The equivalent solvable geometry comprises a 'closed' half-space bounded by a flat conducting screen extending over the entire x–y plane at $z = 0$. Hence, applying the boundary conditions arising from Fig. 4.3, we obtain:

$$\mathbf{J_{ms}} = -2\hat{\mathbf{n}} \times \hat{\mathbf{a}}_\mathbf{y} E_a = 2\hat{\mathbf{a}}_\mathbf{x} E_0 \quad \begin{cases} -a/2 \le x' \le a/2 \\ -b/2 \le y' \le b/2 \end{cases} \tag{4.40a}$$

$$\mathbf{J_{ms}} = 0 \quad \{\text{everywhere outside aperture} \tag{4.40b}$$

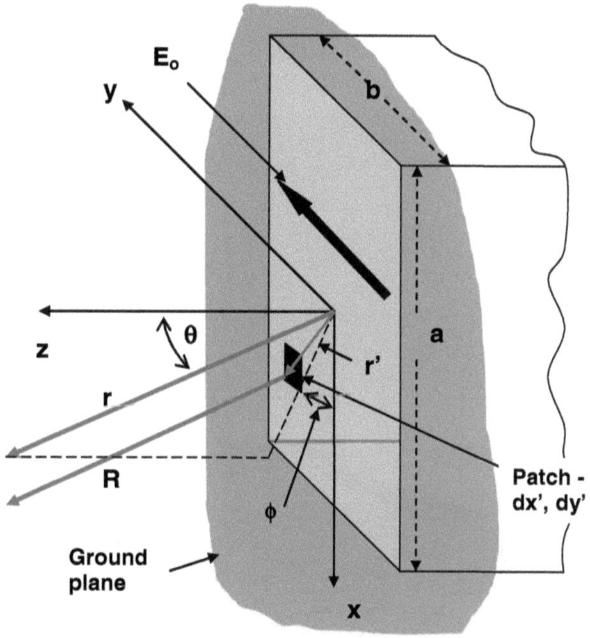

Fig. 4.9 *Rectangular aperture antenna* embedded in a conducting ground plane of infinite extent

The far fields radiated by this magnetic current distribution located in a rectangular aperture can be found by inserting definitions (4.40a) and (4.40b) into Eqs. (4.31) and (4.32) to evolve the potential function components and, hence, into Eqs. (4.33–4.38) to generate the radiated fields. The result is:

$$E_r = 0 \tag{4.41}$$

$$E_\theta = \frac{jabk_0 E_0 \exp(-jk_0 r)}{2\pi r}\left\{\sin\phi\left(\frac{\sin X}{X}\right)\left(\frac{\sin Y}{Y}\right)\right\} \tag{4.42}$$

$$E_\phi = \frac{jabk_0 E_0 \exp(-jk_0 r)}{2\pi r}\left\{\cos\theta\cos\phi\left(\frac{\sin X}{X}\right)\left(\frac{\sin Y}{Y}\right)\right\} \tag{4.43}$$

$$H_r = 0 \tag{4.44}$$

$$H_\theta = -\frac{E_\phi}{\eta_0} \tag{4.45}$$

$$H_\phi = -\frac{E_\theta}{\eta_0} \tag{4.46}$$

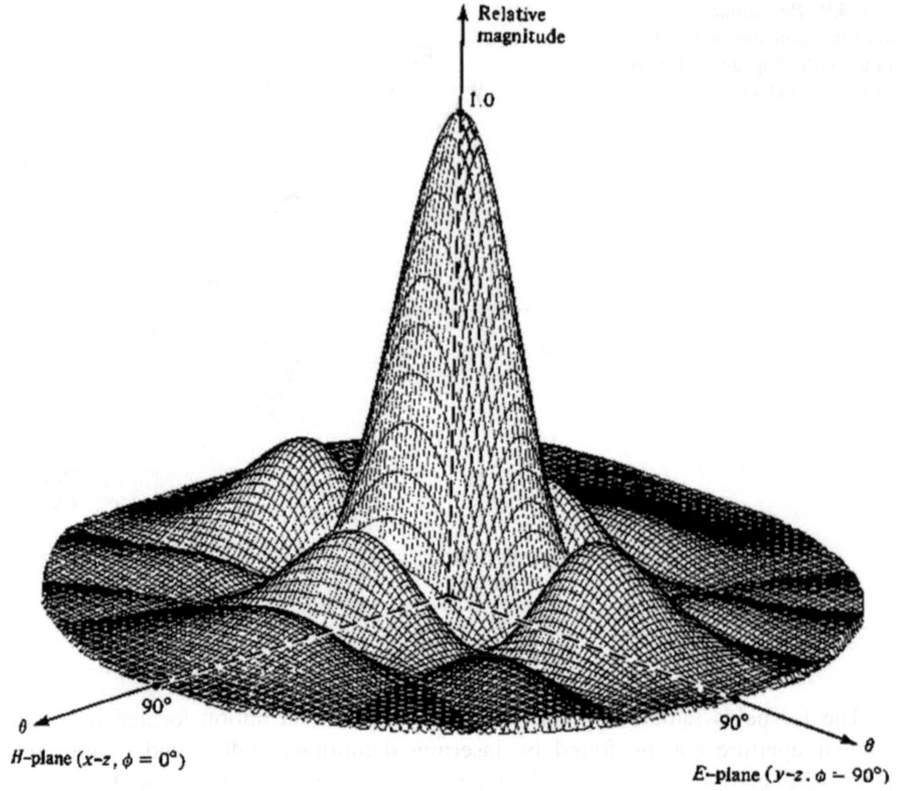

Fig. 4.10 Three-dimensional E-field pattern for a *rectangular aperture* in an infinite conducting plane—constant field case (*Courtesy* of John Wiley & Sons Inc.)

X and *Y* in these equations are given by:

$$X = \frac{k_0 a}{2} \sin\theta \cos\phi \quad \text{and } Y = \frac{k_0 b}{2} \sin\theta \sin\phi$$

The normalised E-field distributions represented by the above equations are plotted using standard mathematical software in Fig. 4.10 for a rectangular aperture with $a = 3\lambda_0$ and $b = 2\lambda_o$, and in Fig. 4.11 for a square aperture with $a = b = 3\lambda_0$.

Figures 4.10 and 4.11 provide a good pictorial representation of the penetration of the aperture radiation pattern into the forward half-space of the antenna, provided it is fully realised that the depicted lobe surfaces are actually constant magnitude power density contours. They could be viewed as 'isodensity' surfaces. Lower power density contours would push much further out into the half-space, while higher power densities produce surfaces closer to the antenna. Actually, in practice, this form of presentation is seldom used, particularly in experimental scenarios, because the volume of measurement involved in generating such images is much

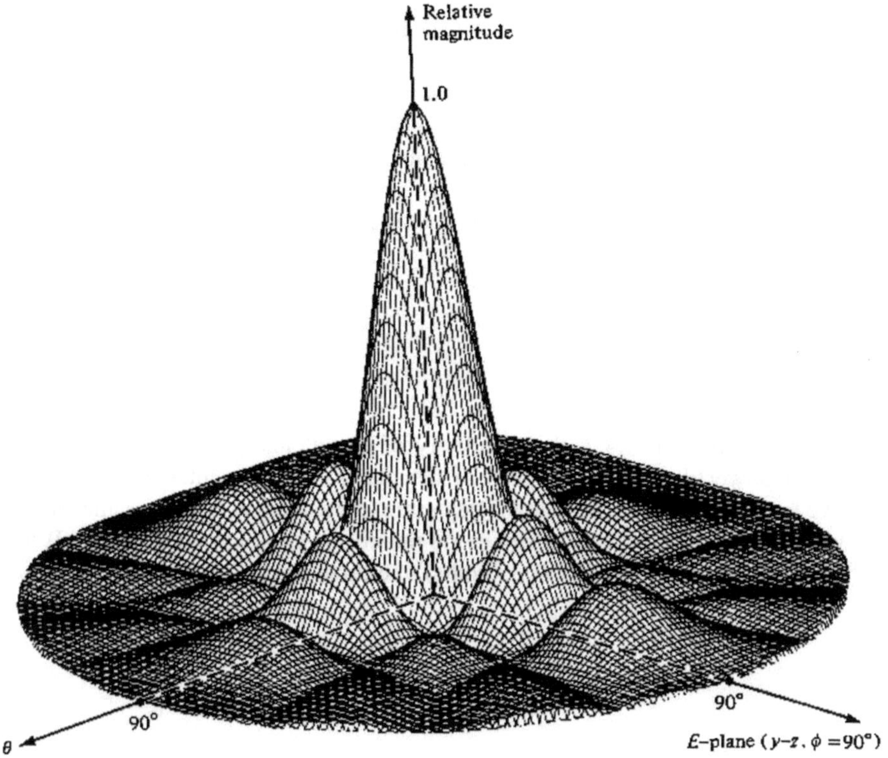

Fig. 4.11 Three-dimensional E-field pattern for a *square aperture* in an infinite conducting plane —constant field case (*Courtesy* of John Wiley & Sons Inc.)

too high to be cost-effective. So, as we have intimated earlier, the convention is to present antenna patterns as two primary 'cuts'—generally the *E*-plane and the *H*-plane (see Fig. 4.10). As the figure demonstrates, the *E*-plane for the aperture is the *x–z* plane or the $\phi = \pi/2$ plane. The E-field distribution on this plane is therefore:

$$E_r = E_\phi = 0 \tag{4.47}$$

$$E_\theta = \frac{jabk_0E_0\exp(-jk_0r)}{2\pi r}\left[\frac{\sin\left(\frac{k_0b}{2}\sin\theta\right)}{\frac{k_0b}{2}\sin\theta}\right] \tag{4.48}$$

On the *H*-plane (*y–z* plane or $\phi = 0$ plane), the E-field has the form:

$$E_r = E_\theta = 0 \tag{4.49}$$

$$E_\phi = \frac{jabk_0E_0\exp(-jk_0r)}{2\pi r}\cos\theta\left[\frac{\sin\left(\frac{k_0a}{2}\sin\theta\right)}{\frac{k_0a}{2}\sin\theta}\right] \tag{4.50}$$

Interestingly, the term inside the square brackets in Eqs. (4.48) and (4.50) has reduced to the spectrum functions (see Table 4.2) of the uniformly illuminated linear apertures of length b and of length a, respectively. The important conclusion here is that the radiation patterns of two-dimensional apertures can usually be adequately represented on the two major planes, or 'cuts', by recruiting the spectrum functions. These functions generally provide good first approximations to mainlobe beamwidths, the locations of the sidelobe zeros and maxima, as well as the magnitudes of the sidelobes.

4.5.2 Directivity

As discussed in Sect. 4.4.2, the directivity of an antenna gives a measure of the enhanced power density at the peak of the mainlobe of an antenna of finite size relative to that of an isotropic antenna delivering the same power at the same far-field radius. If the antenna power density at the mainbeam peak at radius r is p_a (W/m^2) and the total power radiated is P_{rad} (Watts), the directivity of the antenna is defined as:

$$D = \frac{p_a}{P_{\text{rad}}/4\pi r^2} = \frac{4\pi r^2 p_a}{P_{\text{rad}}} \tag{4.51}$$

For an aperture antenna, as described above, if the electric and magnetic fields in the far field have already been determined, as detailed in the previous section, then the radial power density flow at any angle in the forward half-space is given by:

$$U(\theta, \phi) = \frac{1}{2}\text{Re}\left[(\hat{\mathbf{a}}_\theta E_\theta + \hat{\mathbf{a}}_\phi E_\phi) \times (\hat{\mathbf{a}}_\theta H_\theta + \hat{\mathbf{a}}_\phi H_\phi)^*\right] = \frac{1}{2\eta_0}\left(|E_\theta|^2 + |E_\phi|^2\right) \tag{4.52}$$

If at the 'nose' of the mainlobe, $E_\theta = E_\theta^0$ and $E_\phi = E_\phi^0$, then the power flow density there we can computed from

$$p_a = \frac{1}{2\eta_0}\left(\left|E_\theta^0\right|^2 + \left|E_\phi^0\right|^2\right) \text{W/m}^2 \tag{4.53}$$

The total power P_{rad} is more difficult to evaluate. It involves a surface integration of the general power density term $U(\theta, \phi)$ over a hemisphere in the forward half-space of the radiating aperture (see for example Balanis).

However, since the chosen aperture model is embedded in a plane conducting surface of infinite extent, a simpler procedure exists to compute the total radiated power. If the electric field ($E_0(x',y')$) in the aperture is available, which it usually is for known excitation methods, the power density in the aperture can be calculated.

Thence, the total power entering the forward half-space, namely P_{rad}, can be computed by integrating the power density over the aperture area. That is:

$$P_{\mathrm{rad}} = \frac{1}{2\eta_0} \iint_S |E_0(x', y')|^2 \mathrm{d}x' \mathrm{d}y' \tag{4.54}$$

where S is the aperture surface. For the uniformly illuminated rectangular aperture in Fig. 4.9, this gives:

$$P_{\mathrm{rad}} = \mathrm{ab} \frac{|E_0|^2}{2\eta_0} \mathrm{W} \tag{4.55}$$

Also, for this aperture example, Eq. (4.53) in combination with Eqs. (4.48) and (4.50), evaluated at $\theta = 0$, yields:

$$p_a = \left(\frac{\mathrm{ab}}{\lambda} \right)^2 \frac{|E_0|^2}{2\eta_0 r^2} \mathrm{W/m^2} \tag{4.56}$$

Thus, for this particular case, we obtain:

$$D = \frac{4\pi p_a}{P_{\mathrm{rad}}} = \frac{4\pi \mathrm{ab}}{\lambda_0^2} = \frac{4\pi A_{\mathrm{eff}}}{\lambda_0^2} \tag{4.57}$$

where A_{eff} is the effective area of the aperture. It is worth stressing that, Eq. (4.57) is actually a very useful relationship for gauging the directivity (and hence gain) of any aperture antenna, not just for the cited example.

References

1. Collin RE, Zucker FJ (1969) Antenna theory. McGraw-Hill Book Co., New York
2. Harrington RF (1961) time harmonic electromagnetic fields. McGraw-Hill Book Co., New York
3. Balanis CA (1997) Antenna theory. Wiley, New York

Chapter 5
Array Antennas for Solar Systems

Engineering... to define rudely but not inaptly, is the art of doing that well with one dollar, which any bungler can do with two after a fashion.

Arthur Mellen Wellington

Abstract Antenna arrays and the devices, components and techniques, which have evolved to implement them, form a very large subdivision of the antenna literature. In Chap. 5, we provide merely a summary of those aspects of the technology that are appropriate to solar power generation and in particular to satellite-based systems as discussed in Chap. 9. Fortunately, the essence of array behaviour can be deduced by examining linear arrays of isotropic radiators, since planar arrays, which are more difficult to analyse directly, can largely be viewed as combinations of linear arrays. The majority of the chapter is engaged in developing and using well-established polynomial techniques which now underpin array analysis and synthesis procedures. Array radiation patterns and their dependence on element numbers and element spacing are discussed mainly with reference to pattern directivity, pattern gain, beamwidth, sidelobes and grating lobes. The optimising of element spacings to maximise gain with the fewest number of possible elements is considered as a precursor to a brief consideration of synthesising arrays to procure specified sidelobe behaviours. Of course, commercial array simulation and design software are readily available to do this on the Internet. Such modelling tools generally employ some form of synthesis procedure, of which there are many as is outline in this chapter.

5.1 Introduction

A cursory glance at any picture of a solar farm makes it clear that solar power collection systems employ array structures. For example, they can be seen in solar voltaic panels (Chap. 7), in concentrated solar power farms (Chap. 8) and in antenna systems (Chap. 10). However, these arrays differ fundamentally from the conventional antenna arrays employed in radar and communications, in that 'signal' addition is performed at the level of power where phase is irrelevant, whereas in conventional interference-dependent arrays, as we shall see, signal combining is

A.J. Sangster, *Electromagnetic Foundations of Solar Radiation Collection*,
Green Energy and Technology, DOI 10.1007/978-3-319-08512-8_5

performed at the voltage or current level whereupon phase becomes critical. This distinction will be explored further in Chap. 10.

So why do we need to examine conventional array behaviour? The primary reason here, in the context of solar power collection, is to provide a platform for understanding the potential evolution of systems on space platforms (Chap. 9). Such systems require the conversion of the power from collected sunlight into microwave power so that it can then be beamed to the earth's surface with little attenuation. The success, or otherwise, of this concept is highly dependent on creating microwave beams which automatically track and stay focused on the receiving antennas on the ground. Such tracking and target acquisition can only be done safely and reliably by resorting to phased array technology for the satellite mounted microwave antenna.

At microwave frequencies, array antennas generally comprise a linear or planar arrangement of periodically spaced aperture antennas. These may be of the slot, open-ended waveguide, or horn type, as described in Chap. 4. To simulate the electromagnetic wave nature of such periodic distributions of radiators, it is usual, firstly, to model an equivalent periodic array of isotropic radiators, and secondly, to employ the principle of pattern multiplication to determine the radiation pattern of the original array. As the name implies, this principle suggests that if the far-field pattern of the isotropic array is multiplied with that of an individual, or typical, aperture antenna which forms an element of the actual array, the resultant pattern will be that of the original array. However, care must be exercised in following this procedure. Unless the element pattern has been obtained in the array environment, this process ignores element-to-element interaction (mutual coupling) and the resultant pattern becomes an approximation to the required pattern. Nevertheless, the technique is a powerful one and is widely used to assess the properties of potential array designs.

Array antennas will also be examined using the now conventional z-transform method, since it gives access to the study of elementary array synthesis procedures. Although the z-transform method again neglects the effects of mutual coupling between closely spaced neighbouring radiating elements of the target array, for many microwave antennas of the array type, the method remains tolerably effective because mutual coupling is often too small to cause significant deviation from the modelled result.

5.2 Uniform Linear Array of Isotropic Elements

A linear array is one that is formed from a single line of multiple radiating sources. When these individual sources radiate equally in all directions, the array is said to comprise isotropic elements. In addition, the array is described as uniform when the elements are equally spaced and fed with equal signal levels. Such an array is depicted in Fig. 5.1. Although planar array (an array of linear arrays) antennas are generally used in solar collection systems, it is advantageous to study the radiation

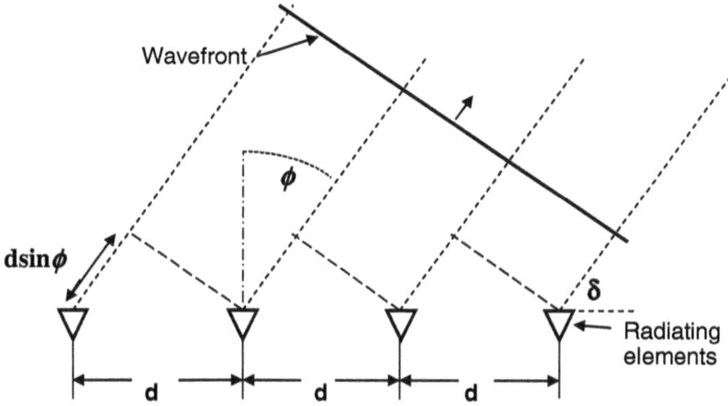

Fig. 5.1 Geometrical relationships for linear array with element spacing *d*

patterns of linear arrays which are mathematically less complex. Nevertheless, the knowledge gained remains entirely relevant to, and representative of, planar arrays.

At a distant point from the antenna, the total *E*-field due to *n* elements can, by superposition, be expressed as:

$$E_T = E_0\left(1 + e^{j\psi} + e^{2j\psi} + \cdots + e^{j(n-1)\psi}\right) \tag{5.1}$$

where E_0 is the field magnitude at the distant point due to a single element, and very importantly:

$$\psi = k_0 d \sin\phi + \alpha \tag{5.2}$$

Here, d is the element-to-element spacing, k_0 is the free-space phase coefficient, and α is the progressive element-to-element phase shift impressed on the array by the feed structure. If we normalise Eq. (5.1) to the element field E_0, we obtain:

$$\frac{E_T}{E_0} = 1 + e^{j\psi} + e^{2j\psi} + \cdots + e^{j(n-1)\psi} \tag{5.3}$$

By replacing the exponential terms by the variable $z(=e^{j\psi})$, the far-field pattern of the linear array can be expressed mathematically as a polynomial, as follows:

$$\varsigma(z) = 1 + z + z^2 + z^3 + \cdots + z^{n-1} \tag{5.4}$$

This polynomial form has been shown [1] to be a powerful mathematical tool in the analysis and synthesis of array antennas. However, before advancing to the treatment of arrays in general, it is perhaps instructive to consider some special cases.

When $\zeta(z)$ is multiplied by z in Eq. (5.4), we obtain:

$$z\varsigma(z) = z + z^2 + z^3 + \cdots + z^n \tag{5.5}$$

Hence, subtracting Eq. (5.4) from Eq. (5.5) gives:

$$(z - 1)\varsigma(z) = z^n - 1$$

giving

$$\varsigma(z) = \frac{z^n - 1}{z - 1} \tag{5.6}$$

With a little manipulation, Eq. (5.6) can be written in the form:

$$\varsigma(z) = \frac{z^{n/2}\left(z^{n/2} - z^{-n/2}\right)}{z^{1/2}\left(z^{1/2} - z^{-1/2}\right)} \tag{5.7}$$

If this equation is recast in its exponential form, with $z = e^{j\psi}$, while recognising the identity:

$$\frac{e^{j\theta} - e^{-j\theta}}{2} = \sin\theta$$

then, we discover that:

$$E_T = \frac{E_0 \sin\frac{n\psi}{2}}{\sin\frac{\psi}{2}} \tag{5.8}$$

Element phase is referred to the array mid-point to secure this simple form. Functionally, Eq. (5.8) is not too dissimilar to Eq. (4.23), the spectrum function for a uniformly illuminated linear aperture. The primary difference is the oscillatory nature of the denominator of Eq. (5.8) which strongly influences the pattern distribution. The result is that the linear array pattern and the linear aperture pattern are comparable only for a limited range of ψ values, and hence element spacings d, as illustrated in Sect. 5.2.1.

5.2.1 Radiation Patterns

Equation (5.8) represents the functional form of the far-field radiation pattern, or polar diagram, of the uniformly excited linear array of isotropic elements. Since the radiation in the forward half-plane $-\pi/2 < \phi < \pi/2$ is clearly identical to the radiation into the reverse half-plane $\pi/2 < \phi < 3\pi/2$, we need consider only the former in the following discussion.

The ratio of sines in Eq. (5.8) implies that the pattern must be multi-lobed. The primary or main lobe occurs, where simultaneously $\psi \to 0$ and $\sin(\psi/2) \to \psi/2$.

Fig. 5.2 Principal lobe—broadside radiation

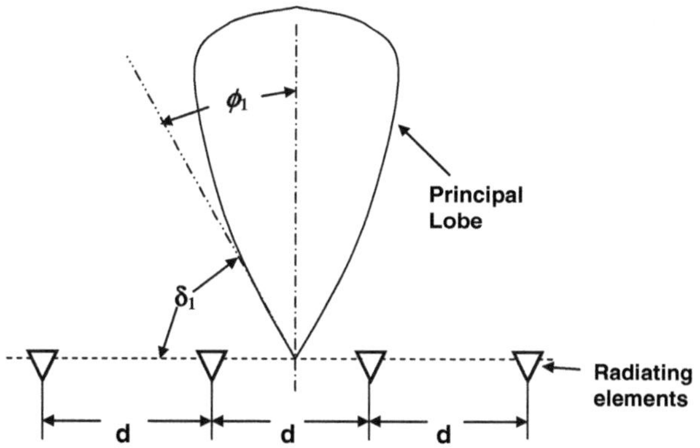

For this lobe, $E_T/E_0 = n$, that is, at the peak of the primary lobe, the E-field at a given radius for the array is n times the E-field at the same radius for a single isotropic element. However, note that if the element spacing d is equal to or greater than λ_0, there may be more than one principal maximum—these secondary maxima are termed grating lobes and appear where $\psi = 0, \pi, 2\pi, \ldots$. It is apparent from Eq. (5.2) that the principal maximum at $\psi = 0$ occurs where:

$$\sin \phi = -\frac{\alpha}{k_0 d} \tag{5.9}$$

When this principal maximum appears in a direction normal to the line of the array, i.e. at $\phi = 0°$, the array is termed a broadside array (see Fig. 5.2). Clearly, for this case, $\alpha = 0°$, which means that the element-to-element phase shift is set to 0°. In the diagram, ϕ_1 denotes the angle to the first null measured from the array normal, while δ_1 provides essentially the same information, but measured from the array baseline—$\delta_1 = 90° - \phi_1$.

When $\phi = 90°$, the principal maximum lies along the line of the array (Fig. 5.3) and the array, in this case, is termed end-fire. The angle from the lobe maximum to the first null is denoted by δ_2 in the figure. From Eq. (5.9), it is clear that for an end-fire array, we require $\alpha = -k_0 d$.

5.2.2 *Broadside Array*

For a broadside array of equally space and equally excited radiators (uniform array), the nulls of the radiation pattern occur where the sine ratio in Eq. (5.8) exhibits zero values, i.e. where:

$$\frac{n\psi}{2} = \pm m\pi \tag{5.10}$$

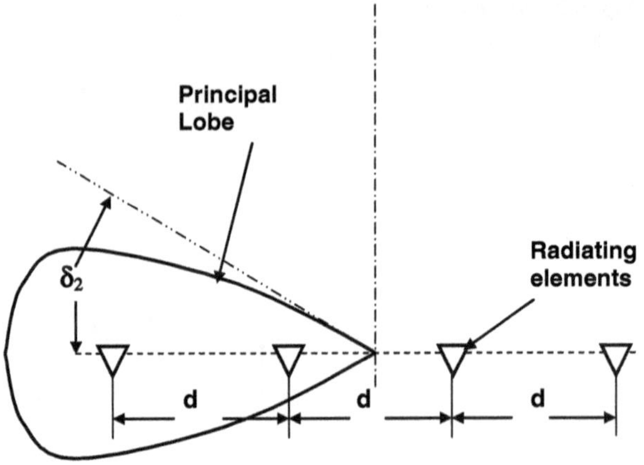

Fig. 5.3 Principal lobe—end-fire array

Hence, substituting for ψ using Eq. (5.2) and noting that for a broadside array, $\alpha = 0$,

$$\frac{nk_0 d \sin\phi}{2} = \pm m\pi \tag{5.11}$$

which implies that:

$$\sin\phi = \pm\frac{2m\pi}{nk_0 d} \tag{5.12}$$

Therefore, the nulls in the pattern appear at angles given by:

$$\phi_m = \sin^{-1}\left(\frac{m\lambda_0}{nd}\right) \tag{5.13}$$

If the array is large, the most likely scenario in solar collection applications, such that $nd \gg m\lambda_0$, then ϕ_m is small for the first few nulls nearest to the main lobe and is given by:

$$\phi_m = \frac{m\lambda_0}{nd} \approx \frac{m}{L/\lambda_0} \tag{5.14}$$

In this relationship, L ($\approx nd$) is the total length of the array. Strictly, $L = (n-1)d$, but the error is small for long arrays. The first nulls which define the principal lobe occur where $m = 1$, and for this broadside case, the beamwidth between first nulls (BWFN) is given by:

$$\text{BWFN} = 2\phi_1 \approx \frac{2}{L/\lambda_0}\ \text{radians} = \frac{114.6^\circ}{L/\lambda_0} \tag{5.15}$$

For reasons intimated earlier, it is much more usual for antenna engineers to employ half-power beamwidth (HPBW) in defining the radiated main beam and for the broadside uniform array, it is not difficult to show that

$$\text{HPBW} = 0.44 \times \text{BWFN} \tag{5.16}$$

So, for this array:

$$\text{HPBW} \approx \frac{2 \times 0.44}{L/\lambda_0} = \frac{50.4^\circ}{L/\lambda_0} \tag{5.17}$$

For this uniform broadside array of isotropic elements with $d < \lambda_0/2$, the radiated power density in the principal lobe can be approximated by:

$$p_{\text{array}} \approx \frac{P_T}{2\pi r^2 \theta_{3\text{dB}}} \ \text{W/m}^2 \tag{5.18}$$

where P_T denotes power transmitted and r is a representative far-field distance from the array. For a single isotropic radiator delivering the same power P_T, the power density at r is:

$$p_{\text{iso}} = \frac{P_T}{4\pi r^2} \ \text{W/m}^2 \tag{5.19}$$

Hence, the directivity (D) of the array has the simple form:

$$D = \frac{p_{\text{array}}}{p_{\text{iso}}} \approx \frac{2}{\theta_{3\text{dB}}} \tag{5.20}$$

In the initial process of assessing array requirements in any given application, simple formulae such as Eqs. (5.17) and (5.20) are very useful for acquiring a quick appreciation of the trade-off between the beamwidth desired and antenna size.

5.2.3 End-Fire Array

The only difference between a uniform end-fire array and a uniform broadside array is in the value of the progressive phase shift α, so the form of Eq. (5.8) is unaltered by the change. Consequently, the condition for pattern nulls remains as in Eq. (5.10). However, in the end-fire case, with $\alpha = -k_0 d$, the equation relating ψ and ϕ becomes:

$$\psi = k_0 d \ \sin\phi - k_0 d \tag{5.21}$$

This can be written in a more convenient form if we note that $\delta = 90° - \phi$, namely

$$\psi = k_0 d \cos\delta - k_0 d \tag{5.22}$$

Hence, combining Eq. (5.22) with Eq. (5.10), we get:

$$\frac{n}{2} k_0 d(\cos\delta - 1) = \pm m\pi$$

or

$$\cos\delta - 1 = \pm\frac{2m\pi}{nk_0 d} = \pm\frac{m\lambda_0}{nd} \tag{5.23}$$

Thus, in terms of δ, the null angles are located at:

$$\frac{\delta_m}{2} = \sin^{-1}\left(\pm\sqrt{\frac{m\lambda_0}{2nd}}\right) \tag{5.24}$$

Once again, directing our attention to a long array with $nd \gg m\lambda_0$, the nearest pattern nulls occur at angles given by:

$$\frac{\delta_m}{2} \approx \sqrt{\frac{m\lambda_0}{2nd}} = \sqrt{\frac{m}{2L/\lambda_0}}$$

i.e.

$$\delta_m \approx \sqrt{\frac{2m}{L/\lambda_0}} \tag{5.23}$$

So, finally, we obtain the useful relations:

$$\text{BWFN} = 2\sqrt{\frac{2}{L/\lambda_0}} \text{ radians} = 114.6\sqrt{\frac{2}{L/\lambda_0}} \text{ degrees} \tag{5.24}$$

For the end-fire array, it is not difficult to show that HPBW = 0.66 × BWFN, and therefore, we can write:

$$\text{HPBW} \approx 76.3\sqrt{\frac{2}{L/\lambda_0}} \text{ degrees} \tag{5.25}$$

Note that the half-power beamwidth for the end-fire array is very much larger than for the corresponding broadside array. For example, when the ratio $L/\lambda_0 = 50$, the end-fire array beamwidth is approximately 15 times greater than that of the equivalent broadside format. Some improvements in end-fire gain can be achieved without increasing the number of radiating elements by judicious adjustment of the element spacings d. The optimum spacing is termed the Hansen–Woodyard condition, which will be examined below.

5.2.4 Scanned Array

The above examination and discussion of broadside and end-fire arrays underlines a major property of array antennas, namely their ability to provide electronic beam scanning. The two cases represent the scanning extremes of zero degrees (broadside) and ninety degrees (end-fire) when the progressive phase shift α is set to, respectively, zero and $-k_0d$. At the $\phi = 0°$ ($\delta = 90°$) scan condition, the principal beam exhibits minimum beamwidth and optimum directivity. As the beam is scanned away from broadside towards the end-fire condition, both the beamwidth and the directivity deteriorate, slowly at first, and then increasingly rapidly as end-fire is approached. This behaviour is illustrated graphically in Fig. 5.4 which presents 3-dB beamwidth as a function of array length for a uniform linear array. These beamwidths were generated by forming the following relationship:

$$\frac{\sin\frac{n\psi}{2}}{\sin\frac{\psi}{2}} = 0.707 \tag{5.26}$$

for uniform linear arrays with $d = \lambda_0/2$. The curves in Fig. 5.4 represent varying scan angles (ϕ) and the scan limit (see dashed line) defines the angle to which the principal beam can be scanned unmodified by the proximity of end-fire. For example, a very long array can be scanned much closer to end-fire than a short one. This figure makes it very clear that array length is the major determinant of the principal lobe beamwidth and hence of the array gain.

This phase scanning ability of array antennas is crucial to any notion of using space platforms to collect solar power and thence beam it to Earth using microwave beams (Chap. 9). As we shall see the proposed power density levels in these beams is potentially lethal to life forms on the ground, if such a beam were to stray away from the ground receiving antenna. Automatic and reliable tracking using phase scanning at the space array will be mandatory for these systems to achieve approval by regulatory bodies.

It is pertinent to ask here whether or not, for any given number of elements n, there is an element spacing d and an array length $L = (n - 1)d$ which procures optimum array gain and minimum beamwidth. For a broadside array calculation based on Eq. (5.8), it can be demonstrated that minimum beamwidth and maximum gain (allowing for power losses in sidelobes and grating lobes) occur where the

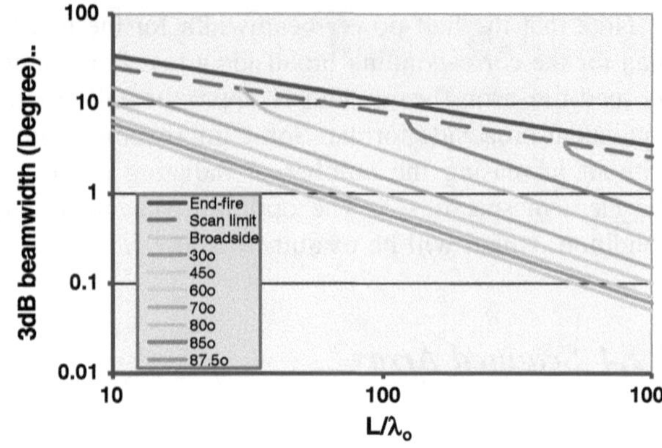

Fig. 5.4 Beamwidth variation with scanning (adapted from [3])

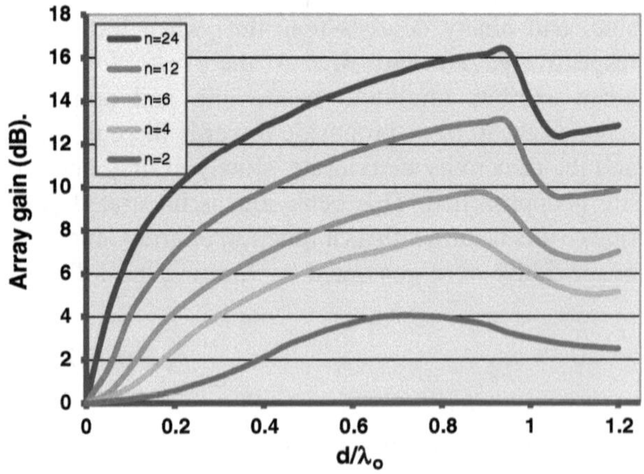

Fig. 5.5 Gain as a function of d/λ_0 for a uniform broadside array (adapted from [4])

element spacing is just less than one wavelength. This is shown clearly in Fig. 5.5 where gain is presented as a function of d/λ_0 for a range of array options with values of n ranging from 2 to 24. The pronounced dip in the value of the gain at $d/\lambda_0 \approx 0.95$ particularly where n is large is produced by the appearance of the first pair of grating lobes in the end-fire directions.

5.3 Array Design Using Theory of Polynomials

In this short section, the intention is to introduce the reader to antenna synthesis, rather than analysis, which we have been directing our attention towards, up to this juncture. The method which will be outlined applies strictly to uniform arrays. Nevertheless, the procedures developed here provide a good general introduction to antenna synthesis concepts without embarking on an overly complex mathematical journey. Actually, in practice, antenna design engineers generally use very sophisticated commercial software to perform the complex operations involved in synthesis, but to do this efficiently, some awareness of the basic process is essential.

It has already been shown (Eq. 5.4) that the far-field pattern of an n-element uniform linear array can be expressed in the polynomial form:

$$\varsigma(z) = 1 + z + z^2 + z^3 + \cdots + z^{n-1}$$

where $z = e^{j\psi}$ and ψ is the difference in phase of waves radiated from any pair of adjacent elements. By a fundamental theory of algebra, a polynomial of order $(n-1)$ has $(n-1)$ zeros (some of which may be multiple zeros) and can be factored into $(n-1)$ binomials [1]. Thus,

$$\varsigma(z) = (z - r_1)(z - r_2)(z - r_3)\cdots(z - r_{n-1}) \quad (5.27)$$

The quantities $r_1, r_2, \ldots, r_{n-1}$ are termed the roots of the $(n-1)$th order polynomial. Since $z = e^{j\psi}$ has magnitude unity, these roots can be represented graphically as located on a unit circle in the complex z-plane. The magnitude of ζ (essentially the far-field pattern) can then be determined by multiplication of the vector magnitudes $(z - r)$ as illustrated in Fig. 5.6, which shows $z = e^{j\psi}$ with ψ measured clockwise from the real axis denoted by Re. The imaginary axis is denoted by Im. Complex vector $r_1 = e^{j\psi_1}$ is also represented on the diagram, and the vector $(z - r_1)$ is just the vector difference between z and r_1.

For arrays formed from isotropic elements, radiation will be symmetrically distributed around the axis of the array, and consequently only radiation angles ϕ in the range $-\pi/2$ to $\pi/2$ need to be considered. From Eq. (5.2), this means that ψ, the argument of z, increases from $-k_0d + \alpha$ to $k_0d + \alpha$ as ϕ varies from $-\pi/2$ to $\pi/2$. Thus, the range of ψ is $2k_0d$. For example, for a broadside array for which $\alpha = 0$, if the element spacing is set at half the free-space wavelength ($d = \lambda_0/2$), then ψ

Fig. 5.6 Unit circle diagram

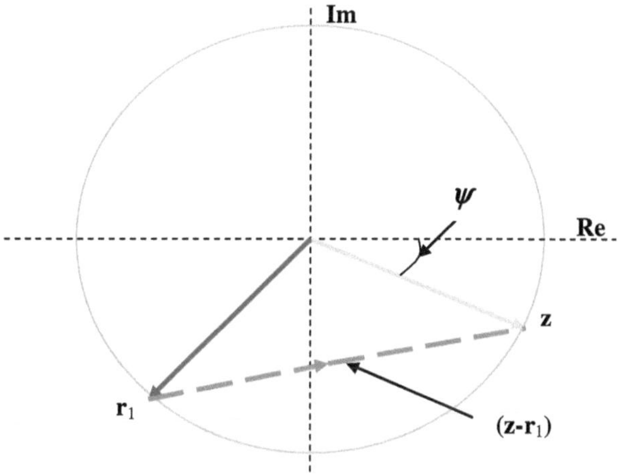

varies from $-\pi$ to $+\pi$, as shown in Fig. 5.7a. Note that z is presumed to travel in the clockwise direction on the unit circle. For $d = \lambda_0/4$ (Fig. 5.7b), the range of ψ is from $-\pi/2$ to $+\pi/2$, while the range is from $+\pi/2$ to $-\pi/2$ when $d = 3\lambda_0/4$, with z traversing the unit circle 1.5 times. Given that main beams occur when $\psi = 0°, 360°, \ldots 2m\pi$, arrays with large element spacings ($d \geq \lambda_0$) generate a grating lobe each time the locus of z passes zero on the circle diagram.

For a uniform linear array antenna, the roots of its polynomial representation (Eq. 5.27) are not restricted as to where on the unit circle they lie and embedded in this observation resides the basis of array synthesis. The roots r_n can be located at

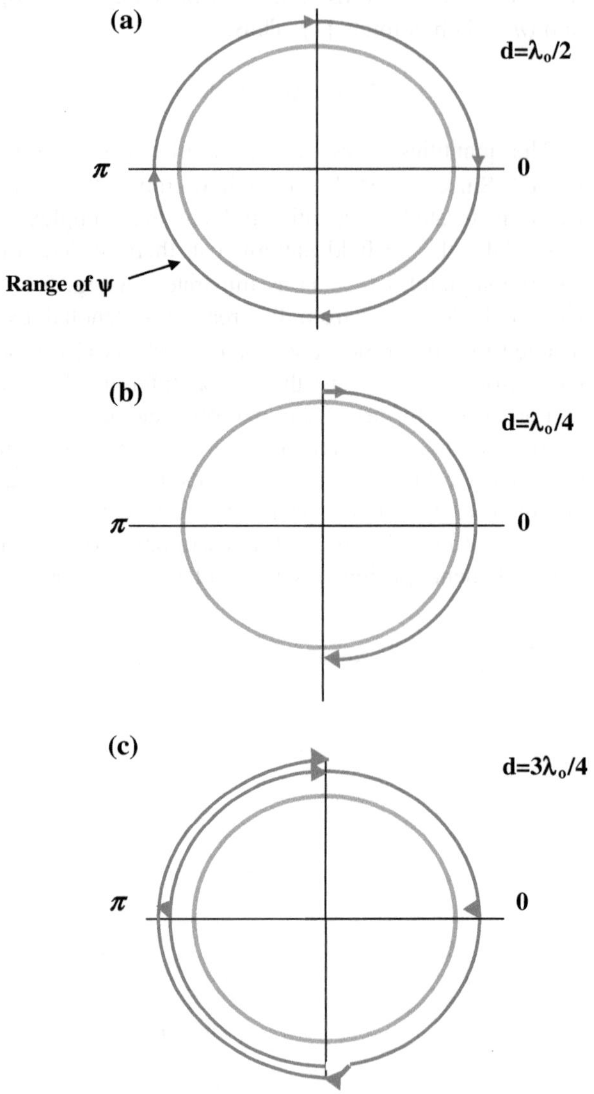

Fig. 5.7 Range of ψ for broadside array ($\alpha = 0$) for varying element separation d: **a** $d = \lambda_0/2$, **b** $d = \lambda_0/4$, **c** $d = 3\lambda_0/4$

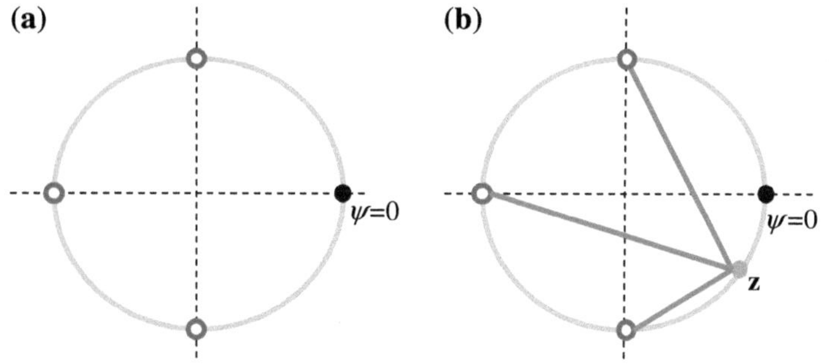

Fig. 5.8 Location of the roots on the circle diagram for a four element uniform array: **a** root positions, **b** showing vectors $(z - r_i)$

any angle ψ except $\psi = 0$ where the primary beam is always located. For the uniform array case, the roots, plus the pole at $\psi = 0$ point (black dot), divide the circle into n equal arcs. This is illustrated in Fig. 5.8a for a four-element array.

The far-field pattern magnitude is obtained by forming the product of the vectors $(z - r_i)$. This can be done graphically by multiplying the lengths of the blue vectors in Fig. 5.8b for all positions of z (red dot) as it moves around the circle. It is not difficult to observe that the principal maximum occurs at $\psi = 0$ where all three vectors exhibit their maximum length. Subsidiary, less pronounced maxima also occur mid-way between the zeros, where the pattern nulls occur. These features can be located in 'real space' (denoted by ϕ) by replacing ψ with ϕ, using Eq. (5.2). For example, the nulls of the four-element array occur at the ψ values $-\pi/2$, $-\pi$, $-3\pi/2$, or more generally at $\psi_m = -2m\pi/n$ with $m = 1, 2, 3$. Therefore, the nulls in the 'real space' radiation pattern occur at angle ϕ_m given by:

$$\sin \phi_m = -\frac{\alpha}{k_0 d} - \frac{2m\pi}{nk_0 d} \tag{5.28}$$

The 'real space' far-field pattern of the four-element array can now be constructed either graphically, or by plotting the function $\zeta(z)$, which has the form:

$$\begin{aligned} \zeta(z) &= (z - \mathrm{e}^{-j\pi/2})(z - \mathrm{e}^{-j\pi})(z - \mathrm{e}^{-j3\pi/2}) \\ &= z^3 + z^2 + z + 1 \end{aligned} \tag{5.29}$$

For $\alpha = 0$ and $d = \lambda_0/2$, an approximation to the resultant pattern as a function of ϕ is shown in Fig. 5.9. This technique for array pattern synthesis is generally attributed to Schelkunoff [1]. The range of ψ, in this case, is 2π.

For an end-fire array with four elements, the root diagram remains unchanged, but with $\alpha = -k_0 d$, this repositions the nulls in 'real space'. If, in addition, we reduce d to $\lambda_0/4$, the range of ψ ranges from $\psi = 0$ to $\psi = \pi$ around the lower half of the circle diagram as shown in Fig. 5.10. The root at $3\pi/2$ is now no longer contributing to the shape of the pattern in real space.

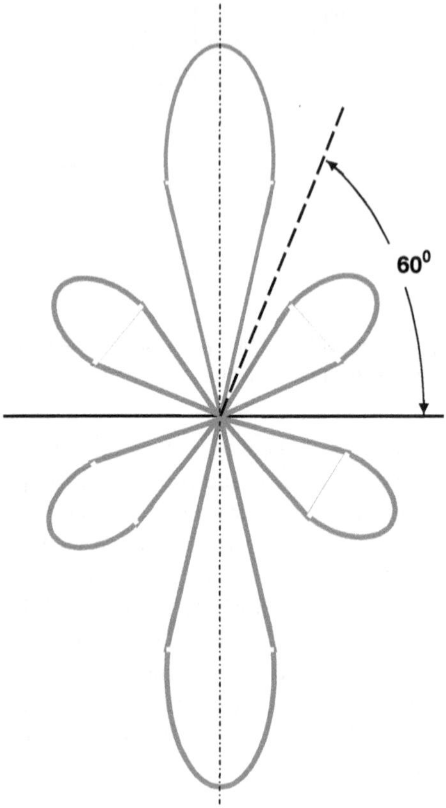

Fig. 5.9 Far-field pattern for a four-element uniform broadside array with $d = \lambda_0/2$

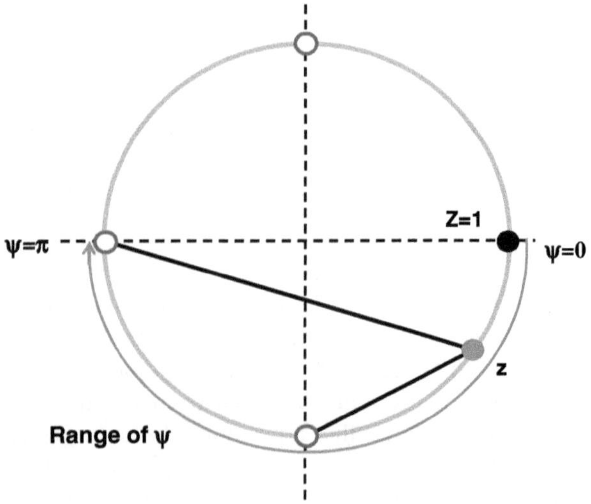

Fig. 5.10 Root location for four-element end-fire array

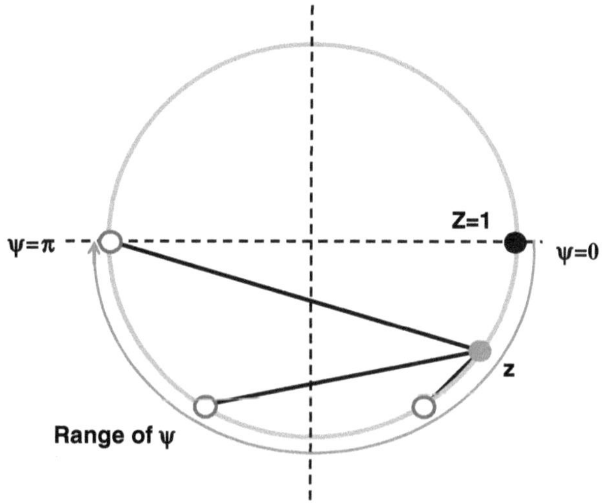

Fig. 5.11 Root locations for four-element end-fire array with improved far-field pattern

The resultant pattern for this case is shown in Fig. 5.12 (curve A). It has a very wide primary lobe and high sidelobes, because of the non-contributing zero. However, if we move all of the roots into the range of ψ as suggested in Fig. 5.11, a much improved pattern emerges for the same number of elements. This arrangement of roots on the circle diagram is generated by a polynomial of the form:

$$\begin{aligned}\varsigma(z) &= \left(z - e^{-j\pi/3}\right)\left(z - e^{-j2\pi/3}\right)\left(z - e^{-j\pi}\right) \\ &= 1 + 2ze^{-j\pi/3} + 2z^2e^{-j2\pi/3} + z^3e^{-j\pi}\end{aligned} \tag{5.30}$$

By making the relation $z_1 = ze^{-j\pi/3}$, Eq. (5.30) can be written in the more convenient form:

$$\varsigma(z_1) = 1 + 2z_1 + 2z_1^2 + z_1^3 \tag{5.31}$$

This represents an array having radiating elements distributed in amplitude as 1:2:2:1 and with a progressive phase shift of

$$\alpha = -\frac{\pi}{2} - \frac{\pi}{3} = -\frac{5\pi}{6} \tag{5.32}$$

This modified end-fire array pattern is sketched in Fig. 5.12 (curve B). Note the improvement in directivity.

If the overall length of the array is fixed but the number of elements it contains is allowed to increase, it becomes possible to improve the directivity still further if the nulls are optimally spaced within the range of ψ on the circle diagram. Curve C in Fig. 5.12 shows the pattern that results when the number of elements is increased to seven by reducing the inter-element spacing to one-eighth of a wavelength, so that

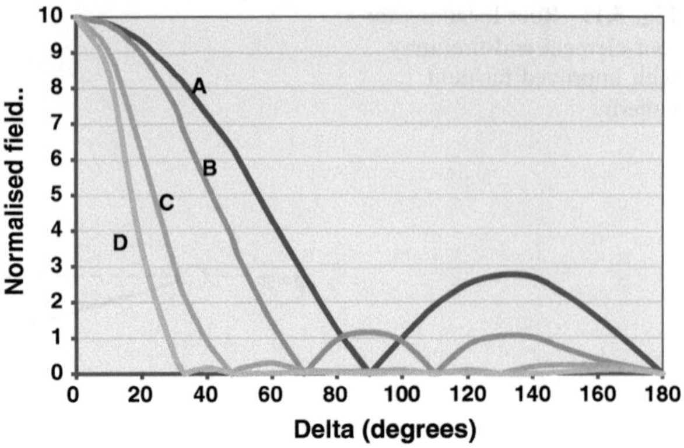

Fig. 5.12 End-fire array patterns: *Curve A (blue)*: *Curve B (red)*: *Curve C (green)*: *Curve D (mauve)* (Color figure online)

the overall array length is still $3\lambda_0/4$. To obtain this result, the nulls are equispaced in the range $\psi = 2k_0d = \pi/2$. Curve D demonstrates the pattern which results when the number of elements is increased even further (13) by reducing the inter-element spacing to $\lambda_0/16$. Again, the nulls are equispaced in the range of ψ. However, these improvements in directivity as will be shown later tends to be achieved at the expense of other important factors of antenna performance.

For the uniform array, as we have seen, the maximum directivity and gain are directly related to the array length. In contrast, the above null adjustment procedures, which impinge on the element excitement levels and phase relations, appear to offer the possibility of arbitrarily improving directivity with an array of fixed length by simply using a sufficiently large number of elements. However, as always in technology, there is a balancing consequence. Closely spaced elements, phased and level controlled to achieve high directivity, present very low impedances to the input feed lines. This means very high currents in the feed lines—or in micro-strip terms, the feed lines become unfeasibly wide. In other words, with 'real' antennas, the practical requirements of efficient delivery of power to each element impose a limit to the directivity improvement which can be extracted from a fixed-length array.

5.3.1 *Optimum Element Spacing*

Having discovered that the location of the roots on the circle diagram for the array polynomial has a major influence on the radiated pattern, it becomes possible to use this graphical technique to determine the optimum element spacing for a uniform array. For the broadside case, the optimum pattern results from including the maximum number of nulls in the range of ψ. Therefore, for a large number of elements n, the maximum number of nulls is active when the range of ψ is 4π as shown in Fig. 5.13a. Since the range of ψ for a broadside array is $2k_0d$, this implies that $4\pi d/\lambda_0 = 4\pi$, and hence, the optimum element spacing $d = \lambda_0$.

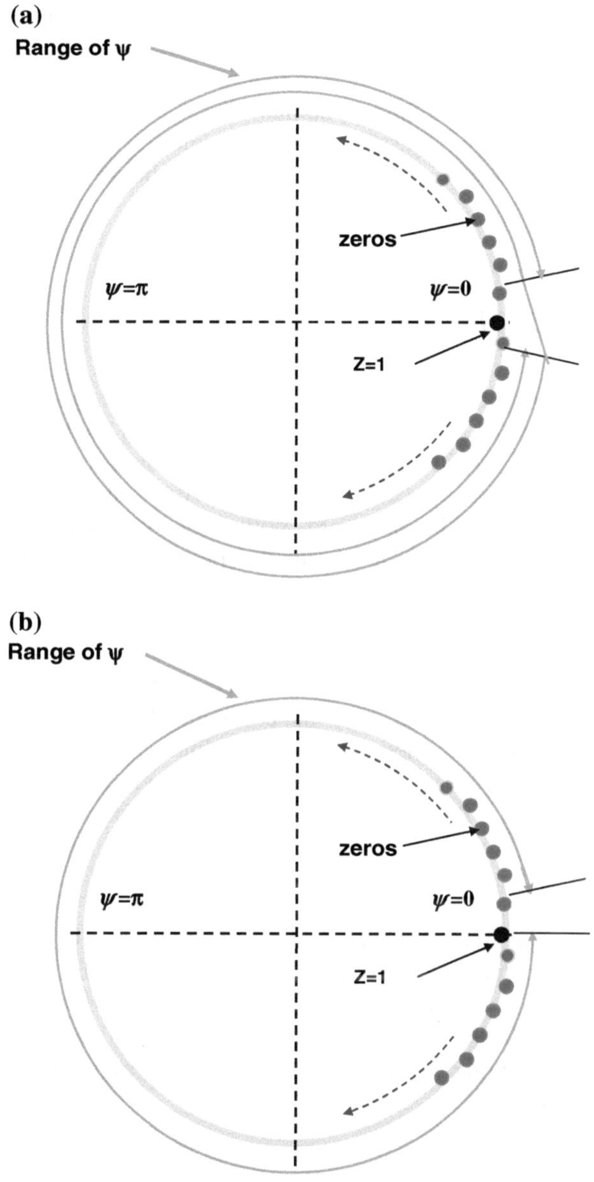

Fig. 5.13 Diagrams showing the range of ψ for optimum element spacing **a** uniform broadside array, **b** uniform end-fire array (the *dotted arrows* imply that the zeros occupy the entire circle)

In the case of the end-fire array, the maximum number of nulls becomes active when the range of ψ approaches 2π, as suggested in Fig. 5.13b, which means that in this case, the optimum element spacing d is almost $\lambda_0/2$. Note that in the limit where $d = \lambda_0/2$, there are two opposed end-fire mainbeams, and the array is bidirectional. Also, for the end-fire array, if it is uniform, the roots are dispersed in an equally spaced format around the unit circle, and so their spacing is $2\pi/n$, except on either side of the pole at $z = 1$. Actually, to further improve directivity, the width of the principal lobe can be reduced further, than suggested in Fig. 5.13b, by simply adjusting the range of ψ so that for $\delta = 0$, the pole lies midway between $\psi = 0$ and the first zero of the polynomial. This is termed the Hansen–Woodyard condition. To achieve this adjustment, an extra inter-element phase shift $-\pi/n$ has to be introduced, or equivalently a phase shift of almost $-\pi$ distributed along the length of the array.

5.3.2 The Binomial Array

A space-based microwave array antenna designed to transfer solar power to the surface of the planet by means of a very high power density microwave beam (Chap. 9) will very likely have to demonstrate exceptionally low sidelobe performance to meet health and safety criteria and to ensure that the system does not present a radiation hazard. With fully adjustable arrays, in both element phase and in element power level, detailed radiation pattern control, encompassing even far-out sidelobes, is certainly feasible. To emphasise this capability, here, we will consider the formation of radiation patterns with no sidelobes.

If the roots of the polynomial $\zeta(z)$ are all equal and can be co-located at the π position on the unit circle, as suggest in Fig. 5.14, then the resultant pattern has only one null or zero at $\psi = \pi$ and consequently should possess no sidelobes. So does this root location scheme lead to a sensible polynomial which can be transformed into a realisable array? For the end-fire array with the unit circle representation shown in Fig. 5.14, the polynomial must have the form:

$$\varsigma(z) = (z + 1)^{n-1} \tag{5.33}$$

The levels to which the individual elements of the array must be driven to secure this solution can be deduced by expanding Eq. (5.33) as a binomial series. For example, for a four-element array, the relative magnitudes are in the ratios 1:2:2:1. This element excitation scheme will result in no sidelobes for the four-element array provided the element spacing d and the progressive phase shift α are also properly imposed.

5.3.3 Supergain Array

Although of limited relevance to solar power collection antennas, it is perhaps interesting to observe that the technique of forming a polynomial with co-located zeros on the circle diagram, employed for the binomial array, can also lead to what is termed a supergain array. If within the fixed-array length, the number of elements is increased, then d and hence the range of ψ are reduced as dictated by Eq. (5.2). In

Fig. 5.14 Roots' locations on unit circle for binomial array

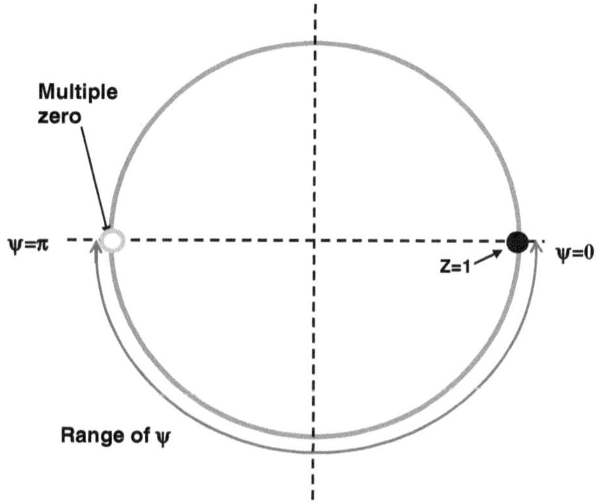

principle, for a small range of ψ, the roots of $\zeta(z)$ can be located very close together, thus yielding a small sidelobe level and a high primary beam directivity. In theory, n can be increased towards infinity with a steady improvement in directional performance being the result. The roots are gradually pushed towards $z = 1$ on the unit circle, and when all approach this point so that all are co-located at $z = 1$, the resultant polynomial converges to:

$$\varsigma(z) = (z - 1)^{n-1} \tag{5.34}$$

When this equation is expanded to polynomial form, it is discovered that the coefficients of the terms in the polynomial, representing the array element drive levels, vary hugely in magnitude and alternate in sign. Generally, this means that supergain arrays with more than a few elements exhibit too low an efficiency to be considered practical. Some success in realising supergain performance with small arrays is reported in Ref. [2].

5.4 Radiation Pattern Synthesis

It is probably clear by now that the radiation pattern of any linear array antenna can be shaped by controlling the positioning and the number of roots on the unit circle. Note that this is also true of any planar array which can be viewed as an array of linear arrays, particularly if element-to-element coupling is weak. In previous examples, it has generally been assumed that the root spacings around the unit circle are constant since the arrays are presumed to be uniform. However, this means that the sidelobe levels are uncontrolled and thus vary in magnitude with

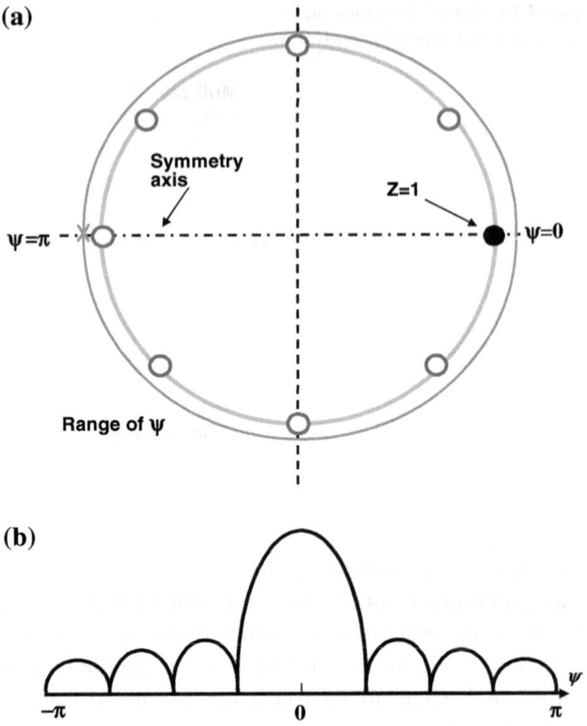

Fig. 5.15 Unequal sidelobes for uniform linear array: **a** root locations for $d = \lambda_0/2$, and 8 elements, **b** resultant graphically generated pattern

their remoteness from the pole or main beam. This is illustrated graphically in Fig. 5.15 where the unit circle and the associated radiation pattern for an eight-element uniform broadside array are depicted. The unit circle (Fig. 5.15a) in this case has a pole at $\psi = 0$ and seven nulls which are symmetrically distributed on either side of the horizontal axis (axis of symmetry). The approximate pattern shows diminishing sidelobe levels as ψ increases on either side of the main lobe.

5.4.1 Tschebyscheff Technique

In antenna synthesis scenarios, the design engineer is usually required to create an antenna with a radiation pattern which accords with the requirements of some communications or radar system into which it will be inserted. Generally, the systems' engineer will specify a desired minimum efficiency, minimum primary beam gain, maximum beamwidth, and maximum sidelobe level. Several other requirements will usually be demanded but these are not relevant here. With the desired sidelobe level expressed as a single number, it has become common to employ techniques which ensure all sidelobes meet the specified level by synthesising a radiation pattern with 'flat' sidelobes. Evidently, the given radiation

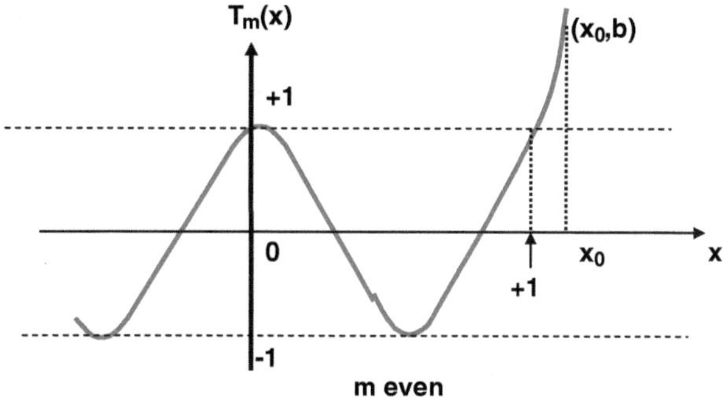

Fig. 5.16 Tschebyscheff function

pattern shape (Fig. 5.15) generated by a uniform array enforcing 'flat' sidelobes entails the introduction of non-uniformity in the element spacings.

For small arrays, the element positioning on the unit circle required to equalize the sidelobe levels can be done empirically. For large arrays, a more controlled procedure is required which automatically establishes the appropriate root locations. Generally, this entails finding a polynomial which naturally exhibits a functional shape which is oscillatory with equal magnitude oscillations on either side of a pole. There are many such functions, but the most popular in array design are the Tschebyscheff functions. These have a characteristic shape (Fig. 5.16), which yields the 'flat' sidelobes. Also, a relatively simple transformation exists to relate this function to unit circle root locations and hence to the polynomial required to define the array. The procedure is described below.

Consider the design of a broadside array antenna ($\alpha = 0, \psi = k_0 d \ \cos\delta$):

As the pattern angle in real space δ increases as follows: $0 \to \pi/2 \to \pi$ then we must also have ψ proceeding as follows: $k_0 d \to 0 \to -k_0 d$ with the main lobe occurring at $\psi = 0$. A suitable transformation between ψ and the x parameter of the Tschebyscheff function is:

$$x = x_0 \cos\left(\frac{\psi}{2}\right) \tag{5.35}$$

Without losing any meaningful content, we can simplify the derivation by setting $d = \lambda_0/2$, in which case $k_0 d = \pi$. So as ψ ranges from $k_0 d \to 0 \to -k_0 d$ on the unit circle, then using Eq. (5.35), the Tschebyscheff parameter x must range from

$$x_0 \cos\left(\frac{\pi}{2}\right) \to x_0 \to x_0 \cos\left(\frac{-\pi}{2}\right)$$

Transferring this requirement to the chosen Tschebyscheff function diagram (Fig. 5.16), the range of x becomes $0 \to x_0 \to 0$.

This forced correspondence between the range of ψ on the unit circle and the range of x on the Tschebyscheff function permits the pole and the zeros in Fig. 5.16 to be appropriately transferred to the unit circle. In the case of Fig. 5.16, there is one pole at x_0 ($\psi = 0$) and four zeros.

Mathematically, the Tschebyscheff function has the form:

$$T_m(x) = \cos(m \cos^{-1} x) = \cos(m\beta) \tag{5.36}$$

where for convenience, we have introduced the equation $\cos\beta = x$. The pattern nulls clearly occur where $\cos(m\beta) = 0$, i.e. where:

$$\beta_k^0 = \frac{(2k-1)\pi}{2m} \qquad k = 1, 2, \ldots, m \tag{5.37}$$

Hence, in terms of the x parameter of the Tschebyscheff function (Fig. 5.16), we must have the zeros located at:

$$x_k^0 = \cos \beta_k^0 \tag{5.38}$$

and using Eq. (5.35) to link x and ψ, we obtain:

$$x_k^0 = x_0 \cos\left(\frac{\psi_k^0}{2}\right) \tag{5.39}$$

On combining Eqs. (5.38) and (5.39), the zeros on the unit circle, representing this Tschebyscheff based array, are located at:

$$\psi_k^0 = 2\cos^{-1}\left(\frac{x_k^0}{x_0}\right) = 2\cos^{-1}\left[\frac{\cos\beta_k^0}{x_0}\right] \tag{5.40}$$

The polynomial, and hence the array format representative of these unit circle root locations, can now be generated using the procedure described in relation to Fig. 5.11. The resultant pattern will possess flat sidelobes in accordance with the original specification, and a pattern synthesis has been accomplished. However, while useful, this is limited pattern synthesis in so far as the specified goals provide no more than a simple envelope or template within which the pattern must lie.

5.4.2 Fourier Series Method

True, or full pattern, synthesis involves the generation of an array antenna whose radiation pattern matches to within specified error criteria a designated 'ideal' pattern. In this case, an iterative mathematical procedure testing the synthesised pattern against the 'ideal' pattern is normally involved in the synthesising process. One such procedure is based on Fourier series. Again, we shall illustrate the method by synthesising a linear array, and for convenience, we assume that the array has an odd number of elements (n). This means that n is best expressed as: $n = 2m + 1$ with $m = 1, 2, \ldots$. As we have already seen the array pattern can be constructed from an n term polynomial $\zeta(z)$, with the form:

$$\varsigma(z) = A_0 + A_1 z + A_2 z^2 + \cdots + A_{2m} z^{2m} \tag{5.41}$$

Noting that $|z^m| = |e^{jm\psi}| = 1$, we can alternatively express the magnitude of $\zeta(z)$ as

$$|\varsigma(z)| = \left|A_0 z^{-m} + A_1 z^{-m+1} + \cdots + A_m + \cdots + A_{2m} z^m\right| \tag{5.42}$$

by dividing Eq. (5.41) by z^m. The polynomial form for $\zeta(z)$ employed in Eq. (5.42) implies that the amplitude distribution (the A_m factors) forms an even function about the central element and the phases of the z terms present an odd function. Hence, we can write:

$$\begin{aligned} A_m &= a_0 \\ A_{m-k} &= a_k - jb_k \\ A_{m+k} &= a_k + jb_k \end{aligned}$$

And so, by pairing terms in Eq. (5.42), that is z^{-k} with z^k, we can reform $|\varsigma(z)|$ as a trigonometric series as follows:

$$\begin{aligned} A_{m-k} z^{-k} + A_{m+k} z^k &= a_k (z^k + z^{-k}) + jb_k (z^k - z^{-k}) \\ &= 2a_k \cos k\psi - 2b_k \sin k\psi \end{aligned} \tag{5.43}$$

But $z^k = e^{jk\psi}$ and consequently combining Eqs. (5.43) and (5.42) yields:

$$|\varsigma(z)| = \left| a_0 + 2\sum_{k=1}^{m} [a_k \cos k\psi - b_k \sin k\psi] \right| \tag{5.44}$$

This representation of the array pattern is in the form of a Fourier series with harmonic coefficients a_k and b_k. Following the mathematical rules of Fourier analysis, we therefore can state that:

$$a_k = \frac{1}{2\pi} \int_0^{2\pi} \varsigma(\psi) \cos(k\psi) d\psi \tag{5.45}$$

$$b_k = \frac{1}{2\pi} \int_0^{2\pi} \varsigma(\psi) \sin(k\psi) d\psi \tag{5.46}$$

In principle, this Fourier representation allows the desired shape of an array radiation pattern to be constructed computationally, by iteratively adding harmonics (and array elements) until the designated pattern has been simulated to a specified degree of accuracy.

References

1. Schelkunoff SA, Friis WT (1952) Antennas, theory and practice. Wiley, New York
2. Dawoud, MM et al (1978) Realization of superdirectivity from active and passive array antennas. In: IEE Conference Publication, No. 169
3. Elliott RS (1963) Beamwidth and directivity of large scanning arrays. Microwave J 7:74
4. The Microwave Engineers Handbook, Page T-136, 1963

Bibliography

5. Collin RE, Zucker FJ (1969) Antenna theory. McGraw-Hill, New York
6. Harrington RF (1961) Time harmonic electromagnetic fields. McGraw-Hill, New York
7. Balanis CA (1997) Antenna theory. Wiley, New York
8. Elliott RS (1981) Antenna theory and design. Prentice-Hall Ltd, Englewood Cliffs
9. Hansen RC (1964) Microwave scanning antennas. Academic Press Ltd., New York
10. Jordan EC (1968) Electromagnetic wave and radiating systems. Prentice-Hall Ltd, Englewood Cliffs
11. Rudge AW, Milne K, Olver AD, Knight P (eds) (1986) The handbook of antenna design: volumes i and ii. Peter Peregrinus Ltd., London

Chapter 6
Solar Radiation and Scattering: Waves or Particles?

'Quantization is a familiar phenomenon in systems where the boundary conditions give rise to standing waves'
'To most non-specialists, quantum mechanics is a baffling mixture of waves, statistics, and arbitrary rules, ossified in a matrix of impenetrable formalism'

Carver A. Mead

Abstract There is some evidence to be found in the literature that modern quantum electrodynamics is inconsistent with classical electromagnetism. The difficulty has mainly evolved from the work of Richard Feynman but, for electrical engineers, it is perhaps more imagined than real. Provided we exclude the esoteric behaviours, both of electrons and photons within atoms embedded in materials, and of exotic particles in the rarefied world of high-energy physics, cumulative electrodynamics provides a smooth link between classical electromagnetism and a version of electromagnetism sympathetic to developments in quantum electrodynamics. In the formulation of this modern electromagnetism, it is likely to be most instructive to adjudge the topic as the study of coherent electron wave functions whose interactions are essentially continuous throughout space.

In this chapter, we demonstrate that electromagnetic theory, including Maxwell's equations, is fully in accord with quantum electrodynamics, provided properly formulated boundary conditions are applied at all interfaces between material regions (of differing composition) and air or vacuum regions. For electrical engineers, wholesale adoption of probabilistic quantum electrodynamics is unnecessary except in energetic processes where interactions between isolated particles occur at the microscopic level, or where power levels are so low that isolated photons can appear. However, in this latter case, the level would normally be judged to be zero and, therefore, insignificant in engineering terms. The chapter suggests that the use of Feynman diagrams, emanating from QED, and applying them to photons in order to model reflection, transmission, refraction and diffraction at the macroscopic level, seems to be delving into metaphysics much more than does resorting to fields. For engineering students, the classical approach, based as it is on the application of

A.J. Sangster, *Electromagnetic Foundations of Solar Radiation Collection*,
Green Energy and Technology, DOI 10.1007/978-3-319-08512-8_6

electric and magnetic fields and of electromagnetic waves, can be confidently assumed to correctly illuminate and fully explain electromagnetic behaviours in macroscopic electrical devices and systems.

6.1 Introduction: What Is Really Being Collected?

Depending on the textbooks, you choose to read on optical topics, and in particular, whether or not they are aimed at physics or electrical engineering students, it is inevitable that you will have to grapple with the 'waves or particles' dilemma—the so called 'uncertainty principle'. Textbooks in engineering electromagnetism almost invariably make no reference to the photon as a particle, and the theory chapters in this book (Chaps. 2–5) have followed the same course. Physics texts on optics, on the other hand, tend to emphasise the importance of photons as particles. This dichotomy is true of even the most modern texts [1, 2]. Nevertheless, it is probably fair to say that in only a very few isolated areas of the science of electromagnetic engineering does the need to acknowledge the fact that action at a distance may not be explicable by classical fields and waves. (As we shall see, the notion of 'distance' can differ considerably between an engineer and a physicist!) Since one of these areas of doubt is optics, and in particular photovoltaics, topics that are relevant to solar collection, it is important to establish a dependable engineering treatment of the photon, which comfortably accommodates and hopefully dispels the conundrum of particle/wave duality.

Recent developments in particle physics, and in particular, quantum electrodynamics, have been subtly pressurising electrical engineering practitioners to move away from traditional, or classical, field concepts, as encapsulated in Maxwell's equations, towards the more 'physically correct' operations (as physicists see it) involving the exchange of energy to represent action at a distance. In fact, some have suggested that the electric field **E** and the magnetic flux density **B** should really be viewed as belonging in the realms of metaphysics [3].

Solar voltaic collectors, discussed in Chap. 7, are commonly analysed and modelled from a quantum mechanical standpoint. Consequently, the aim of this chapter is to focus on examining the influence of quantum electrodynamics on electrical science, and in particular, on classical electromagnetism, from an engineering perspective, in an effort to establish a sound base for theoretical modelling and to ensure that best practice is brought to the design of solar collectors. Newer notions of collective electrodynamics actually provide strong support for continued application of classical electromagnetic theory even in photonic devices, with perhaps, an increased role for electrodynamics than has hitherto been the case. It is perhaps pertinent to note that Stratton [1] was of the opinion that:

> Whatever form the equations of quantum electrodynamics ultimately assume, their statistical average over large numbers of atoms must lead to Maxwell's equations

Wave/particle duality increasingly enforces engineers to consider the following questions. Are electric and magnetic fields best represented by tracing the exchange of photons? Are radio and optical transmissions carried by photons in the form of particles, or waves? What is the most appropriate way to model electromagnetic boundary-value problems? Do we use electromagnetic wave (EMW) theory, or quantum electrodynamic (QED) theory? These questions have huge importance to the design, modelling and simulation of devices and systems ranging from radio wave antennas to computer 'chips' to X-ray machines.

Classical deterministic electromagnetic wave theory, as we have seen in earlier chapters, is most commonly linked to James Clerk Maxwell; on the other hand, it is probably fair to say that modern probabilistic quantum electrodynamics is generally identified with Richard P. Feynman. Of course, Maxwell is not the originator of the electromagnetic wave equations, which carry his name, nor is Feynman—the creator of the equations and techniques, which are now termed quantum electrodynamics. But both were hugely influential figures within these disciplines.

6.2 Classical Electromagnetism

Nature appears to be governed by five forces, namely electric, magnetic, gravitational, plus those responsible for holding together the atomic nucleus, which are termed the 'weak' and the 'strong' nuclear forces. In engineering, only the first three are of any significance, except perhaps, for the handful of engineers employed in the rarefied world of particle physics. In electromagnetism, the forces of interest are, of course, the electric field and the magnetic field, and the topic significantly underpins what can be termed electrical engineering science. This is the branch of the discipline that moulds the electric and magnetic forces and their associated electrical effects, by means of electrical components, circuits, machines, devices and systems generally to the benefit, hopefully, of society and civilisation. Electromagnetic engineering creates the 'real estate' of the world of electricity.

The subject is not new. In about 600 BC, so the story goes, a Greek mathematician Thales of Miletus apparently observed that when amber (*elektron* in Greek) is rubbed against silk, sparks are produced, and he further noted that the treated amber appeared to attract materials such as feathers and straw. Thales is also attributed with the discovery of the attractive influences displayed by a natural magnetic rock called loadstone, found in the locality of a place called Magnesia, in ancient Greece. Apart from being the source of the names for the electric and magnetic forces, more than twenty-two centuries were to pass before serious studies of these forces were initiated.

The history of the development of the science of electromagnetics is generally considered to begin with William Gilbert (1540–1603) from Colchester in England, who was the first to carry out rational experiments on electrical effects. His book 'De Magnete' is viewed as a landmark publication in electrical science, and as its title suggests, it was mainly directed at understanding magnetism. It is clear that in

Gilbert's time, scientists would have been well aware of electrical phenomena such as friction-induced charge and lightning, but it took another century before Benjamin Franklin (1706–1790) living in North America, determined the basic nature of electric charge; namely that it existed in positive and negative forms, and perhaps more significantly, that charge is conserved. Franklin is also the originator of the lightning conductor that he devised to protect buildings 'from mischief by thunder and lightning'.

6.2.1 Influence of QED

Is it possible that a major tool of electrical engineering used extensively for research into low frequency, high frequency, microwave, millimetre wave and optical devices, namely classical electromagnetic theory, including Maxwell's equations, is in the process of being superseded by a newer theory? During 50 years of immersion in electromagnetism, ranging over a wide spectrum of activities [4, 5], the author has found it difficult to ignore the niggling doubts about the rigour of classical electromagnetism emanating from the physical sciences propelled by the competing claims for an alternative analytical approach based on quantum electrodynamics. Over this period, quantum mechanical developments have evolved steadily in order to explain electrical forces in atoms through the mechanism of energy transitions. Electric and magnetic field concepts have been found wanting. However, it has always seemed safe to assume that probabilistic quantum electrodynamics (QED), which tends to view electric and magnetic fields as metaphysics, does not seriously impinge on electrical science at the macroscopic level. It is well documented that the probabilistic feature of quantum mechanics has been controversial from its inception and remains so even today. This is perhaps hardly surprising given Einstein's disdain for the idea of light quanta as probabilistic little 'bullets', as expressed in the following quotation from his 1905 paper [6]:

> Would it not be possible to replace the hypothesis of light quanta by another assumption that would also fit the known phenomena? If it is necessary to modify the elements of the theory, would it not be possible to retain at least the equations for the propagation of radiation and conceive only the elementary processes of emission and absorption differently than they have been until now?

Or, in engineering language, Einstein considered that quanta of light or more generally electromagnetic energy quanta should be of concern only where electromagnetism interacts with atoms at the microscopic level. This view from one of our greatest scientists would seem, on the face of it, to provide more than sufficient justification for continuing both to practice classical electromagnetism and to inculcate it in new generations of electrical engineers. However, according to Richard Feynman, in his short text entitled 'QED' [3], this is not so. Classical electromagnetic theory is deemed passé, and quantum electrodynamics is claimed to more correctly describe all electrical phenomena.

It is perhaps relevant to note that Maxwell has been quoted as observing:

The mind of man has perplexed itself with many hard questions. Is space infinite, and in what sense? Is the material world infinite in extent, and are all places within that extent equally full of matter? Do atoms exist or is matter infinitely divisible?

So, undoubtedly, he would have been perplexed by recent developments in modern particle physics, which are undermining field concepts. He would probably have been dismayed at the possible retreat by electrical engineers from deterministic electromagnetic theory to accommodate probabilistic quantum electrodynamics, which asserts that all modern (non-nuclear) physics can be traced back to the interaction between atoms and electrons. As Feynman himself [3] has forcefully suggested, quantum electrodynamics:

describes *all* the phenomena of the physical world except the gravitational effect

So, should the electrical engineering profession be preparing to supplant electromagnetic theory, in science and engineering teaching, by quantum electrodynamics or is the *status quo* acceptable? This question is investigated and hopefully answered below.

6.3 Photon Scattering

Before we answer the above question, it is perhaps pertinent to consider why it is necessary to ask it at all. Until 100 years ago, the accepted and settled understanding of scientists on the nature of electromagnetic transmission through space was that the process was undoubtedly achieved by means of waves. The nature of electromagnetic wave propagation was thoroughly established by James Clerk Maxwell by 1873, and just before the turn of the century Oliver Lodge and Heinrich Hertz demonstrated their existence. The first doubts, that all was not well with the theory, arose from measurements of blackbody radiations, which clearly did not fit the classical models that then existed. The conundrum was resolved in 1900 by Max Planck who reluctantly suggested that perhaps the radiation is absorbed and emitted by the blackbody in discrete packets, or quanta, of energy E, given by:

$$E = hf \tag{6.1}$$

where f is the frequency in Hertz and h is Planck's constant (6.626×10^{-34} J s). The adjustment was so successful in modelling the experimentally procured results that the idea of quantization had to be taken seriously by dubious scientists.

Nevertheless, it was still possible to argue that the energy quanta were possibly a property of the atomic nature of matter rather than the light itself. However, the debate was farther consolidated in favour of quanta by Albert Einstein in 1905 in a paper on the photoelectric effect in metals. Photoelectricity occurs when a metal or alkali is immersed in electromagnetic waves at optical frequencies, the result being the emission of an electron from the irradiated surface. Previous research had shown that there is a threshold frequency below which the process is unobserved. This is difficult to explain using classical electromagnetic theory, which fails to predict a frequency dependency for such emissions. Again, the answer to this dilemma was quantisation. Theory and measurement become aligned if one of Planck's quanta of light assumed to displace an electron in the targeted metal surface. According to Einstein, if it requires an energy level of ϕ (the work function) to dislodge a surface electron, then for a metal irradiated by light at frequency f Hz, the released electrons will emerge with kinetic energies U_{KE} equal to, or less than:

$$(U_{KE})_{\max} = hf - \phi \tag{6.2}$$

So, the cut-off frequency for photoelectric emission when no electron acquires kinetic energy is:

$$f_{co} = \frac{\phi}{h}\,\text{Hz} \tag{6.3}$$

However, even Einstein himself continued to have doubts that this meant light, and hence, electromagnetic waves were quantised—see quotation in Sect. 6.2.1.

6.3.1 Compton Effect

The experiment which largely confirmed that light (actually X-rays) was quantised is attributed to Arthur H. Compton and was reported in 1923. Compton originated the use of the term photon to describe a quantum of light. The essence of his measurement is pictorially summarised in Fig. 6.1.

In the experiment represented by Fig. 6.1, Compton irradiated a graphite target with high-energy X-rays from an X-ray tube. By employing X-rays, he could be sure that the energy incident on the target was much higher than electron vibration energies and their binding energies and so the dislodged electrons could be considered to have been originally free and at rest, as suggested in the figure. What he observed was scattering much more indicative of particle collisions, than of wave scattering. The target electron was reported to recoil at relativistic speed while a 'photon' was scattered with reduced energy (and hence frequency) in a complementary direction. Classically, light of sufficient intensity for the electric field to accelerate a charged particle to a relativistic speed will cause radiation–pressure recoil and an associated Doppler shift of the scattered light but the effect would become arbitrarily small at sufficiently low light intensities regardless of wavelength, and this was not observed.

If the light (and X-rays) is assumed to consist of particles, an explanation for the low-intensity Compton scattering becomes readily explicable. And so, Compton's experiment convinced physicists that light can behave as a stream of particle-like objects (quanta) whose energy is proportional to the frequency. In the figure that represents particle-like scattering, the following definitions apply:

Fig. 6.1 Compton scattering

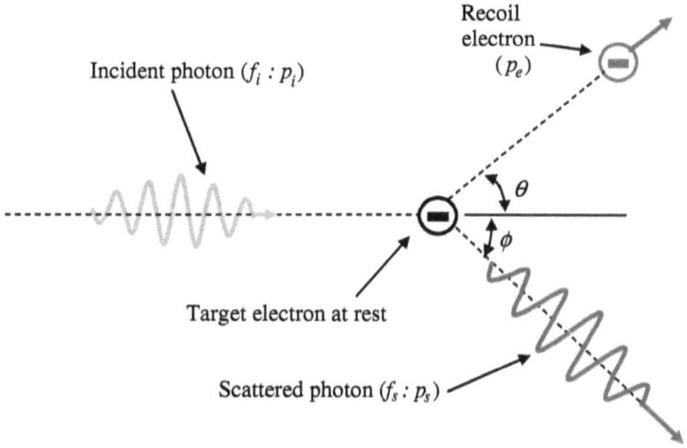

For the input photon:
$$\lambda_i = \frac{c}{f_i} = \text{wavelength}$$
$$p_{mi} = \text{momentum}$$
$$E_i = hf_i = \text{energy}$$
$$c = \text{velocity of light}$$

For the scattered electron:
$$m_0 = \text{electron mass}$$
$$p_{me} = \frac{\sqrt{E^2-(m_0c^2)^2}}{c}\ \text{electron momentum}$$
$$E = \text{kinetic energy}$$
$$m_0c^2 = \text{rest energy}$$

For the scattered photon:
$$\lambda_s = \frac{c}{f_s} = \text{wavelength}$$
$$p_{ms} = \text{momentum}$$
$$E_s = hf_s = \text{energy}$$
$$\Delta\lambda = \text{wavelength shift due to scattering} = \frac{h}{m_0c}[1 - \cos\phi]$$

Because the energy and momentum of a system must *both* be conserved, it is not generally possible for the electron simply to move in the direction of the incident photon, as it would in the wave scenario where radiation pressure acts upon the electron. The interaction between electrons and high-energy photons (when comparable to the rest energy of the electron, 511 keV) results in the electron being given part of the original energy (making it recoil in direction θ), while a photon contains the remaining energy. It is emitted in a different direction (ϕ) and at a lower frequency from the original, so that the overall momentum of the system is conserved.

6.3.2 Young's Experiment

While the idea that light was quantised into photons and that these could display both wave and particle behaviour was largely accepted by the late 1920's the probabilistic nature of the photon wave function that has remained troublesome, particularly for engineers. The answer according to Richard Feynman [3] is an enhanced role for quantum electrodynamics, at the expense of classical electromagnetism, in the practice of electrical science. This is rather ironic given that his comprehensive textbook entitled 'Lectures in Physics Part II' [7] has become recommended reading for many engineering electromagnetics courses, in UK universities and colleges, which continue to follow the classical approach. At about the time 'QED' [3] was being published in 1990, physicists were still struggling with particles, fields and point sources, wave functions, self-action, infinities, quantum transitions, etc., as Feynman admits in his 'Lectures in Physics'. So, quantum electrodynamics seemed at the time to provide the most likely route out of these difficulties. In fact, he appears to have been so convinced of this that he strongly advocated the corpuscular view of light in the form of photons.

> Thus light is something like raindrops—each little lump is called a photon—and if the light is all one colour, all the 'raindrops' are the same size.

To clarify the issue of the probabilistic behaviour of the photon, and also of its wave/particle duality, it may be helpful to revisit a well-rehearsed light interference experiment. This experiment is examined in detail in QED [3] where the concept of photon scattering is consolidated through the use of Feynman diagrams. These were originally devised to analyse graphically, particle interactions in atoms in matter. Here, we will concentrate on one particular version, namely diffraction by parallel slits in a perfectly conducting infinitesimally thin screen (Fig. 6.2)—usually termed Young's two-slit experiment. This case has considerable relevance to antenna theory (see Chap. 5), from low radio wave frequencies to optical frequencies, as well as many other engineering devices that employ diffraction gratings.

For a screen with infinitely long slits in the *y*-direction, the scattering problem reduces to a two-dimensional one with all variations confined to the *x*-*z* plane. The parallel slits are very narrow such that the width (w) of each can be presumed to accord with the inequality, $w << \lambda$. Further, we assume that the separation between the slits (d) is governed by the requirement $d \geq \lambda/2$. A routine solution of Maxwell's equations in an unbounded space, when applied to the semi-infinite region to the left of the screen dictates that the incoming wave is 'plane'. In mathematical terms, it can be expressed quite simply as:

$$\begin{aligned} E_x &= Ae^{-jk_z z}e^{j\omega t} \\ H_y &= \frac{A}{\eta}e^{-jk_z z}e^{j\omega t} \end{aligned} \tag{6.4}$$

where $k_z = {}^{\omega}/_{v_z} = {}^{\omega}/_{c}$ and $\eta = {}^{E_x}/_{H_y}$. An almost equal magnitude reflected wave is also formed to the left of the screen, which combines with the incoming wave to form a dominant standing wave.

At the slit, the incoming plane wave induces an electric field that becomes a source of Huygen's waves to the left and right of the screen as suggested in Fig. 6.2.

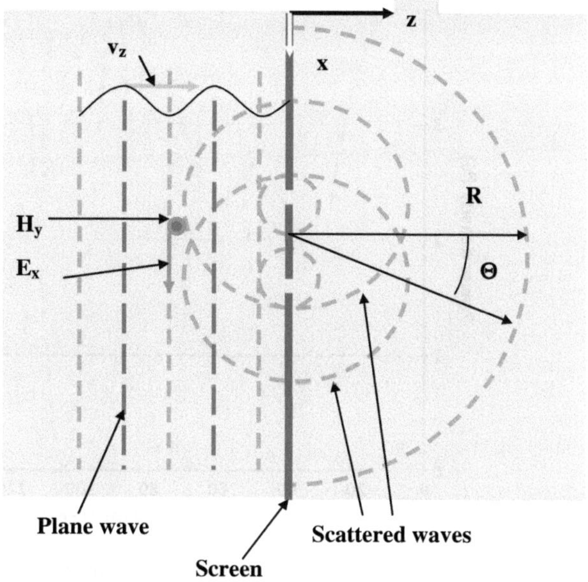

Fig. 6.2 Parallel slit diffraction pattern construction by cylindrical Huygen's waves

For this two-dimensional problem these waves, which are solutions of the Maxwell equations, are cylindrical waves. They can be expressed in the exponential form:

$$E_s = \frac{C}{\sqrt{R}} e^{-jkR} e^{j\omega t} \tag{6.5}$$

with a similar expression for H_s. The constant C represents the peak magnitude of the scattered wave at the slit. If R ≫ d, then the waves from the two slits can easily be shown to combine mathematically to give:

$$E_T = \frac{2C}{\sqrt{R}} \cos\left[\frac{\pi d}{\lambda} \sin\Theta\right] e^{j\omega t} \tag{6.6}$$

If we normalise the system such that a single slit produces a power density at R that is (say) 1 % of the incident level, then the power density at R for the slit pair, namely p W/m^2, can be expressed as:

$$p = 4\cos^2\left[\frac{\pi d}{\lambda} \sin\Theta\right] \tag{6.7}$$

The transmitted power density p as expressed through Eq. (6.7) is depicted graphically in Fig. 6.3. The red solid line represents $d = \lambda/2$, while the mauve chain dotted curve applies to a spacing $d = 2 \times \lambda$. The blue dashed line represents the single-slit level. These oscillatory curves are variously termed diffraction patterns, interference patterns and radiation patterns. They represent the changing phase

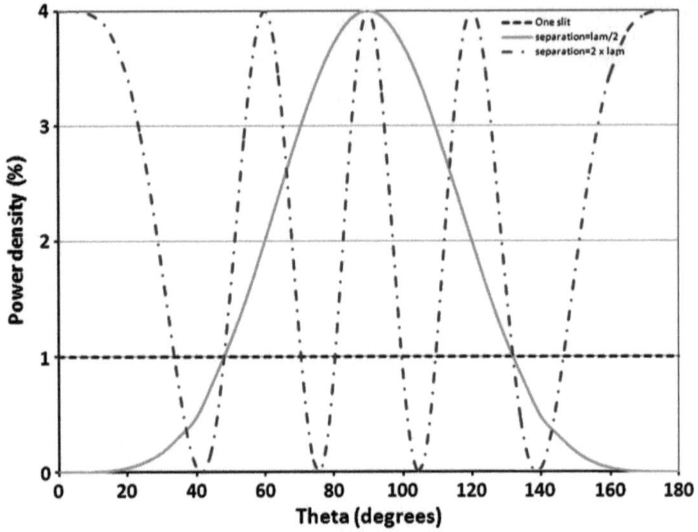

Fig. 6.3 Diffraction patterns for parallel slit examples: **a** single slit *dashed line*, **b** two slits separated by $\lambda/2$ *solid curve* and **c** two slits separated by 2λ *chain dotted curve*

deviation between the waves emanating from the slits as the observation angle Θ progresses from 0° to 180°.

This seemingly complex interference example is relatively easily modelled using wave theory. Any competent senior school pupil if presented with the exponential representation of electromagnetic waves, could perform the above derivation with little difficulty. So how would this analysis proceed if one took a particle view of electromagnetic 'action at a distance' rather than a wave view?

6.3.3 Photons and Interference

The quantum electrodynamic representation of the problem of diffraction by slits in an infinite flat conducting screen is shown in Fig. 6.4. Now a photon source on the left of the screen emits light (or radio waves) uniformly in all directions, with some photons reaching slit 1 and others appearing at slit 2. In accordance with Feynman [3], for analysis purposes, a photon is a vector possessing a probability amplitude. As the photon travels through space, the vector amplitude rotates (phase delays) with the passage of time. The square of the vector amplitude expresses the probability of an event for a photon intercepted by a photon collector.

In the figure, the source is located so that the distance S to both slits is the same. Consequently, the photons passing through to the screen will be in time phase as they enter the right-hand domain. Photons will stream in all directions from each slit. A photon collector is located at radius R and at angle Θ to 'count' the number

Fig. 6.4 Parallel slit diffraction pattern constructed using probability amplitude vectors for photons. (The thought experiment requires the light level to be so low that single photons arrive at the screen—this is 10^{-12} μW!!!)

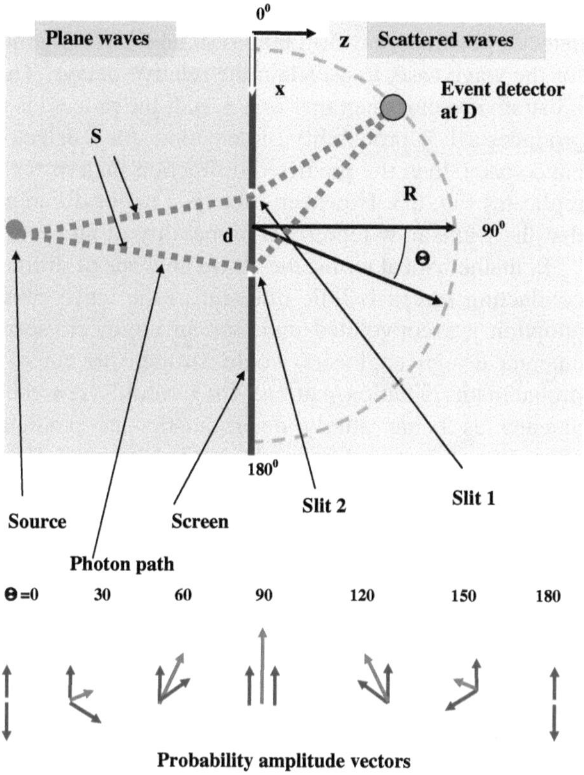

of collection events at that position (D). A stream of photons from slit 1 and a stream from slit 2 travel to the event detector over slightly different distances, which changes with Θ. The method of determining the probability of detection at a given event detector location is illustrated at the foot of Fig. 6.4. The method is graphical and depends on tracking the probability amplitude vector as a photon travels to the detector. Each vector rotates in a clockwise direction (like a stopwatch hand), and at the destination, the vectors are added following the usual vector rules. So, if we arbitrarily assume that each vector is aligned parallel to the screen and pointing towards—x (see 90° case in Fig. 6.4), then for an event collector at 90°, the distance travelled by a photon from slit 1 matches that from slit 2, so each probability vector will rotate clockwise be the same amount and remain parallel. The number of rotations is irrelevant, so we can represent the position at the event detector as shown in the central vector summation diagram. The resultant amplitude of the vector at 90° is thus twice the contributing components. At other positions of the event detector, the time of travel differs for a photon from slit 1 as compared to one from slit 2. If we take the 0° case, with $d = \lambda/2$ say, relative phase change depends only on the delay associated with the photon from slit 2 as it travels distance d to slit 1. For $d = \lambda/2$, this results in an additional delay of 180°. The vector position at the detector is shown in the first diagram in Fig. 6.4 and yields a zero magnitude resultant. This is also true for a detector at $\Theta = 180^{o}$. At intermediate detector positions, say 30°, 60°, 120° and 150°, trigonometry has to be employed, as for the wave case, to ascertain the relative delays. This is shown in Fig. 6.4 by the illustrative vector diagrams 2, 3, 5, 6. If the process is normalised so that a single slit produces a 1 % probability of detection, for a detector at distance R into the right-hand space, then the predicted diffraction pattern for the slit pair, not surprisingly, replicates Fig. 6.3. However, there is a major difference from the Huygen's case in that the *y*-axis now represents probability of an event rather than power density.

In mathematical terms, the above analysis of diffraction by two slits in an infinite conducting screen is little different to the wave case, but the justifications for its adoption are convoluted and, for an engineer, seem remote from reality. Most antenna design engineers would struggle to come to terms with the idea of a probabilistic radiation pattern. They would view the pattern of radiation from an antenna as being utterly deterministic—as predictable and dependable as the chemistry of the metal from which it is fabricated. Fortunately, as Fig. 6.5 illustrates this scenario virtually never arises. The interference pattern is probabilistic only at power levels that are so low they are largely irrelevant to the practice of electrical engineering. At measurable power density levels, from an electrical engineering perspective, photons evidently coalesce into waves [10], and the need to consider 'probabilities' disappears.

6.3.4 Coherence

This simple two-slit diffraction problem is clearly much more tedious and arduous to solve by employing probabilistic vectors rather than Huygen's waves. However, 'QED' justifies the approach by considering the effect of introducing an event detector within each of the slits. As we have just seen, when light (photons) are generated by a single source (*S*), the Feynman method predicts the presence of a probabilistic interference pattern at the event detector (see Fig. 6.3) due to screen apertures at slit 1 and slit 2. The same pattern is also easily generated by wave theory as the apertures become *coherent* point sources of Huygen's waves. Crucially, however, 'QED' suggests that when photon detectors are inserted at slit 1 and slit 2 to detect the presence of photons in the aperture, the interference pattern disappears and the signal at the event detector is constant with 2 % probability. The electrodynamic argument for this change is convoluted. When no event detectors exist in the slits, it is not possible to know which of the two slits any given photon leaving the source will pass through. Therefore, there is interference for an event at D. With event detectors in the slits, the nature of the process is completely changed. Photons passing through slit 1 and detected that there are no longer in phase with those detected at slit 2, since according to the 'uncertainty principle', if photon positions are known, their momenta (frequency or phase) are not (i.e. we can observe the interference pattern or the presence of a photon in a slit but not both). So, there are now two completely separate events. Probabilities should now be

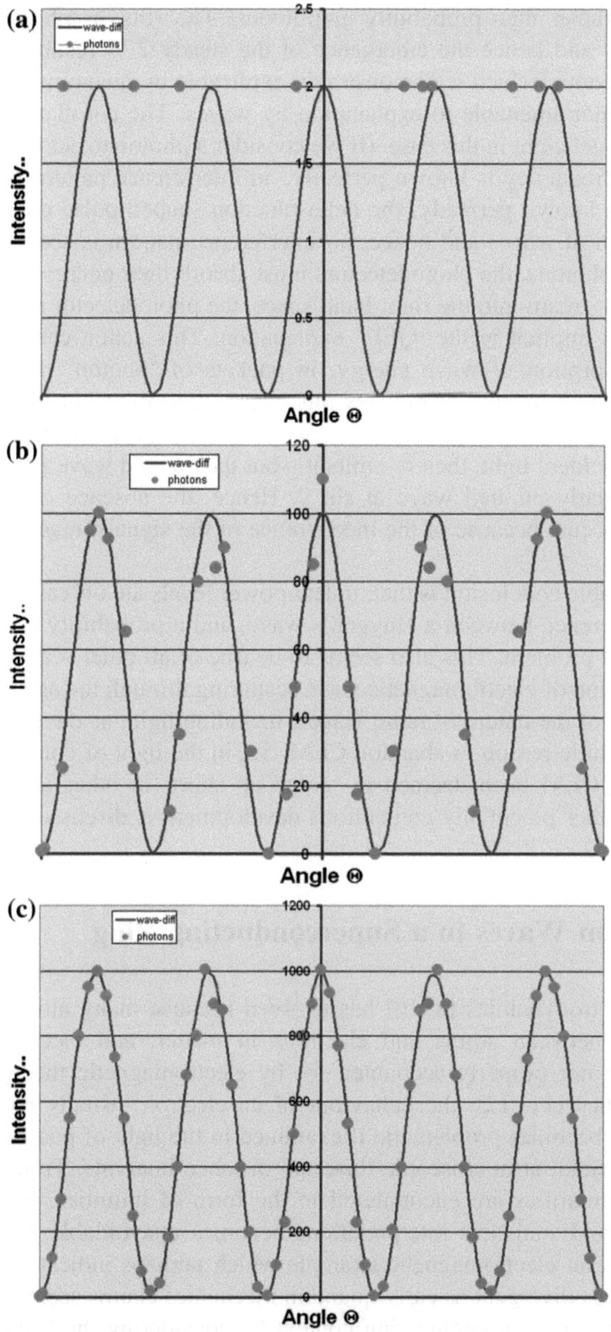

Fig. 6.5 Young's two-slit experiment showing photon distributions for 10, 500 and 5,000 photons superimposed on classical interference pattern (N.B. 5,000 photons = 3×10^{-3} pico-Watts/m^2—in engineering terms, this power level is, to all intents and purposes, zero)

added at D, rather than probability amplitudes! The slits, as photon sources, act independently and hence the emergence of the steady 2 % result. It is postulated that this behaviour, which is demonstrably explicable by quantum electrodynamics (photons), is not amenable to explanation by waves. The corollary is that electromagnetism is deficient in this case. (If we consider a photon to act not unlike a radar pulse—if its frequency is known perfectly, an interference pattern is observed—if its position is known perfectly, the delta-function-shaped pulse contains a continuous spectrum of waves and hence, no interference pattern is seen).

To detect photons, the photodetectors must absorb their energy. But for photons to continue to stream into the right-hand space, the photodetector must then re-emit photons, as is implicit in the 'QED' explanation. This action can surely be interpreted as absorption of wave energy, in packets of 'photon' magnitude, at the detector that subsequently re-radiates this energy into the right-hand space? If so wave theory can easily explain the phenomenon. The detector at slit 1 absorbs a quantum of incident light, then re-emits it—but the emitted wave is now *incoherent* with the similarly emitted wave at slit 2. Hence, the absence of an interference pattern at D occurs because of the incoherence of the signals, reaching the detector there.

The inevitable conclusion is that, unless power levels are unfeasibly low, there is really no difference between a Huygen's wave, and a probability vector, treatment of the two-slit problem. This also seems to be true of all other scattering problems. The examination of electromagnetic wave scattering through the agency of a photon interpretation of the nature of radio waves, including light, as developed in 'QED', seem to give little reason to abandon CEM. So, in the light of continuing advances in QED, has CEM been deemed to 'come up short' in other areas of electrical science? Another potentially contentious development is discussed below.

6.4 Electron Waves in a Superconducting Ring

Quantum electrodynamics [8–10] has evolved because many atomic interactions, for example between atoms and electrons in matter, and electrons with other electrons, are not properly accounted for by electromagnetic theory. In classical electrodynamics [11, 12], the behaviour of an electron, usually represented as a point charge, becomes problematic if examined in the light of possible interactions with its own field, at microscopic time and distance intervals. Troublesome mathematical irregularities are encountered in the form of infinities. So, the need for quantisation and statistical interpretations becomes unavoidable. However, let us take a look at an electromagnetic example which perhaps indicates that QED and CEM are not as divergent as early quantum mechanics claims seemed to suggest. It can be illustrated in an engineering context by considering the behaviour of electrons in a superconducting loop [10] as indicated in Fig. 6.6.

In Fig. 6.6a, the superconducting loop represents a simple inductor that any second-year undergraduate on an electrical engineering degree course could handle.

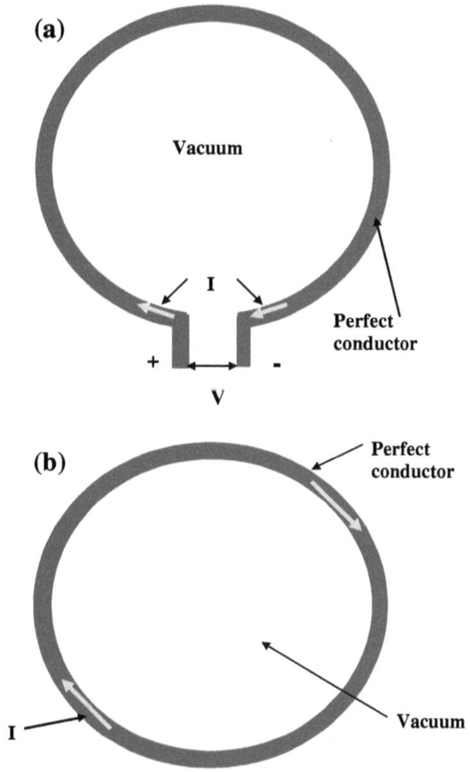

Fig. 6.6 Superconducting loop in vacuum **a** open, **b** closed

The inductance L (Henries) is dependent on the loop dimensions. If the current is initially zero and rises to a final value I $(0 < i < I)$ in time T, then familiar circuit relations yield:

$$V = \frac{\mathrm{d}\Phi}{\mathrm{d}t} = L\frac{\mathrm{d}i}{\mathrm{d}t} \tag{6.8}$$

Or at time T:

$$\Phi = LI = \int_0^T V\,\mathrm{d}t \tag{6.9}$$

so that at time T, the energy stored in the loop (W) is:

$$W = \int_0^T Vi\,\mathrm{d}t = \frac{1}{2}LI^2 \text{ Joules} \tag{6.10}$$

In classical electromagnetism (CEM), we would infer that this energy is stored in the magnetic field. In electrodynamics, on the other hand, each moving electron in the loop couples with every other electron so storing energy in the process.

Now, if the perfectly conducting and isolated loop is closed as suggested in Fig. 6.6b with the current I continuing to flow, an energy W can be stored in the loop in perpetuity with nothing to impede the moving electrons. In practice, it is possible to close the loop without quenching the current. A persistent current is formed. But on the basis of a CEM analysis of this macroscopic problem, the expectation arising from the Faraday law (Eq. (6.8)) would be that if $V = 0$ for the closed loop, then $\frac{\mathrm{d}\Phi}{\mathrm{d}_t} = 0$. In contradiction to practice, CEM predicts that it should not be possible to develop a current in, or a flux through, the perfectly conducting ring. This is because CEM tends to view current flow in a metal as a stream of electronic charge drifting through the material. As a consequence, it fails to predict the above superconducting ring phenomenon despite the macroscopic context. For an answer to this dilemma, we need to look elsewhere.

Measurements [10] on superconducting loops of varying dimensions reveal that the flux is quantized and that equation (6.9) should more correctly read:

$$\Phi = \int V \, \mathrm{d}t = n\Phi_0 \tag{6.11}$$

This equation suggests that at the microscopic level, the electric potential (V) within the loop material is not zero. Experiments, accurate to one part in 10^9, indicate that the flux quantum $\Phi_0 = 2.06783461 \times 10^{-15}$ V sec. Measurements also reveal that changing n to $n + 1$ or $n - 1$ is very difficult—presumably because electrons come in essentially one form, with a tightly specified energy level and, hence, a limited frequency range. The explanation for these behaviours can be deduced from QED, which focuses on 'real' physical parameters such as energy, momentum, frequency and phase. Einstein has provided the relation:

$$W = \hbar\omega \text{ Joules} \tag{6.12}$$

while de Broglie has determined that momentum

$$\vec{p}_{om} = \hbar\vec{k} \text{ kg m/s} \tag{6.13}$$

The constant of proportionality $\hbar$ is just Plank's constant divided by 2π. These equations dictate that at the quantum (atomic)-level energy is synonymous with frequency, while momentum is synonymous with phase. The quantity $\vec{k}$ is defined as the propagation vector of a wave in rad/m. The arrow-topped vector notation [10] is used here to be consistent with QED formulations. When applied to an electron, this can only mean that the electron has wave characteristics, so that a vast number of interacting and coupled electrons in the superconducting ring form a coherent wave such that the cumulative phase around the loop is

$$\phi = \oint_{\text{loop}} \vec{k} \cdot \mathrm{d}\vec{\ell} \text{ radians} \tag{6.14}$$

Consequently, the loop is in essence a ring resonator for the cumulative electron wave which dictates that the phase is quantised such that $\phi = 2n\pi$. The reason for the quantum behaviour is thus apparent. Given that accumulated phase equates to the time-integrated frequency difference between two points in the loop and that frequency is synonymous with energy, it can be shown [10] that

$$\phi = \frac{q_0}{\hbar} \int V \, \mathrm{d}t \tag{6.15}$$

The non-zero voltage is evidence of an electron wave. Hence, combining Eqs. (6.15) and (6.9), we get

$$\phi = \frac{q_0}{\hbar} \Phi = \frac{q_0}{\hbar} n \Phi_0 \tag{6.16}$$

So magnetic flux can be equated to phase accumulation of the electron wave function, which is usually represented by the magnetic vector potential $\vec{A}$ and by analogy with Eq. (6.14), this leads to the equation:

$$\Phi = \oint_{\text{loop}} \vec{A} \cdot \mathrm{d}\vec{\ell} \tag{6.17}$$

Furthermore, mutual coupling experiments using adjacent loops of the form suggested in Fig. 6.6 indicate that the interaction between the loops associated with the action at a distance of the electrons in each loop can for any loop size, shape orientation and separation, be represented by the vector potential $\vec{A}$. It can be calculated by (see Chap. 3, Eq. (3.44)):

$$\vec{A} = \frac{\mu_0}{4\pi} \oint_{\text{loop}} \frac{\vec{I}}{r} \mathrm{d}\ell \tag{6.18}$$

where $\vec{I}$ is the current in the loop and μ_0 is the permeability of vacuum. Equations (6.14), (6.16) and (6.17), and a generalised form of (6.18), are the fundamental relations governing quantum electrodynamics in cumulative form [10], which recognises that electrons have wave characteristics and do not exist as isolated 'point-form' particles.

Does this evidence mean that we should prefer the above quantum electrodynamic equations to those of Maxwell, given the inability of CEM to provide a meaningful link between fields and charge at the atomic level [10], and more pertinently, its failure to properly model the closed superconducting loop—a plainly macroscopic problem—of engineering relevance?

6.5 Electromagnetic Ring Resonator

We can possibly answer the above question from an engineering perspective by examining a not too dissimilar structure comprising the electromagnetic waveguide ring resonator, which is well modelled by CEM. Our aim is to attempt to determine whether or not its modal (quantised) behaviour can be further illuminated by QED and the electron wave function. The rectangular waveguide ring (Fig. 6.7) is assumed to be formed from infinitesimally thin perfectly conducting walls, and to trap a vacuum in the interior space. In Fig. 6.7a, the ring of average length L metres, which is presumed to be much larger than guide width a, is fed at one port by a pulsed source of frequency f_0, which can be varied continuously. The second port is terminated in a matched load. In Fig. 6.7b, the ring is closed to form a ring resonator.

To model this geometry, classical electromagnetic theory [13–15] would employ Maxwell's equations to form a second-order differential equation, in terms of the electric field vector or the magnetic field vector (**E** or **H**), which is applicable to the interior region. For example:

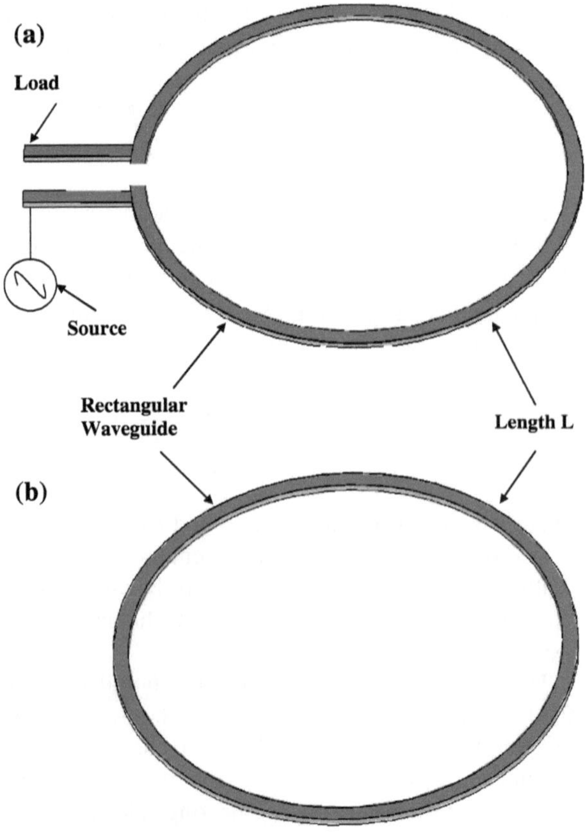

Fig. 6.7 **a** Waveguide loop with and matched load and **b** waveguide ring resonator. In both cases, the waveguide has width a, height b, length L with perfectly conducting walls and evacuated interior

$$\nabla^2 \mathbf{E} = \frac{1}{c^2}\frac{\partial^2 \mathbf{E}}{\partial t^2}$$
$$\nabla^2 \mathbf{H} = \frac{1}{c^2}\frac{\partial^2 \mathbf{H}}{\partial t^2} \quad (6.19)$$

When one or other of these is solved in conjunction with the relevant tangential field boundary conditions at the perfectly conducting walls, namely:

$$\mathbf{n} \times \mathbf{E} = 0 \quad (6.20)$$

$$\mathbf{n} \times \nabla \times \mathbf{H} = 0 \quad (6.21)$$

Here, $\mathbf{n}$ is the unit vector normal to the interior walls. The result for TE modes can be evolved as follows [13]. The wave Eq. (6.19) in a rectangular waveguide defined by means of a Cartesian coordinate frame can be expanded as follows for the H_z field component:

$$\frac{\partial^2 H_z}{\partial x^2} + \frac{\partial^2 H_z}{\partial y^2} + \frac{\partial^2 H_z}{\partial z^2} = \frac{1}{c^2}\frac{\partial^2 H_z}{\partial t^2} \quad (6.22)$$

If we limit our horizon to seeking sinusoidal wave solutions at frequency ω rad/s, then z-dependence of the field can be expressed as $\exp(-\bar{\gamma}z)$, where $\bar{\gamma}$ may be complex, while time dependence has the form $\exp(j\omega t)$. Hence, Eq. (6.22) becomes:

$$\frac{\partial^2 H_z}{\partial x^2} + \frac{\partial^2 H_z}{\partial y^2} + k_c^2 H_z = 0 \quad (6.23)$$

where

$$k_c^2 = \bar{\gamma}^2 + \frac{\omega^2}{c^2} \quad (6.24)$$

The solution is:

$$H_{zmn} = C\cos\frac{m\pi x}{a}\cos\frac{n\pi y}{b}\exp -\bar{\gamma}z \exp j\omega t \quad (6.25)$$

In this equation, m and n are integers such that $m = 0, 1, 2 \ldots, \infty$ $(m = n \neq 0)$, and $n = 0, 1, 2 \ldots, \infty$ $(n = m \neq 0)$. This solution also requires that:

$$k_c^2 = k_{cmn}^2 = \left(\frac{m\pi}{a}\right)^2 + \left(\frac{n\pi}{b}\right)^2 \quad (6.26)$$

Hence, from Eq. (6.24), we must have

$$\bar{\gamma}^2 = \bar{\gamma}_{mn}^2 = \left(\frac{m\pi}{a}\right)^2 + \left(\frac{n\pi}{b}\right)^2 - k^2 = k_{cmn}^2 - k^2 \tag{6.27}$$

where $k = \omega/c$. In this equation, k is the free-space propagation coefficient, k_{cmn} is the cut-off coefficient, while γ_{mn} is the waveguide propagation coefficient for the mnth solution or mode. The other field components of the mode are derived from H_z using the following relations derived from the Maxwell equations:

$$\begin{aligned} H_{xmn} &= -\frac{\bar{\gamma}_{mn}}{k_{cmn}^2}\frac{\partial H_{zmn}}{\partial x} \\ H_{ymn} &= -\frac{\bar{\gamma}_{mn}}{k_{cmn}^2}\frac{\partial H_{zmn}}{\partial y} \\ E_{xmn} &= -\frac{j\omega\mu_0}{k_{cmn}^2}\frac{\partial H_{zmn}}{\partial y} \\ E_{ymn} &= \frac{j\omega\mu_0}{k_{cmn}^2}\frac{\partial H_{zmn}}{\partial x} \end{aligned} \tag{6.28}$$

In the terminated waveguide example (Fig. 6.7a), the field solutions are modal associated with a double infinity of transverse standing wave solutions. These are termed the TE_{mn} modes. The rectangular waveguide also supports a double infinity of TM_{mn} modes. While the system exhibits modal solutions (see Chap. 2), it is important to note that the frequency of any given solution is solely dictated and controlled by the source. If the source coupler is designed to excite only the TE_{10} mode (the most common arrangement), the excitation by a short pulse, of time duration much less than the propagation delay through the waveguide, will be represented by a spectrum of frequencies around the carrier frequency, all in the TE_{10} mode, as the pulse passes through the guide. On the other hand, a very long pulse, of width much greater than the propagation delay, will be represented by a TE_{10} mode at the carrier frequency only, and this frequency is wholly controlled by the source. Mathematically, such solutions are complete, and the solutions are fully in accordance with practice. This completeness has paved the way for a widespread development of sophisticated simulation software packages, based on moment methods and finite element methods. These modelling tools permit the user to confidently address highly complex boundary-value problems that emanate from engineering research.

If we direct our attention now to Fig. 6.7b, here, we have closed the open ends of the perfectly conducting waveguide loop to form a ring resonator. This is not unlike the superconducting ring, except that the 'action' takes place within the vacuum space rather than within the perfectly conducting material. The EM solution proceeds as described above, but with the additional boundary condition that the phase around the ring for both clockwise and anticlockwise waves must be an integer number (p) of cycles. The standing wave solution thus has the form:

$$H_{zmn} = C \cos \frac{m\pi x}{a} \cos \frac{n\pi y}{b} \cos \beta_{mn} z \exp j\omega t \tag{6.29}$$

where

$$\beta_{mn} = \frac{2p\pi}{L} \tag{6.30}$$

While the phase coefficient for the ring waveguide modes is clearly quantised (Eq. (6.30)), this is not in the quantum electrodynamic sense (c.f. Eq. (6.16)). Quantisation comes purely from the ring resonator geometry and not from any properties of possible EM particles: that is photons.

So does QED provide a better insight into the behaviour of this EM resonator. There are two reasons why we might wish to question the CEM solution. Firstly, there is the issue of adopting electric and magnetic fields as the primary variables. Feynman [16] has actually observed that:

> there is no such concept as 'the' field, an independent entity with degrees of freedom of its own

Modern physics tends to view the idea of fields as an outdated stratagem to help visualise 'action at a distance'. Secondly, there is the allocation of boundary conditions that remain troublesome. What do the boundary conditions really imply, when in reality, at the interface between two materials, there are electrons loosely or tightly attached to atoms? Microscopically, a boundary layer is granular.

6.6 EM Waves and QED

In QED, the ring resonator analysis might proceed as follows. If the trapped energy is not so low that only isolated photons are detectable, the coupled electron waves occupying an infinitesimally thin layer on the interior surfaces of the superconducting waveguide walls form current flows as yet unknown. Each element of current can be represented by a current density vector **J** (A/m^2). The electron interactions within the conductor and across the vacuum space can be represented by the vector potential **A** for each **J** in volume dv. Hence, generalising Eq. (6.18), we obtain:

$$\mathbf{A} = \frac{\mu_0}{4\pi} \underset{\text{allspace}}{\oiint \int} \frac{\mathbf{J}}{r} \mathrm{d}v \tag{6.31}$$

We also have an electric potential (V) associated with the charge density (ρ) within dv. That is:

$$V = \frac{\eta_0}{4\pi} \underset{\text{allspace}}{\oiint \int} \frac{\rho}{r} \mathrm{d}v \tag{6.32}$$

Here, $\eta_0 = \sqrt{\mu_0/\varepsilon_0}$ and ε_0 is the permittivity of vacuum. Strictly speaking, **A** and V can be viewed as representing the possibility of interaction between electrons in motion. Mindful of maintaining a healthy distance from the notion of fields, Mead [10] is quoted as suggesting that they provide: 'a book-keeping device for evaluating the effect at a particular point of all currents everywhere in space'. So a QED waveguide computation would commence with a crude 'trial' assumption for the waveguide wall currents and for each element dv, it will be necessary to compute (with accurate book-keeping) the **A** and V at every other element, producing changes in **J** and ρ at these elements which in turn influences the original elements. A largely intractable, complex, non-linear, iterative process is the result. Such a direct QED computation, if attempted, would hopefully converge on a steady state current distribution in the waveguide walls, which is recognisably a mode of the rectangular waveguide as outlined above. Actually, in practice, the linearity, at the isolated current element level, between **A** and **J**, and V and ρ, is used to invert the problem. For example, Eq. (6.31) inverts to become

$$\nabla^2 \mathbf{A} = -\mu_0 \mathbf{J} \tag{6.33}$$

a second-order differential equation, which is readily solvable if the boundary conditions on **A** can be established.

For the waveguide problem, therefore, a QED solution implies the employment of the magnetic vector potential **A** and the electric potential V and formulating the differential equations that these variables must satisfy for time-varying currents. These are:

$$\nabla^2 \mathbf{A} - \frac{1}{c^2}\frac{\partial^2 \mathbf{A}}{\partial t^2} = -\mu_0 \mathbf{J} \tag{6.34}$$

$$\nabla^2 V - \frac{1}{c^2}\frac{\partial^2 V}{\partial t^2} = -\eta_0 \rho \tag{6.35}$$

But, within the interior of the vacuum filled waveguide, **J** and ρ are zero. Therefore, QED requires the solution of wave equations identical in form to the classical electromagnetic versions (Eq. (6.19)). These can be solved if the boundary conditions on the perfectly conducting walls can be formulated. Such boundary conditions have been provided by Mead [10] for a perfectly conducting surface supporting an infinitesimal current layer denoted by $\mathbf{J}_s$:

$$\begin{aligned} \left.\frac{\partial \mathbf{A}}{\partial s}\right|_{\text{surf}} &= -\mu_0 \mathbf{J}_s \\ \left.\frac{\partial V}{\partial s}\right|_{\text{surf}} &= -\mu_0 c^2 \rho \end{aligned} \tag{6.36}$$

As Mead observes, in relation to this use of potentials in solving QED problems:

> This view provides the conceptual basis for the historic success of the boundary-value approach to electromagnetic problems.

In other words, we have come full circle. When applied to classical EM problems, QED essentially ends up solving the same differential equations together with appropriate boundary conditions. So provided **B** and **E** can be related mathematically to **A** and *V*, and electromagnetic boundary conditions can be equated to those applied to **A** and *V*, CEM is in accord with QED. Consistency between the two approaches is secured if:

$$\begin{aligned} \mathbf{B} &= \nabla \times \mathbf{A} \\ \mathbf{E} &= -\nabla V \end{aligned} \qquad (6.37)$$

It is ironic that in the early history of electromagnetism, Maxwell actually developed his now famous equations in terms of the vector potential **A,** which he termed the electromagnetic momentum. Given today's controversy over the significance of fields, the fact that Oliver Heaviside and Heinrich Hertz disagreed with Maxwell's use of **A** is perhaps rather unfortunate, since the result was that they reworked his equations into the format widely used today, in which **E** and **B** are the principal variables [17].

In summary, modern electromagnetism should retain field equations but should sensibly reflect appropriate aspects of quantum electrodynamics. To do so, it should embrace the following statements:

1. A photon is a measure of the energy that is exchanged in an atomic or particle transition within a material. Its magnitude is proportional to frequency.
2. Photons as discrete particles only occur in free space, or in macroscopic spaces containing EM waves, at power densities that are several orders below practical engineering levels.
3. In the 'real' world, 'actions at a distance' for non-isolated particles are transmitted by wave mechanisms that are deterministic. The probabilistic aspects of the photon wave function can be deemed to be insignificant in the engineering context.
4. The fundamental variables in electromagnetic theory are the electronic charges and their motions. The electron can have wave characteristics.
5. Quantum electrodynamic equations relating energy and momentum to electron charge, current and potential are a fundamental aspect of electromagnetism and are as important as the Maxwell's relations.
6. 'Action at a distance' is expressed, in QED, through a vector potential **A** and a scalar potential *V*.
7. **B** and **E** are auxiliary variables in electromagnetic theory, derived from **A** and *V*.

So in conclusion, the practice of electromagnetism for electrical engineers is predominantly a deterministic field-based discipline. It is of course necessary to acknowledge that **E** and **B** are related to the quantum electrodynamic variables **A** and *V*, and that in turn, these variables are properly related through electrodynamics to electron current and charge.

References

1. Stratton JA (2007) Electromagnetic theory. Wiley, New Jersey
2. Kenyon IR (2008) The light fantastic. Oxford University Press, Oxford
3. Feynman RP (1990) QED. Penguin Books Ltd., London
4. Sangster AJ (1965) A method for the analysis of waveguide coupling. Proc IEE 112:2171–2179
5. Sangster AJ et al (2008) Open-ended microwave oven for flip-chip assembly. IEE Proc (MAP) 2(1):53–58
6. Einstein A (1905) Ann Phys 17:891–921
7. Feynman RP, Leighton RB, Sands M (1964) Lectures on physics II. Addison-Wesley, Reading
8. Yariv A (1989) Quantum electronics. Wiley, Singapore
9. Peskin ME, Schroeder DV (1995) An introduction to quantum field theory. Westview, Boulder
10. Mead CA (2000) Collective electrodynamics. MIT Press, Cambridge
11. Vanderlinde J (1993) Classical electromagnetic theory. Wiley, New York
12. Jackson J (1998) Classical electrodynamics. Wiley, New York
13. Kraus JD (1984) Electromagnetics. McGraw-Hill, New York
14. Rao NN (1994) Elements of engineering electromagnetics. Prentice-Hall, New Jersey
15. Sadiku MNO (1995) Elements of electromagnetics. Oxford University Press, New York
16. Wheeler JA, Feynman RP (1945) Rev Mod Phys 17:157–181
17. Hertz H (1900) Electric waves. MacMillan, London

Chapter 7
Solar Photovoltaics

In atomic theory we have fields and we have particles. The fields and the particles are not two different things. They are two ways of describing the same thing—two different points of view

P.A.M. Dirac

Abstract In this chapter, it is demonstrated that the photovoltaic cell which forms the basis of one class of solar power system is essentially a semiconductor junction diode formed from doped crystalline silicon. The diode action is developed here in electrical engineering terms. Consequently, when it is illuminated with light, the photovoltaic mechanism can readily be explained by viewing the device as immersed in an electromagnetic wave which has the effect of modifying the quiescent state of the diode, thus causing a photo-current to flow. The basic photodiode equation, which forms the 'bedrock' of solar PV module and array design and development, is key to array operation and control. It can thus be evolved, as the chapter demonstrates, from the laws of engineering electromagnetism supported by fundamental thermodynamics. For regular solar modules and arrays, employing identical cells throughout their structures, it is shown that the cell's theoretical model is formally applicable to the whole module or even the entire array. Measured *I–V* characteristics are available from manufacturers for photovoltaic diode cells, and it is observed that the efficient and effective incorporation of such cells into modules and arrays necessitates the matching of theoretical parameters to experimental data. There is a plethora of mathematical techniques to do this, but here, we have concentrated on the Newton tangent method which cogently illustrates the process. The chapter concludes with a brief examination of array power collection and efficiency from a theoretical perspective. The need for power level control, and panel directional tracking of the sun, in order to maximise these important parameters is given some consideration in the penultimate section, while in the final section, the implications of technology advances on array efficiency are perused.

7.1 Introduction—Photovoltaic Collectors

The influence on modern consumer electronics of photovoltaic (PV) technology has been advancing steadily in the past ten to fifteen years, notwithstanding its rather quiet and unheralded start. Now, however, progress is overt and rapid. Today, solar

A.J. Sangster, *Electromagnetic Foundations of Solar Radiation Collection*,
Green Energy and Technology, DOI 10.1007/978-3-319-08512-8_7

cells are commonly incorporated into calculators, watches, radios and toys, and are increasingly to be found powering street signs, parking meters and traffic lights. Its use in electrical power engineering to provide a sustainable source of energy for modern communities, which are transitioning away from fossil fuels, is already well established as we shall see in this chapter.

For students of electrical science, PV can actually be a difficult subject because it is commonly taught from a quantum mechanical perspective. However, if it is assumed (see Chap. 6) that the light incident on the PV junction has sufficient intensity to validly presume that the incident solar power comprises electromagnetic waves, then many of the conceptual challenges faced by students can be circumvented. It is perhaps worth noting that at lower than light frequencies, the photovoltaic phenomenon is still present, but in this case, it is interpreted as semiconductor diode detection.

7.1.1 Solar Cell Electronics

Photovoltaic cells are in essence large-area (typically 10 cm in diameter) semiconductor diodes which are primarily formed from doped silicon. It has a crystalline structure in which the atoms bind together by a generous and powerful sharing of electrons. In its pure form, silicon is a poor conductor at normal temperatures, since few of these electrons are 'free' as in a conductor. However, when silicon is doped with a small amount of arsenic, atoms of the dopant get bound up in the silicon crystal lattice. But arsenic, or antimony, has one more electrons in the outer shell of its atom than silicon, and this electron does not get used in the binding and sharing process associated with the formation of covalent bonds. Consequently, the impurity atoms (donor atoms) contribute 'free' electrons which can drift through the crystal (Fig. 7.1). The material, termed N-type silicon, now conducts although not as well as a metal. A similar process occurs when silicon is doped with a material like boron, gallium or indium, each of which has fewer electrons in the outer ring of its atom (acceptor atom). In this case, at the positions within the silicon lattice where the indium has taken the place of a silicon atom, a binding electron is missing, forming a 'hole' in the lattice (see Fig. 7.1). Any free electron entering this P-type material will be attracted into the 'holes' and conduction again results. In generating this explanation, significant liberties have been taken with the solid-state physics. Nevertheless, the explanation retains sufficient depth to impart a reliable appreciation of the topic.

The really interesting aspect of this silicon doping exercise arises when N-type material is in contact with P-type material [1]. Note that electrons in an *N*-doped silicon crystal are termed majority carriers (of charge) as are holes in a *P*-doped region. On the other hand, electrons which manage to diffuse into the P region become minority carriers as are holes which migrate into the N layer. Minority carriers also exist in the P and N regions due to thermal agitation of the silicon atoms. If the thermal energy is strong enough, covalent bonds can be broken

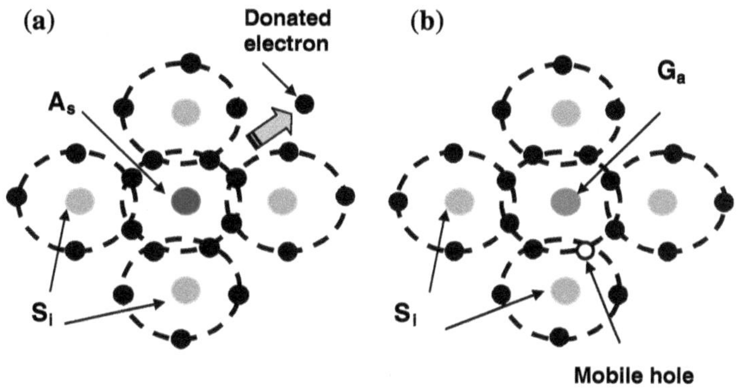

Fig. 7.1 **a** *N*-doped silicon with a donor atom (arsenic) and, **b** *P*-doped silicon with an acceptor atom (gallium) embedded in the silicon lattice

releasing, respectively, minority electrons and holes. The rate of thermal generation of minority carriers is exponentially dependent on temperature which means that semiconductors are notoriously temperature sensitive.

At the instant, when the N-type and P-type materials are brought together, the second law of thermodynamics comes into action, encouraging thermally agitated 'free' electrons in the N-type material to diffuse across the junction, additionally so, because of the P-type holes which are waiting to be filled. The process of diffusion is quite common in nature; it is picturesquely present when a smoke ring spreads inexorably into the still surrounding air, dispersing in accordance with the second law. In the PN junction, the process will continue until the negative charge on the P side of the junction and the positive charge (due to electron deficiency) on the N side result in a residual barrier voltage, and a concomitant electric field (see Fig. 7.2) across the junction, which prevents further charge movement by diffusion. This electric field is a charge separation field, as described in Chap. 2. It is important to be clear that only charge within the semiconductor is involved in forming the voltage barrier and the electrically neutral silicon crystal which existed at the outset remains so. The barrier voltage does not appear at the terminals of the device, and no current flows if it is 'shorted' as shown in Fig. 7.2.

The depletion layer is so described because the charge separation results in narrow regions which are relatively charge free on either side of the junction. The width of this layer is dependent on the crystal preparation, and it need not extend equally on the P and N sides of the junction. The relative widths are related to the number of impurity atoms per unit volume in the P and N materials.

The flow of charge across the depletion layer as the barrier builds comprises two components—the majority carriers from the P layer (holes) which recombine with electrons in the N layer resulting in current I_{rp} (r = recombination; p = P layer) and majority carriers from the N layer (electrons) which recombine with holes in the P layer producing a current I_{rn}. These currents continue even after the barrier voltage reaches maximum height (typically 0.3–0.7 V) because statistically, in a 'warm'

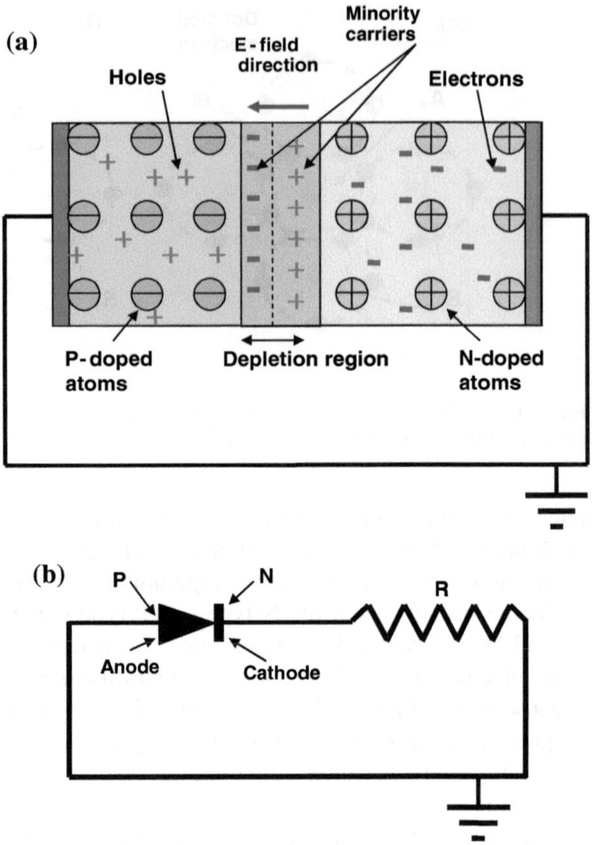

Fig. 7.2 PN junction showing **a** formation of depletion or barrier layer in unbiased arrangement and, **b** its equivalent electrical circuit

device, there is always the possibility of a hole or an electron having enough energy to 'vault' the barrier. That the barrier does not keep building is due to the minority carriers in the P and N regions which 'escape' atoms due to thermal agitation. If these appear near the depletion layer, they are 'swept' across it by the electric field there. Again there are two components of current—I_{tp} associated with thermally agitated minority carriers travelling from the N layer to the P layer and I_{tn} associated with minority carriers moving in the opposite direction. Since a hole moving left is equivalent to an electron shifting rightwards, or vice versa, we can state that the total recombination current:

$$I_r = I_{rp} + I_{rn}$$

and the total thermally generated current is

$$I_t = I_{tp} + I_{tn}$$

In equilibrium (unbiased) conditions as depicted in Fig. 7.2

$$I_r = I_t = I_s$$

where I_s is termed the saturation current.

With the above knowledge of the quiescent conditions within an unbiased semiconductor, it is not difficult to see that any external operation which lowers the barrier voltage, or field, will boost I_r but have little effect on I_t. This is described as forward bias and can be procured by incorporating a voltage source in the external circuit or by immersing the junction in an electromagnetic wave with suitable field polarity, as we will see below. Alternatively, any operation which raises the barrier voltage height suppresses I_r relative to I_t which remains essentially unchanged. The current flow thus remains very low. This is termed reverse bias.

7.1.2 PN Junction Basic Equations

The development of practical photovoltaic arrays relies heavily on modelling techniques in which the theoretical junction characteristics, together with solar illumination data (see Sect. 1.3.1), are incorporated into simulation software designed to iteratively optimise solar collection performance of a multi-element photovoltaic array. The equations governing the operation of photovoltaic or semiconductor junctions, when illuminated by an electromagnetic wave at light frequencies, are therefore important. They can be determined, perhaps not unexpectedly, by applying Maxwell's equations to the depletion layer—see Fig. 7.3.

The appropriate Maxwell equation for a volume containing free charge is the Gauss law expressed through Eq. (3.7), that is

$$\nabla \cdot \mathbf{D} = \rho \tag{7.1}$$

The depletion layer is presumed to be located in a Cartesian frame with the x-direction normal to the semiconductor interfaces, and the y–z plane lying parallel to the interfaces. It is clear from Fig. 7.3 that changes in voltage, field strength and charge through the semiconductor junction occur in the x-direction only, with little change occurring in the y- and z-directions. In the Cartesian frame, Eq. (7.1) becomes

$$\frac{\partial D_x}{\partial x} + \frac{\partial D_y}{\partial y} + \frac{\partial D_z}{\partial z} = \rho \tag{7.2}$$

which reduces to

$$\frac{\mathrm{d}D_x}{\mathrm{d}x} = \rho(x) \tag{7.3}$$

Fig. 7.3 Semiconductor junction plus electron barrier potential. **a** Carrier (electrons and holes) concentration on logarithmic scale, **b** charge accumulation in depletion region, **c**, **d** *E*-field distribution (*purple*) in the depletion region and resultant built-in barrier voltage (*green*) (color figure online)

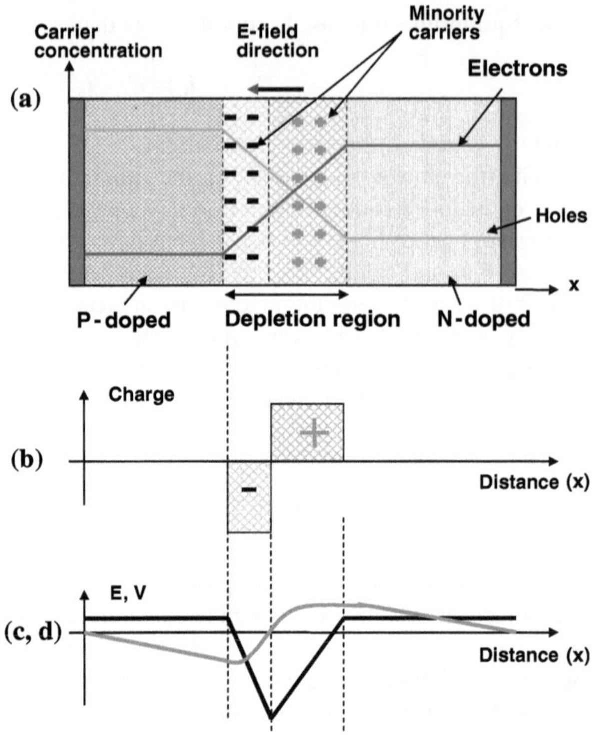

For the *N*-doped portion of the depletion layer, which has a surfeit of minority carriers or positive charge, plus the *P*-doped layer which has a surfeit of negative charge, the charge density can be expressed as

$$\rho(x) = e(N_a - N_d) \tag{7.4}$$

where N_a and N_d are, respectively, the number density of negatively and positively charged minority carriers, each of magnitude equal to the charge of an electron (e). Also $D_x = \varepsilon_r\varepsilon_0 E_x$; hence, we can write

$$\frac{\mathrm{d}E_x}{\mathrm{d}x} = \frac{e(N_a - N_d)}{\varepsilon_r\varepsilon_0} \tag{7.5}$$

But

$$E_x = -\frac{\mathrm{d}V_n}{\mathrm{d}x} \tag{7.6}$$

so in voltage terms Eq. (7.5) becomes

$$\frac{\mathrm{d}^2 V_n}{\mathrm{d}x^2} = \frac{e(N_a - N_d)}{\varepsilon_r \varepsilon_0} \tag{7.7}$$

In principle, this second-order differential equation (Poisson equation) can be solved directly if the depletion layer boundary conditions can be established and the charge flows across the depletion layer junction can be defined.

Nevertheless, it is perhaps more instructive to build the solution by pursuing empirical reasoning. On integrating Eq. (7.5) with respect to x across the N-doped section of the depletion layer, we get

$$E_x = \frac{eN_d x}{\varepsilon_r \varepsilon_0} \tag{7.8}$$

For a constant level of free charge over this portion of the depletion layer, as suggested in Fig. 7.3, this equation dictates a linearly increasing E-field across the layer, as indicated in the figure. The opposite trend develops in the P-doped layer where the minority carriers are negative. If the thickness of the depletion region in the N-doped material is d_n, then the voltage (V_n) across this layer is readily available by integration. Thus

$$V_n = \int E_x \mathrm{d}x = \frac{eN_d d_n^2}{2\varepsilon_r \varepsilon_0} \tag{7.9}$$

and for the P-doped layer with N_a acceptor ions per unit volume:

$$V_p = -\frac{eN_a d_p^2}{2\varepsilon_r \varepsilon_0} \tag{7.10}$$

Now, given that the depletion layer as a whole must be electrically neutral in the quiescent state, the following relation applies:

$$N_a d_p = N_d d_n \tag{7.11}$$

By combining Eqs. (7.9–7.11), it is not difficult to show that

$$\frac{V_p}{V_n} = \frac{N_d}{N_a} \tag{7.12}$$

Consequently, the 'barrier' voltage across the complete depletion layer of thickness $d = d_p + d_n$ can be expressed as

$$V_j = V_n - V_p = V_n\left(1 + \frac{N_d}{N_a}\right) = \left[\frac{eN_d d^2}{\varepsilon_r \varepsilon_0}\right]\left(1 + \frac{N_d}{N_a}\right) \tag{7.13}$$

As noted earlier, this voltage does not appear in the external circuit because the device as a whole remains electrically neutral as suggested in Fig. 7.3d—otherwise, the semiconductor junction would defy power conservation. The charge, *E*-field and voltage magnitudes and trends are not to scale in Fig. 7.3. They are merely illustrative of the junction diffusion process.

Due to diffusion, the density of minority carriers (electron density n_p) in the depletion layer on the P side of the junction (see Fig. 7.3) is related to the density of electrons (n_n) in the *N*-doped region by the Boltzmann equation which applies to all diffusion processes. That is

$$n_p = n_n \exp\left(\frac{-eV_j}{k_B T}\right) \tag{7.14}$$

where T is the absolute temperature in Kelvin and k_B is the Boltzmann constant (1.3807×10^{-23} J/K). A similar equation applies to p_p and p_n the number densities of holes. When a forward bias V is applied to the junction, the minority carrier density increases due to charge injection across the 'lowered barrier' giving

$$n_p + \Delta n_p = n_n \exp\left(\frac{-e(V_j - V)}{k_B T}\right) \tag{7.15}$$

Hence, using the semiconductor equation, namely $n_p p_p = n_n p_n$, it is possible to write

$$n_p + \Delta n_p = n_p \exp\left(\frac{eV}{k_B T}\right) \tag{7.16}$$

and

$$\Delta n_p = n_p\left[\exp\left(\frac{eV}{k_B T}\right) - 1\right] \tag{7.17}$$

As long as V persists and is positive, this excess charge at the boundary leaks steadily into the bulk semiconductor, producing a 'forward bias' current (I_{fb}) proportional to the incremental charge densities. That is

$$I_{fb} = I_s\left[\exp\left(\frac{eV}{k_B T}\right) - 1\right] \tag{7.18}$$

V can also be negative in which case the current represented is the reverse bias current I_{rb}. Equation (7.18) is the desired solution to the Poisson equation in its most basic form.

7.1.3 Photovoltaic Action

From the perspective of radiation detection, this current I_{fb} will flow in the external circuit of the semiconductor junction during the phase of the incoming wave which creates an E-field across the depletion region equivalent to a voltage $+V$. On the other hand, in the following half-cycle when the E-field reverses, the voltage in the depletion layer becomes negative and the junction becomes reverse biased. The current generated (I_{rb}) is tiny (typically in the microamp range) and is essentially equal in magnitude to I_s. In photodiode terms, it is often termed the dark current. When these currents in the external circuit are passed through a rectifying circuit, a DC voltage is generated which is related to the power in the incident electromagnetic wave. The process represents AC to DC conversion.

The mechanism is illustrated in Fig. 7.4. When the instantaneous electric field across the junction due to the electromagnetic wave is in the direction of reducing the impeding charge separation field (lowering the potential barrier, see Fig. 7.3d), electrons will start to find their way across the junction, and a current (I_{fb}) flows (see Fig. 7.4) as discussed in the previous section. The holes drifting across the semiconductor junction are balanced by a current flow in the external circuit. On the other hand, in the half-cycle of the wave when the electric field enhances the charge separation field, electrons continue to be prevented from crossing the junction which is reduced to the saturation value. The averaged charge flowing across the junction thus contributes to a DC current through the semiconductor diode resulting from its immersion in the AC electromagnetic wave—namely light in the case of a PV junction.

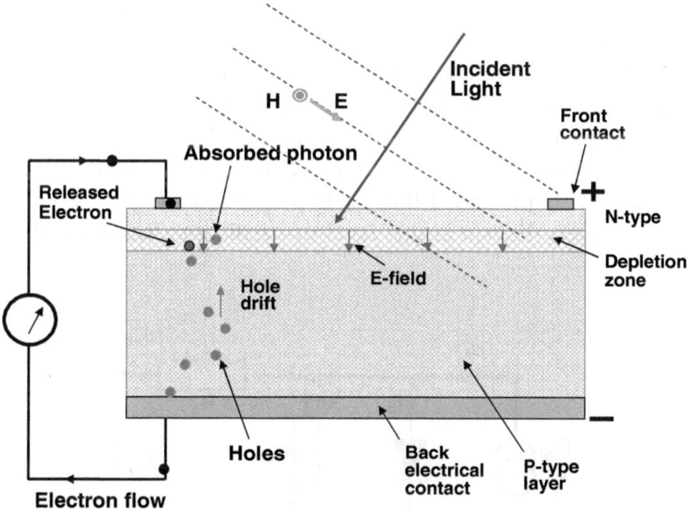

Fig. 7.4 Schematic of photovoltaic crystal showing an electron, released by the EM field, moves to the P layer causing holes to drift towards the N layer

At light frequencies, the above process can alternatively be explained from a quantum mechanics perspective. Photons of the appropriate frequency and energy can, by absorption into the doped depletion layer atoms, dislodge electrons with enough energy to surmount the potential barrier, hence creating an instantaneous current flow as before. In fact, modern solar cells actually have a thin layer of intrinsic material (undoped silicon) between the P-type and N-type semiconductors (PIN diodes) which helps improve photon collection and hence efficiency (see Sect. 7.3.3).

7.2 PV Array Simulation

As intimated earlier in this chapter, solar collection arrays based on photovoltaic cells are largely designed by simulation which requires the existence of reliable mathematical models of practical PN junctions. Such models are now relatively well developed as the literature attests [2–16]. Generally, they rely on an accurate knowledge of the semiconductor junction physics.

The current–voltage relationship for semiconductor junctions and hence PV cells is quite familiar to most electrical engineers, and for any given cell, it can usually be generated by experimental means with very little difficulty. Consequently, the formulation of cell models generally involves converting the measured current–voltage relationship into electrical circuit parameters more appropriate to array design. These are usually the saturation current, series and shunt resistance and quality factor. The literature describes a range of possible models of PV cells, and these are comprehensively summarised in reference [2]. Here, we will follow through representative procedures described in reference [12–14] to generate a model based on the circuit shown in Fig. 7.5.

Arguably, the most common model for a photovoltaic cell comprises a single diode, a current source, and parallel (R_{sh}) and series (R_s) resistances representing electrical and thermal losses, as depicted in Fig. 7.5. The current source (I_{ph}) represents the injection of power into the cell by incident light at power density level p_{sol} (W/m^2). It is presumed that I_{ph} is proportional to p_{sol}. T_j is the junction temperature in degrees Kelvin at this level of irradiation. Basically, the modelling

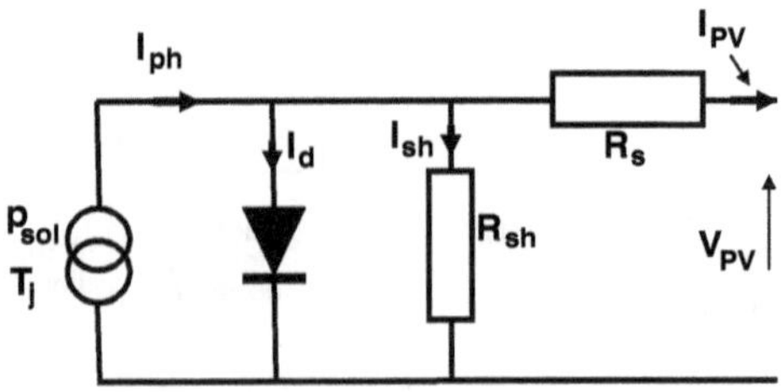

Fig. 7.5 Equivalent electrical circuit for illuminated solar cell

imperative is to determine the unknown parameters in Fig. 7.5 so that the resultant theoretical I_{PV}–V_{PV} characteristic matches that of a manufactured cell intended for use in a solar array panel. Such a characteristic for a solar cell (6 cm diameter) fabricated by R.T.C. in France [12] is shown in Fig. 7.6. In Fig. 7.6a, the I_{PV}–V_{PV} characteristics are shown in the conventional diode form for an unilluminated diode (dark blue circles) and for an illuminated cell generating a 0.76 A 'photon' current (I_{ph}) (orange diamonds). The solar industry tends to present the same data in the form shown in Fig. 7.6b which presumes that the cell is irradiated (typically with p_{sol} = 1,000 W/m^2 at T_j = 33 °C).

The 'dark' curve in Fig. 7.6a depicts the traditional diode characteristic with essentially zero current in reverse bias (V negative) and a rapidly growing forward bias current when the voltage exceeds a threshold—in this case at ~0.4 V. Illumination at typically 1,000 W/m^2 introduces a current flow I_{ph} which is in a direction that imposes a negative current (−0.76 A) bias on the photodiode. Note that at V_{PV} = 0 the short circuit current $I_{sc} = I_{ph}$. On increasing V_{PV} in the positive sense from this point, instead of producing a forwardly biased current in the diode, as in the 'dark' case, it diminishes the cell current (I_{PV}) until the open circuit condition (V_{oc}) occurs, when I_{PV} = 0.

The theoretical I–V characteristic for a solar cell can be developed by applying Kirchhoff's law to the equivalent circuit in Fig. 7.5. This gives

$$I_{PV} = I_{ph} - I_D - I_{sh} \tag{7.19}$$

But $I_D = I_{fb}$, so from Eq. (7.18)

$$I_D = I_0\left[\exp\left(\frac{e(V_{PV} + I_{PV}R_s)}{A_q k_B T_j}\right) - 1\right] \tag{7.20}$$

where I_0 is the reverse saturation current of the diode (=I_s) and A_q is a quality factor for the diode. This is the degree to which the measured diode deviates from an ideal diode, for which A_q = 1. Hence,

$$I_{PV} = I_{ph} - I_0\left[\exp\left(\frac{e(V_{PV} + I_{PV}R_s)}{A_q k_B T_j}\right) - 1\right] - \frac{V_{PV} + I_{PV}R_s}{R_{sh}} \tag{7.21}$$

In this equation, I_{ph} (Amps) is the photocurrent. It is equal to the diode short circuit current (I_{sc} A) where V_{PV} = 0. Hence, we obtain for I_{ph}

$$I_{ph} = I_{sc}\left(1 + \frac{R_s}{R_{sh}}\right) + I_0\left(\exp\frac{eI_{sc}R_s}{A_q k_B T_j} - 1\right) \tag{7.22}$$

Equation (7.21), with I_{ph} replaced by Eq. (7.22), represents a multitude of possible I_{PV}–V_{PV} characteristics depending on the values of the unknowns R_s, R_{sh}, A_q and I_0 and cannot be solved analytically. The generally accepted method of matching the theoretical I–V curve represented by Eq. (7.21) to the measured points

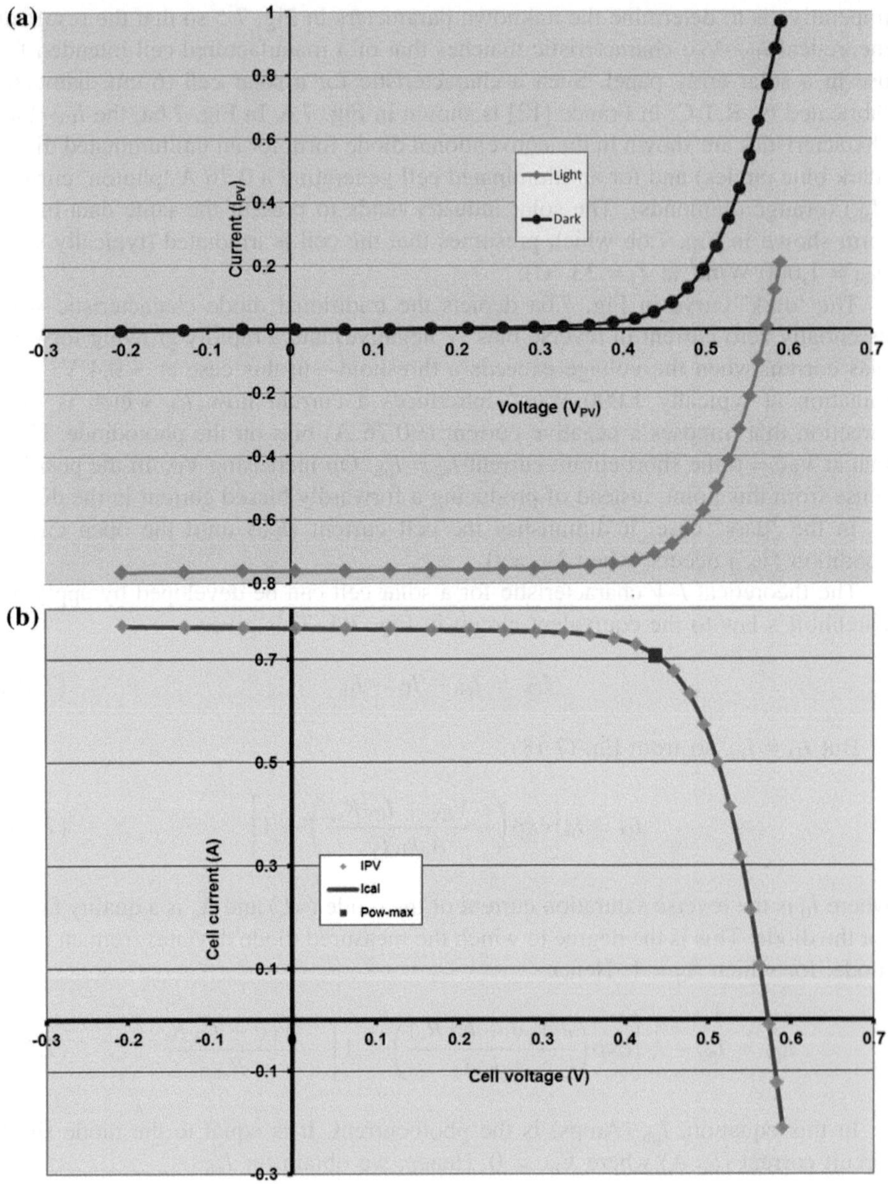

Fig. 7.6 **a** Conventional *I–V* characteristics for non-irradiated cell (*dark*) and for an illuminated cell (*light*) and **b** measured (*red diamonds*) and modelled (*green curve*) *I–V* characteristics for the same cell in solar industry form. The *mauve square* denotes the maximum power point (color figure online)

for a given cell is a nonlinear least-squares error minimisation technique applied at each experimental point. The most attractive of these is Newton's method [17], which is accessible in algorithmic form on software such as MATLAB [18].

7.2.1 Newton Iteration Procedure

The solution method can be summarised as outlined below. In essence, in order to fit Eq. (7.21) to the measured data, it is necessary to minimise a function $S(\theta)$ which expresses the square of the difference between the desired value and the computed value of a quantity of interest, in our case cell currents I_i at i values of this quantity where $i = 1, 2, 3 \ldots N$. θ_k $(k = 1, 2, 3, 4)$ represents, respectively, the set of unknown parameters R_s, R_{sh}, A_q and I_0. To accommodate this spectrum of values mathematically, we require to formulate a matrix **[S]** where

$$S_i(\theta_k) = [I_i - I_{PV}(V_i, I_i, \theta_k)]^2 \tag{7.23}$$

and

$$[\mathbf{S}(\theta)] = \begin{bmatrix} S_1(\theta_1)S_1(\theta_2)S_1(\theta_3)S_1(\theta_4) \\ S_2(\theta_1)S_2(\theta_2)S_2(\theta_3)S_2(\theta_4) \\ \ldots \\ \ldots \\ S_i(\theta_1)S_i(\theta_2)S_i(\theta_3)S_i(\theta_4) \\ \ldots \\ S_N(\theta_1)S_N(\theta_2)S_N(\theta_3)S_N(\theta_4) \end{bmatrix} \tag{7.24}$$

For the matrix element $S_i(\theta_k)$, V_i and I_i are the measured values of the cell voltage and current, respectively, at point i. Examination of Eq. (7.21) makes it clear that the I_{PV} is implicitly bound into the I–V cell equation. However, simplification is obviously possible by replacing V_{PV} and I_{PV} on the right of the equals sign by the measured values at the point of interest i, namely V_i and I_i. Hence

$$I_{PV} = I_{ph} - I_0\left[\exp\left(\frac{e(V_i + I_iR_s)}{A_q k_B T_j}\right) - 1\right] - \frac{V_i + I_iR_s}{R_{sh}} \tag{7.25}$$

To find the minimum of a function such as $[S(\theta)]$, the suggested route is to employ a numerical analysis procedure such as Newton's method of tangents (also known as the Newton–Raphson method), although many other possibilities exist [17]. To explain the numerical process, it is sensible to examine its application to a simple quantity $f(x)$ which is a function of the single variable x. The method then entails a successive search for better approximations to the roots (or zeroes) of the real-valued function (see Fig. 7.7).

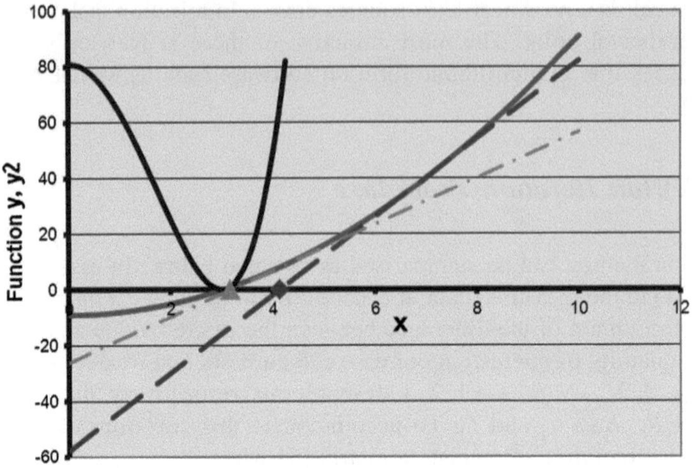

Fig. 7.7 Newton method of tangents numerical iteration procedure applied to *mauve curve* at $x = x_0 = 7$: $x_1 = 4.14$, $x_2 = 3.167$. The root is at $x = 3$

Given a mathematical quantity f, which is a function of the real dimension x, and its derivative $f'(x)$, we commence with a trial value x_0 for a root of the function f. Provided the function satisfies all the assumptions made in the derivation of the formula, a better approximation for the root (zero) of $f(x)$ is x_1 given by

$$x_1 = x_0 - \frac{f(x_0)}{f'(x_0)} \tag{7.26}$$

Graphically, the point (x_1, 0) is the intersection with the x-axis of the tangent to f at the point (x_0, $f(x_0)$). This intersection occurs at $x = 4.14$ in Fig. 7.7 (green diamond).

The process can be repeated as dictated by

$$x_{n+1} = x_n - \frac{f(x_n)}{f'(x_n)} \tag{7.27}$$

until a sufficiently accurate value of x is reached. In Fig. 7.7, the second tangent at $x = 4.14$ (chain dotted straight line) produces an intersection $x_2 = 3.167$ (red triangle) which is in error on the desired value of 3.0 by 5.5 %.

To improve the chances that the iterative procedure will track towards the desired root, or zero, the technique can be applied to a function $F(x)$ which is the square of the original function $f(x)$, i.e. $F(x) = [f(x)]^2$, see Fig. 7.7—U-shaped blue curve. The advantage of this change is that the iteration always progresses 'downhill' independent of the original trial value—that is whether or not it is less or greater than the root value. Of course, the original trial must be within the mathematical 'trough' containing the minimum.

For a multidimensional function such as $[S(\theta)]$, the Newton method can still be applied, in which case the interation involves 'tracking downhill' to a minimum in a multidimensional surface. If we form a matrix $[J(\theta)]$ (termed the Jacobean) whose elements are the partial differentials of the corresponding elements in $[S]$, namely

$$[J(\theta)] = \begin{bmatrix} \frac{\partial S_1}{\partial \theta_1} \frac{\partial S_1}{\partial \theta_2} \frac{\partial S_1}{\partial \theta_3} \frac{\partial S_1}{\partial \theta_4} \\ \frac{\partial S_2}{\partial \theta_1} \frac{\partial S_2}{\partial \theta_2} \frac{\partial S_2}{\partial \theta_3} \frac{\partial S_2}{\partial \theta_4} \\ \cdots \\ \cdots \\ \frac{\partial S_i}{\partial \theta_1} \frac{\partial S_i}{\partial \theta_2} \frac{\partial S_i}{\partial \theta_3} \frac{\partial S_i}{\partial \theta_4} \\ \cdots \\ \frac{\partial S_N}{\partial \theta_1} \frac{\partial S_N}{\partial \theta_2} \frac{\partial S_N}{\partial \theta_3} \frac{\partial S_N}{\partial \theta_4} \end{bmatrix} \tag{7.28}$$

then the Newton procedure can be expressed succinctly as

$$[\theta_n] = [\theta_{n-1}] - \frac{[S(\theta_{n-1})]}{[J(\theta_{n-1})]} \tag{7.29}$$

where $1 < n < N$ the maximum number of iteration steps. When applied to the cell represented by the measured I–V curve in Fig. 7.6, the green solid line trace results, which is patently in excellent agreement with the measured points.

For a four-dimensional surface represented by $[S]$, the Newton method of searching for a root is susceptible [12] to oscillation and overflow. The solution is a minor modification in the iterative equation (7.29) as delineated in Eq. (7.30). The modifier λ_L is termed the Levenberg parameter and is typically of the order of 0.1. Its effect is to 'slow' the convergence process and thereby avoid oscillation.

$$[\theta_k] = [\theta_{k-1}] - \lambda_L \frac{[S(\theta_{k-1})]}{[J(\theta_{k-1})]} \tag{7.30}$$

As has already been intimated, the Newton procedure converges reliably only if the original trial solution is not too inaccurate. The determination of a suitable trial calls for the availability of a computational routine which can produce an approximate solution quickly and reliably. Such a routine is provided by reference [13].

7.2.2 Solar Cell Conductance Method

Referring to Eq. (7.21), it is possible to make some approximations which have little impact on the accuracy of the equation. Firstly, if the reverse bias applied to the cell is large, then evidently

$$\frac{eV_{\mathrm{PV}}}{k_{\mathrm{B}}T_{\mathrm{j}}} \ll 1 \tag{7.31}$$

And if we also make the reasonable assumption that $R_{\mathrm{sh}} \gg R_{\mathrm{s}}$, then the shunt conductance can be deduced from the reverse bias cell characteristic by implementing a simple linear fit. In this case, the calculated value of R_{sh} gives the shunt current $I_{\mathrm{sh}} = V_{\mathrm{PV}}/R_{\mathrm{sh}}$ which can be deducted from the measured cell current to establish the current through the junction itself.

Hence, under forward bias conditions for which $V_{\mathrm{PV}} + R_{\mathrm{s}}I_{\mathrm{PV}} \gg k_{\mathrm{B}}T_{\mathrm{j}}$, Eq. (7.21) reduces to

$$I_{\mathrm{PV}} = I_{\mathrm{ph}} - I_0\left[\exp\left(\frac{e(V_{\mathrm{PV}} + I_{\mathrm{PV}}R_{\mathrm{s}})}{A_{\mathrm{q}}k_{\mathrm{B}}T_{\mathrm{j}}}\right)\right] \tag{7.32}$$

In this case, the -1 within the bracket containing the exponential term is swamped, while the final term is insignificant for $R_{\mathrm{sh}} \rightarrow \infty$. Using this equation, we can now formulate a theoretical conductance

$$G = \frac{\mathrm{d}I}{\mathrm{d}V} = \frac{\mathrm{d}I_{\mathrm{PV}}}{\mathrm{d}V_{\mathrm{PV}}} \tag{7.33}$$

which has the form

$$G = -\frac{e}{A_{\mathrm{q}}k_{\mathrm{B}}T_{\mathrm{j}}}(1 + R_{\mathrm{s}}G)(I_{\mathrm{PH}} - I) \tag{7.34}$$

where $I \equiv I_{\mathrm{PV}}$. If Eq. (7.34) is reformulated as

$$\frac{G}{(I_{\mathrm{ph}} - I)} = -\frac{e}{A_{\mathrm{q}}k_{\mathrm{B}}T_{\mathrm{j}}}(1 + R_{\mathrm{s}}G) \tag{7.35}$$

and plotted on a $G/(I_{\mathrm{ph}} - I)$ versus G orthogonal axis pair, Eq. (7.35) represents a straight line with intercept $-e/k_{\mathrm{B}}T_{\mathrm{j}}$ on the vertical axis and slope $-eR_{\mathrm{s}}/A_{\mathrm{q}}k_{\mathrm{B}}T_{\mathrm{j}}$ as shown in Fig. 7.8. The final approximation involves the not untypical cell property that the reverse saturation current I_0 is very much less than the photon current I_{ph}, and hence, we can equate I_{ph} to the short circuit current I_{SC}. I_0 can be evaluated by standard diode procedures from the I–V characteristic of the given cell. The calculation steps described here can readily be formulated into an algorithm which should be capable of very rapid evaluation on any respectable computer.

For the solar cell depicted in Figs. 7.6 and 7.8, the extracted parameters based on the Newton method and on the conductance method are listed and compared in Table 7.1. It is clear that good agreement pertains and that in this case, the conductance method would present excellent trial values for the Newton iteration.

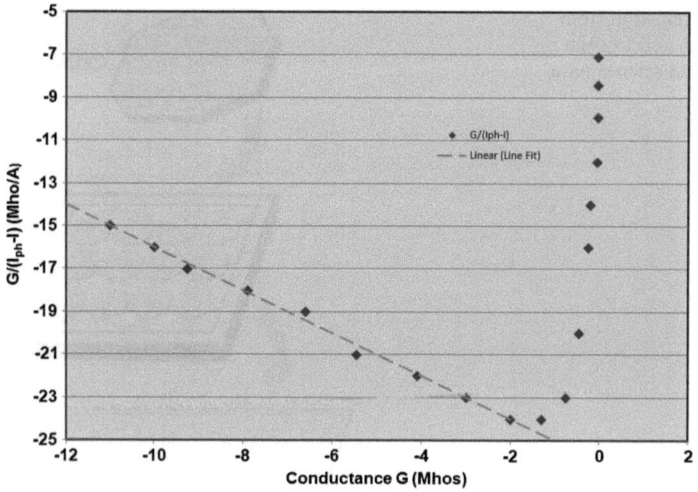

Fig. 7.8 Conductance-to-cell current ratio as a function of conductance G for a commercial solar cell (R.T.C. France). For the line fit (*dashed line*), the y-axis intercept yields $-e/k_BT_j$, while the slope gives $-eR_s/A_qk_BT_j$

Table 7.1 Comparison of parameter extraction methods for solar cell at 33 °C

Parameter	Newton method	Conductance method
R_{sh} Ω	0.0186	0.02386
R_s Ω	0.0364	0.0385
A_q	1.4837	1.456
I_s μA	0.3223	0.46
I_{ph} A	0.7608	0.7603

7.3 Cells, Modules and Arrays

Solar cells are predominantly manufactured from crystalline silicon, which is also known as solar-grade silicon. Without getting sidetracked into chemistry, it is relevant to examine briefly the fabrication of photovoltaic cells. Bulk silicon is available in several categories according to its crystal structure and size and whether or not it exists originally as an ingot, ribbon or wafer. The main categories are monocrystalline silicon, polycrystalline silicon, ribbon silicon and hybridised crystalline silicon.

Monocrystalline silicon wafers are sliced from an approximately 10-cm-diameter p-doped cylindrical ingot. Boron is most commonly used as the dopant. The N layer can be achieved by stacking the wafers in a heated tube through which a carrier gas is passed, laced with, for example, phosphorous. An oxide layer which grows on the

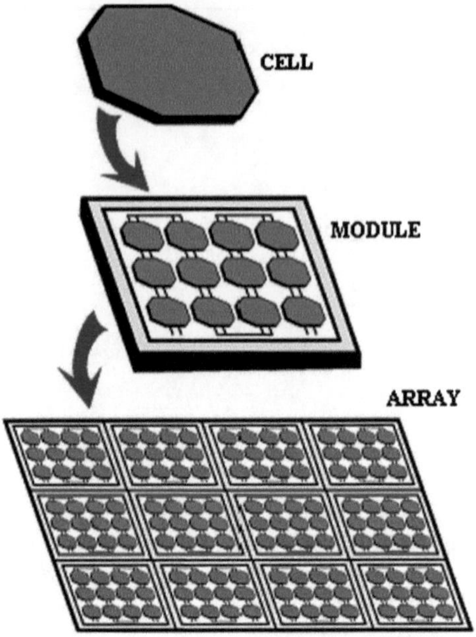

Fig. 7.9 Transition from photovoltaic cell to solar array. *Source* science.nasa.gov

Fig. 7.10 Solar modules with circular monocrystalline cells from Siemens. *Source* google.images.com

wafer surface releases phosphor ions by diffusion into the silicon, displacing the boron. The oxide is later removed by etching. The resultant circularly shaped photovoltaic cells are clearly not ideally suited to mounting in square or rectangular frames to form solar modules and arrays as is evident in Fig. 7.9. Slicing off and wasting portions of expensive silicon to improve the 'fill factor' is generally not recommended, so a poor 'fill factor', and a non-optimal per unit area solar collection performance for the module or array, is usually accepted, as is shown in practical examples from Siemens shown in Fig. 7.10.

Polycrystalline silicon offers cell manufacturers the distinct advantage that wafers can be sliced from square ingots formed by carefully allowing molten silicon in large rectangular moulds to cool slowly. The resultant cells are less costly than comparable monocrystalline varieties, but they are measurably less efficient due to increased electrical resistivity. By extruding thin films from molten silicon, a polycrystalline material is formed termed ribbon silicon, which offers low-cost fabrication of wafers although these tend to be relatively lossy. Introduced into commercial production in around 2009, hybrid silicon uses existing polycrystalline moulds with small ‘seeds’ of monocrystalline material. The result is an ingot with a monocrystalline-like material in the interior with a polycrystalline outer shell. When sliced to form wafers, the inner sections provide high-efficiency monocrystalline-like cells of square cross section, while the pared outer edges of the wafer can be employed in less stringent cell applications. The result is a production line that can fabricate monocrystalline cells at prices more reminiscent of polycrystalline devices.

As usual in engineering, ‘trade-offs’ need to be negotiated, in this case between cell cost, cell efficiency, wafer shape and module ‘fill factor’. Once these basic decisions have been made, the way in which they impinge on the design and development of systems can be estimated through modelling, which includes all aspects of cell and module construction. This process is touched upon in the next section.

7.3.1 Electrical Circuit Representation

In simulation terms, the transition from the mathematical model of a solar PV cell to a model comprising modules in the form of an array (see Fig. 7.11) involves many combinations of identical cells to form the modules and multiple interconnections of matching modules to form arrays. A module for solar array applications typically comprises series-connected cells across which the cumulative voltage meets an externally dictated requirement. A typical number is 72 resulting in a voltage in the 30–40 V range. If a planned system requires a higher operating voltage, then an appropriate number of modules to secure the desired voltage are connected in series. Parallel strings are introduced to generate a specified level of power. The electrical circuit representation of an array is shown in Fig. 7.12, where the circle motif, split vertically, denotes series cells (n_s in number) and when split horizontally, it denotes parallel modules (number = n_p). Other parameters are as defined in Fig. 7.5.

In Fig. 7.12, the series and parallel strings of photodiodes differ from a circuit with a single diode only through the constant multipliers n_s and n_p. This implies that the *I–V* characteristics for a module, and for an array, must be formally the same as for the single cell shown in Fig. 7.6b. This is illustrated in Fig. 7.13a where the I_{PV}–V_{PV} characteristics of a 72-cell module are presented for a range of different irradiance levels. The quantity of major interest in the development of a multi-cell module, or array, is the power delivered to the external circuit due to solar radiation. If we form the equations $I_{\mathrm{PVm}} = n_p I_{\mathrm{PV}}$ and $V_{\mathrm{PVm}} = n_s V_{\mathrm{PV}}$, then the power delivered is given by

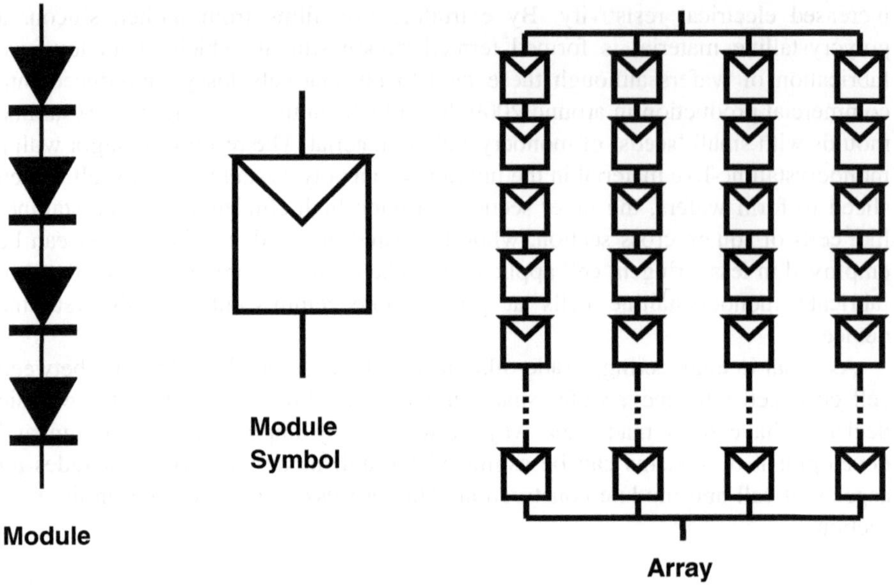

Fig. 7.11 Cell interconnections in solar modules and arrays

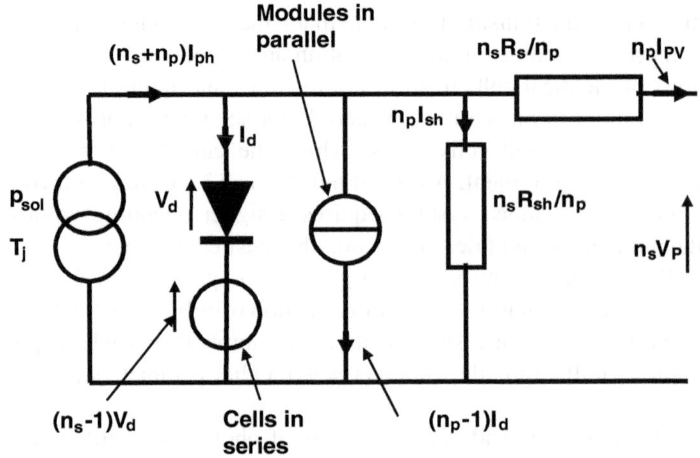

Fig. 7.12 Electrical circuit for solar module comprising series and parallel cell connections

$$P_{out} = V_{PVm} \times I_{PVm} \tag{7.36}$$

It is clear from Fig. 7.13a that P_{out} is positive for $0 < V_{PVm} < V_{oc}$. At the extremes of this range, $P_{out} = 0$ both where $V_{PVm} = V_{oc}$ since $I_{PVm} = 0$, and at $I_{PVm} = I_{sc}$ since $V_{PVm} = 0$. Evidently, therefore, P_{out} must have a maximum value in

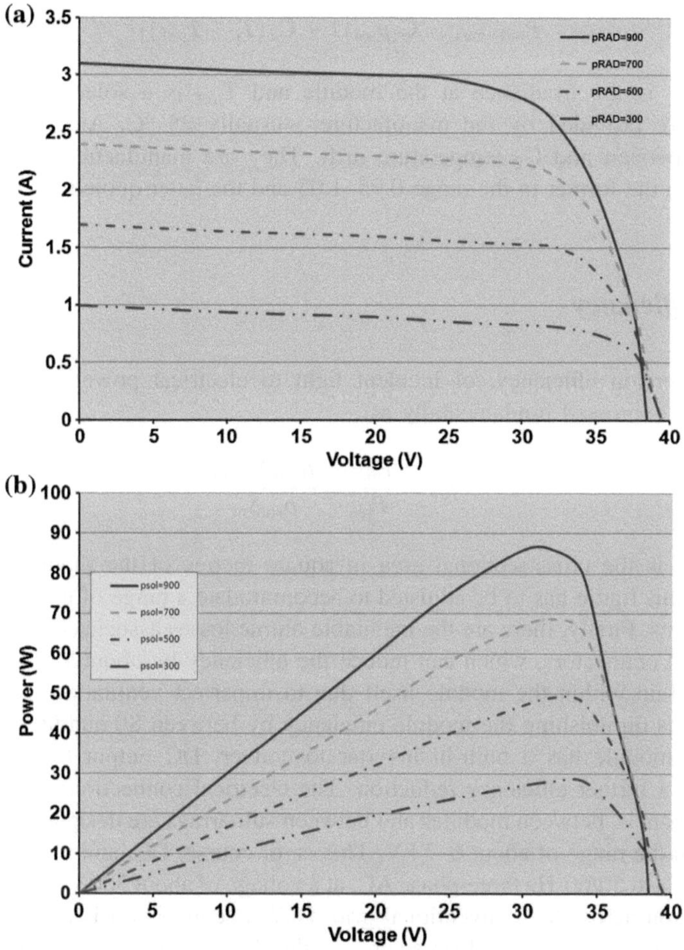

Fig. 7.13 Current (**a**) and power (**b**) responses of 72-cell module for different levels of irradiation (p_{sol})

the above range of *V*. This maximum occurs at the 'knee' of the *I*–*V* curve (see Fig. 7.6b). P_{out} expressed as a function of the terminal voltage (V_{PVm}) is displayed in Fig. 7.13b, for the 72-cell solar module. The power maximum is unmistakable. Close scrutiny of this figure shows that the maximum power point (MPP) moves towards a higher voltage as the irradiance (p_{sol}) decreases. This is a source of efficiency deterioration as is discussed below. The MPP also displays a slight dependence on T_j [2]. These trends are summarised in the equation [19]:

$$P_{\text{out-max}} = C_1 p_{\text{sol}} \left[1 + C_2 \left(T_{\text{j}} - T_{\text{jref}}\right)\right] \tag{7.37}$$

where p_{sol} is the irradiance at the module and T_{jref} is a solar panel reference temperature provided by the manufacturer—usually 25 °C. Also C_1 represents panel dispersion and C_2 temperature drift. They are manufacturer-supplied constants with the former in the range 0.95–1.05 and the latter quoted as −0.47 %/°C.

7.3.2 Efficiency

The conversion efficiency, of incident light to electrical power, of a solar PV module is expressed fundamentally as

$$\eta_{\text{sol}} = \frac{P_{\text{out}}}{P_{\text{in}}} = \frac{I_{\text{PVm}} V_{\text{PVm}}}{p_{\text{sol}} S_{\text{m}}} \tag{7.38}$$

where S_m is the cross-sectional area in square metres of the modular panel. In practice, this figure has to be adjusted to accommodate a range of unavoidable loss mechanisms. Firstly, there are the inevitable ohmic losses associated with electrical cables and connectors, which can reduce the efficiency by about 98 %. Secondly, heat build-up within the module itself due to imperfect ventilation increases cell losses, thus diminishing the module efficiency by between 80 and 90 %.

If the module has a built-in inverter to convert DC output to AC, this can introduce a further efficiency reduction. The electrical connections between cells, between panels, between modules and between sub-arrays are designed to achieve a voltage in the range of about 6–7 kV. This is, of course, DC, and it is necessary to convert this to 50/60-Hz three-phase AC, at a voltage of about 500 kV for long-range transmission across the conventional grid. DC/AC conversion is technically quite simple, and essentially involves switching the direction of the DC current in the primary winding of a step-up transformer. The arrangement is termed an inverter. In high-power, high-voltage systems, solid-state thyristors, or mercury arc valves, can be used to perform the electrical switching. With the help of harmonic filtering, a three-phase sinusoidal AC voltage, at the level of the grid, is thus formed, across the output terminals of the secondary of the transformer. To handle an array power level of 1 GW, the inverter system is large and is likely to require a power house occupying an area of about 300 m × 300 m, on the solar farm site. When switching losses, transformer losses and mismatch losses within the power house equipment are factored in, the efficiency is degraded by a factor of about 0.85.

Modular efficiency can also be downgraded if the module is not operating at the MPP (see Fig. 7.13b) for all solar radiation levels. To achieve this, MPP tracking must be employed. A wide range of tracking methods have evolved in the past 10–15 years from the most common 'perturb and observe' technique to systems based on fuzzy logic [2]. Here, we will illustrate the tracking concept using the incremental conductance technique, which in essence searches iteratively for the maximum of the power curve (Fig. 7.13b) through conductance. The implementation of the method commences with Eq. (7.36) which for convenience we choose to write as

$$P_{\text{out}} = I_{\text{pv}} \times V_{\text{pv}}$$

Hence, differentiating with respect to V_{pv} gives

$$\frac{\mathrm{d}P_{\text{out}}}{\mathrm{d}V_{\text{pv}}} = \frac{\mathrm{d}(V_{\text{pv}}I_{\text{pv}})}{\mathrm{d}V_{\text{pv}}} = I_{\text{pv}} + V_{\text{pv}}\frac{\mathrm{d}I_{\text{pv}}}{\mathrm{d}V_{\text{pv}}} \tag{7.39}$$

This can be written in the alternate form:

$$\frac{1}{V_{\text{pv}}}\frac{\mathrm{d}P_{\text{out}}}{\mathrm{d}V_{\text{pv}}} = \frac{I_{\text{pv}}}{V_{\text{pv}}} + \frac{\mathrm{d}I_{\text{pv}}}{\mathrm{d}V_{\text{pv}}} \tag{7.40}$$

The first term on the right of this equation is module conductance at V_{pv} (a positive quantity), while the second term is the rate of change of the conductance (a negative quantity—see Fig. 7.13a). At MPP, therefore, where the rate of change of P_{out} must be zero, the conductance has to equate to the rate of change of conductance. Furthermore, if

$$\left|\frac{I_{\text{pv}}}{V_{\text{pv}}}\right| > \left|\frac{\mathrm{d}I_{\text{pv}}}{\mathrm{d}V_{\text{pv}}}\right|$$

the tracking algorithm is on the up slope of Fig. 7.13b, and it is on the down slope if

$$\left|\frac{I_{\text{pv}}}{V_{\text{pv}}}\right| < \left|\frac{\mathrm{d}I_{\text{pv}}}{\mathrm{d}V_{\text{pv}}}\right|$$

Formulating such an MPP search algorithm is relatively routine [2] if the modelled I–V characteristic [i.e. Eq. (7.21)] for the cell or module has been established. With such an algorithm incorporated into the solar panel electronics, the efficiency deterioration attributable to changing irradiation strength is of the order of 95 %. In the absence of such tracking, it can be as low as 80 %.

Loss of efficiency is also incurred if the solar PV arrays employ static panels. For example, due to the daily solar movement in the sky, it is estimated that in Germany the efficiency of solar farms changes by as much as 30 % [3] in the summer for a fixed array and by 10 % in the winter. So clearly, there is considerable gain to be secured by incorporating motorised control of the directionality of solar panels. Currently, most systems which have adopted solar tracking are of a single-axis format following only the daily movement of the sun. Double-axis systems which can also track seasonal movement of the sun are relatively uncommon. Where tracking has been applied, the control is open loop based on algorithms which rely

Fig. 7.14 Typical large-area photovoltaic farm (*Courtesy* of Shutterstock)

on the type of equations discussed in Sect. 1.3. Unfortunately, the cost of incorporating tracking into PV array systems is not insubstantial, so that the additional yield in efficiency which this technology obviously offers does not, as yet, compensate for the extra expenditure incurred.

From Eq. (7.38), it is apparent that the DC power generated by a solar cell is given approximately by the product of its efficiency, its area in square metres and the irradiance. A solar cell, typically 100 cm^2 in collection area, with an optimistic efficiency of 20 %, located in the Sahara desert and pointing at the sun with an irradiance of typically 1,000 W/m^2, will generate about 2 W. Of course, the sun shines for only 50 % of the day in near-equatorial regions so we have to assume that we will collect 1 W averaged over time. Clearly, we will need many millions of cells (see Fig. 7.14) of this area to generate significant power, i.e. in excess of 100 MW [20–22].

7.3.3 Array Sizing, Monitoring and Optimisation

That solar photovoltaic systems intended for electrical power supply applications make extensive use of cell, module or array models can be illustrated by inspecting some common developmental and monitoring tasks associated with this technology. These include array design and sizing, conversion of 'field'-monitored performance measurements into a standard reporting format, array performance optimisation, and real-time comparison of measured with theoretically expected performance for fault finding in, and maintenance of, large arrays. These

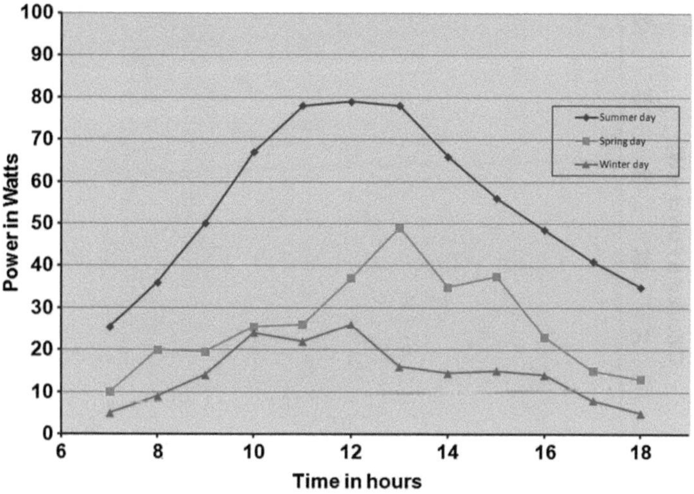

Fig. 7.15 Solar panel power delivery during a single day, at different times of year

applications require models that are both accurate and computationally fast. Two not untypical examples are sufficient to establish this assertion.

Figure 7.15 is based on modelling using measured irradiation levels from a site for a proposed solar farm in France, and taken before array installation. By employing a theoretical model such as one discussed earlier in this chapter, it becomes possible to estimate the potential power delivery from the various locations across the site if covered by an array of solar panels of a predetermined specification. The sizing of the array and the acquisition of optimum location and orientation data for individual panels can thus be assessed before construction and installation begins. The cost savings of such an approach can be very large.

Modelling also permits active monitoring and control of operational solar arrays particularly where the site employs radiation monitors [22]. Performance of the solar array model, when compared to the actual structure, will permit real-time fault detection, such as panel deterioration, contamination of panel surfaces and electrical interconnection faults. This is important for arrays covering tens of thousands of square metres. Figure 7.16 provides an example of real-time monitoring where measured power collection (x-axis), for an array capable of delivering 30 kW, is compared with computationally predicted results (y-axis) from a model based on solar irradiation in clear conditions. Ideally, the graph should be a straight line at 45°, but over the two-hour period of sampling, there was variable cloud cover resulting in intermittent irradiance levels at the real array. Hence, the sampled points are scattered on either side of the 'optimum' line. Intermittent cloud is a serious problem in the optimisation and monitoring of solar arrays and tends to be the primary reason for not siting them in Northerly or Southerly regions of the globe where the weather is unpredictable, rather than simply because of weaker sunlight —although this is obviously a factor.

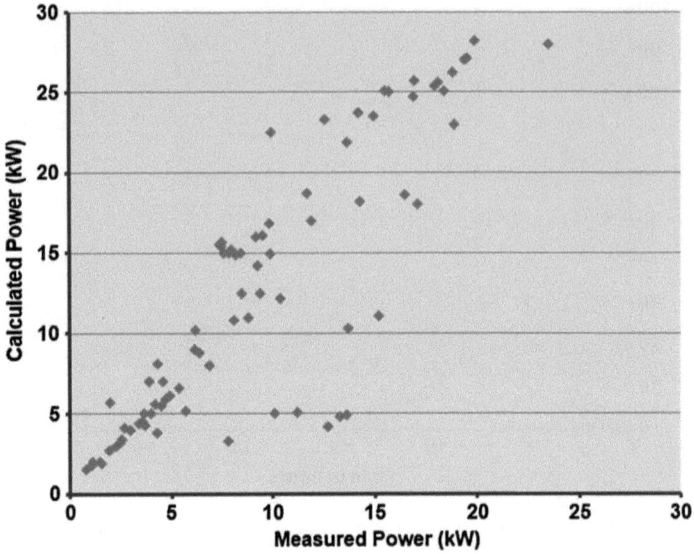

Fig. 7.16 Poorly correlated computed and measured DC power results from an array under cloudy skies

7.3.4 State-of-the-Art Cell Fabrication

Most solar cells in operation today are single crystal silicon cells as we have seen. The silicon is purified and refined by well-established techniques into a single crystal (typically 120 mm in diameter and 2 mm thick). It has been proposed that high-efficiency solar modules could be procured by introducing an additional step into the fabrication technique [23] which involves taking the silicon crystal, and making a multitude of parallel transverse slices across the wafer (rather like finely slicing a round loaf of bread), creating a large number of wafers which are then aligned edge to edge (slices of bread laid out flat to be toasted in the sun) to form a cell comprising 1,000 wafers of dimensions 100 mm × 2 mm × 0.1 mm laid end to end on the 100-μm edges (see Fig, 7.17). A total exposed silicon surface area of about 2,000 cm^2 per side is thus realised.

As a result of this slicing, the electrical doping and contacts that were on the face of the original crystal are located on the edges of the wafer, rather than the front and rear as is the case with conventional cells. This has the interesting effect of making the cell sensitive from both the front and rear (a property known as bifaciality) [23]. Using this technique, one silicon crystal should result in a cell capable of generating 10–12 W of electrical power in bright sunlight. In order to achieve this level of power output from unsliced silicon crystal cells, we would require about 40 crystals. The electrical contacts formed from evenly spaced metal tabs on the wafer edges can be connected (an added technical complication in the fabrication of sliced silicon cells) to a larger 'bus' conductor to transmit the power. It is proposed that

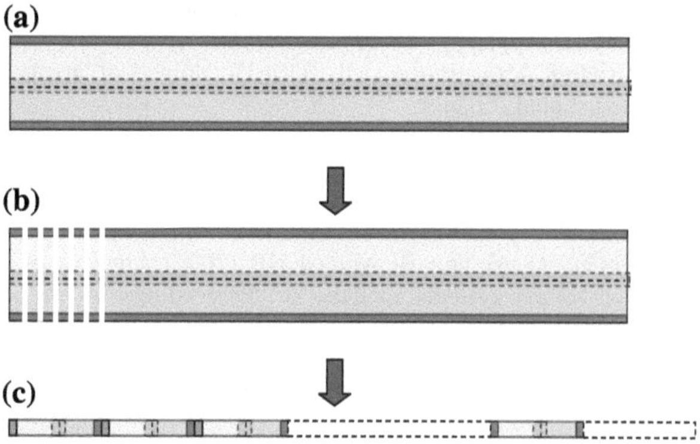

Fig. 7.17 Wafer dicing for high-efficiency application. **a** Conventional doped silicon wafer, **b** dicing concept and **c** laying slices end to end

the cell should be covered with a thin protective layer of dielectric with an anti-reflective coating. Cells of this construction, in addition to boosting power gathered per crystal, should represent an excellent compromise between cost-effectiveness, reliability and efficiency.

References

1. Cutler P (1972) Solid-state device theory. McGraw-Hill Book Co., New York
2. Rekouia D, Matagna E (2012) Optimization of photovoltaic power systems. Springer, London
3. Krauter S (2006) Solar electric power generation. Springer, Berlin
4. Jones AD, Underwood CP (2002) A modelling method for building-integrated photovoltaic power supply. Buil Serv Eng Res Technol 23(3):167–177
5. Jones AD, Underwood CP (2002) A modelling method for building integrated PV systems. Sol Energy 70(4):349
6. Laplaze D, Youm I (1985) Modelling of a photovoltaic cell. Solar Cells 14:167 (in French)
7. Wolf M, Rauschenbach H (1963) Series resistance effects of solar cell measurement. Adv Energy Convers 3(2):455–479
8. Charles JP, Abdelkrim M, Muoy YH, Mialhe P (1981) A practical method of analysis of the current-voltage characteristics of solar cells. Solar Cells 4(2):169–178
9. Agarwal SK, Muralidharan R, Agarwala A, Tewary VK, Jain SC (1981) Solar cell parameter evaluation considering the series and shunt resistance. J Phys 14(11):1634–1638
10. Cabestany J, Castaner L (1983) A simple solar series resistance measurement method. Rev Phys Appl 18(9):565–567
11. Chegaar M, Azzouzi G, Mialhe P (2006) Simple parameter extraction method for illuminated solar cells. Solid-State Electron 50:1234–1237
12. Easwarakhanthan T, Bottin J, Bouhouch I, Boutrit C (1986) Non-linear minimization algorithm for determining the solar cell parameters with microcomputers. Int J Solar Energy 4:1–12
13. Ouennoughi Z, Chegaar M (1999) A simple method for extracting solar cell parameters using the conductance method. Solid-State Electron 43:1985
14. Chegaar M, Ouennoughi Z, Hoffmann A (2001) A new method for evaluating illuminated solar cell parameters. Solid-State Electron 45:293–296
15. Mialhe P, Khoury A, Charles JP (1984) A review of techniques to determine the series resistance of solar cells. Phys Status Solidi A 83(1):403–409

16. Cabestany J, Castaner L (1983) Evaluation of solar cell parameters by non-linear algorithms. J Phys D Appl Phys 16(12):2547–2558
17. Salvadori MG, Baron ML (1961) Numerical methods in engineering. Prentice-Hall International, London
18. MATLAB. www.mathworks.co.uk/products/matlab/
19. Richard Williams J (1977) Solar energy. Ann Arbor Science Publishers Inc., Ann Arbor
20. Lorenzo E et al (1994) Solar energy: engineering of photovoltaic systems. Progensa
21. Kurokawa K, Energy from the desert. http://www.iea-pvps.org/
22. Gergaud O, Multon B, Ahmed HB (2002) Analysis of experimental validation of various photovoltaic systems models. In: Proceedings of 7th international electromacs congress, Montreal, Canada
23. Blakers A, Weber K, Everett E, Franklin E, Deenapanray S (2006) Sliver cells—a complete photovoltaic solution. In: IEEE 4th world conference on photovoltaic energy conversion, Hawaii

Chapter 8
Concentrated Solar Power

Engineers participate in the activities which make the resources of nature available in a form beneficial to man and provide systems which will perform optimally and economically.

L. M. K. Boelter

Abstract In the drive towards providing massive levels of electrical power through the agency of solar ray gathering, concentrated solar power (CSP) techniques offer potentially the most efficient, cost-effective solutions. These techniques are largely based on solar reflectors of one form or another, and several geometries are examined in this chapter. Four CSP options are currently receiving major attention by the power-generating industries. These are systems based on the use of parabolic troughs, linear Fresnel reflectors, heliostats and parabolic dishes. To assist in assessing these alternative schemes, we initially consider in Chap. 8 the relationship between an optical receiver and an antenna, and this leads to the introduction of ray optics as a design tool for large optical reflectors. In power collection terms, an optimum level of ray concentration is shown to exist in theory, and this knowledge provides useful pointers towards the design of efficient collectors. Also, at high-power concentration levels, optical reflectors are subject to thermal effects and the thermodynamics required to comprehend the relevant phenomena are presented in Sect. 8.3.

8.1 Introduction

When they were very young, few readers will not have been fascinated by the 'game' of paper singeing, or of starting a fire from kindling, simply by focusing sunlight through a lens to form a 'hot spot' on the target. Fundamentally, this is the key to concentrated solar power (CSP) systems, except that optical mirrors are preferred to lenses, which are impractical on a large scale. Obviously, the concept is easily expressed, but how do we formulate a mathematical description of the process which will enable CSP systems to be designed and developed in an assured, effective and efficient manner?

A.J. Sangster, *Electromagnetic Foundations of Solar Radiation Collection*,
Green Energy and Technology, DOI 10.1007/978-3-319-08512-8_8

Table 8.1 Relative strengths and weaknesses of four main CSP technologies

Collector type	Thermal efficiency	Op. temp range (°C)	Concentration ratio	Tracking	Tech. maturity	Relative Cost
Parabolic Trough	Low	50–400	15–45	One axis	Advanced	Low
Linear Fresnel	Low	50–300	10–40	One axis	Moderate	Very low
Heliostat	High	300–2000	150–1500	Two axis	Recent development	High
Parabolic Dish	High	150–1500	100–1000	Two axis	New development	Very high

Serious research and development effort have been directed towards CSP for several decades with the aim of focusing sunlight by means of mirrors onto a collector or receiver which is commonly a fluid-filled tube or boiler, thus converting solar energy into thermal energy. The heated fluid is then used to drive a turbine/generator set much as is done in a conventionally powered electricity generation station where the heat is produced by fossil fuel combustion. Stated in this way the concept is, in principle, easily realisable with well-established technology, but progress has been hindered by the wide range of possible routes to efficient execution (see Table 8.1). The chosen implementation method, while necessarily having to meet stringent technical constraints, at each stage of the collection and conversion process, if high efficiency is to be secured, must also meet severe environmental and economic goals. Arguably, these latter demands have been much more difficult to satisfy. The collector types, enunciated in column one of the table, are illustrated schematically in Fig. 8.1.

The primary electromagnetic wave-related descriptions of the light-concentrating processes associated with each category of CSP system, as depicted in Fig. 8.1, will be developed in this chapter. But firstly, we need to establish the mathematical developments which underpin the design equations for devices whose main purpose is to gather and concentrate light by means of large optical reflectors. This will be addressed in Sect. 8.2

8.2 Solar Collectors as Antennas

Self-evidently, simply by comparing Fig. 8.1a, d with Fig. 4.1g, or Fig. 8.1b, c with Fig. 5.1, CSP collectors are antennas of either the aperture or the array type. But in the literature, solar collectors are seldom viewed as electromagnetic wave-based antennas. So at what point does the electromagnetic antenna theory outlined in Chaps. 4 and 5 transmute into the ray optics, which governs the optical theories employed in concentrated solar power publications [1–7].

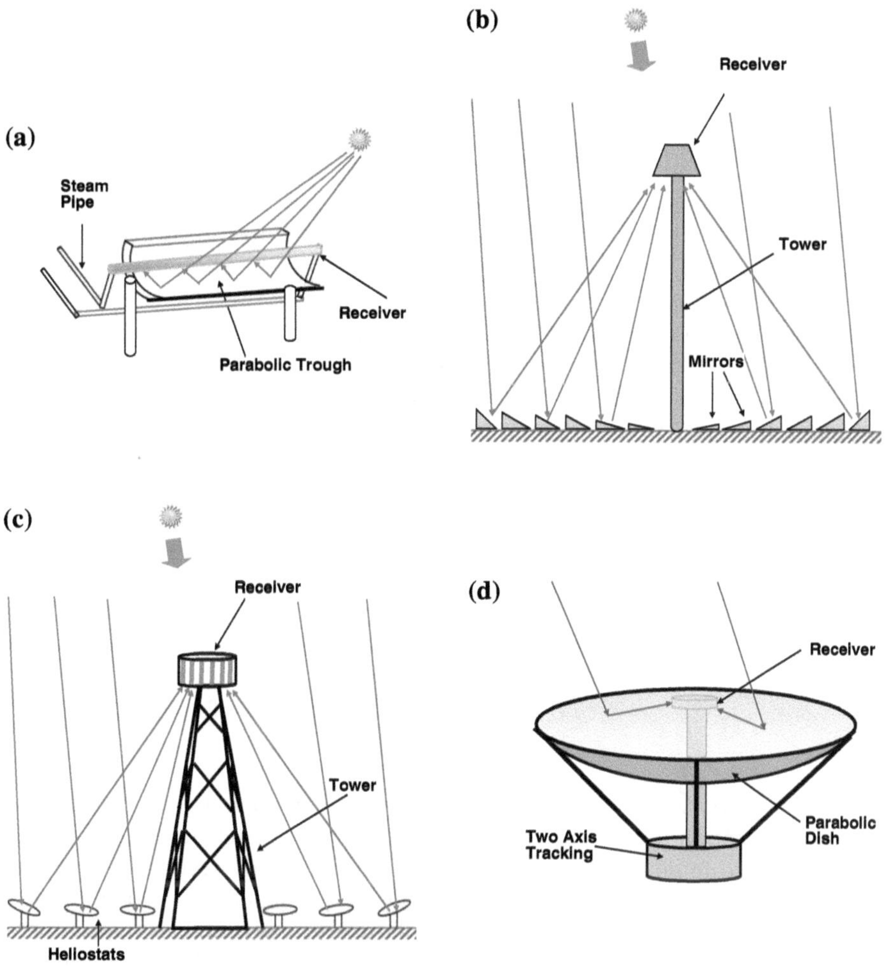

Fig. 8.1 Examples of the main CSP technologies in schematic form: **a** parabolic trough, **b** Fresnel, **c** heliostat, and **d** parabolic dish

The electromagnetic field equations, expressing the transmission and reception characteristics of antennas as a class of boundary value problem, are summarised in Table 4.1. The significant and generally applicable equations are:

$$\mathbf{A}(\mathbf{r},t) = \frac{\mu_0}{4\pi}\iiint_V \frac{\mathbf{J}(\mathbf{r}')}{|\mathbf{r}-\mathbf{r}'|}\mathrm{d}V' \tag{8.1}$$

and

$$\mathbf{A}_m(\mathbf{r},t) = \frac{\varepsilon_0}{4\pi}\iiint_V \frac{\mathbf{J}_m(\mathbf{r}')}{|\mathbf{r}-\mathbf{r}'|}\mathrm{d}V' \tag{8.2}$$

where as we have seen $\mathbf{J}(\mathbf{r}')$ denotes conventional conduction current, usually on a metallic surface, and $\mathbf{J}_m(\mathbf{r}')$ denotes magnetic current (E-field), usually in an aperture contained within V. These equations are in principle applicable to electromagnetic antennas of any description, and they form the kernel of many antenna design software packages. For most antennas, it is seldom possible to implement Eqs. (8.1) and (8.2) directly as we have demonstrated in Chap. 4. The use of equivalence theorems to circumvent this difficulty is discussed there. However, for antennas displaying major dimensions, which are extremely large by comparison with a wavelength, as in CSP systems, even where equivalence techniques are adopted, the computational intensity involved is much too severe to contemplate the direct integration route to a solution. Fortunately, it is possible by using symmetry, imaging and other techniques, to formulate mathematical approximations, which hugely ease the task. The simplifications also use concepts from optics, such as Huygen's principle, which we have already encountered (Sect. 4.3.1).

The source of these simplifying approximations, which can be applied to antennas classed as extremely large in wavelength terms, is not too difficult to establish. As we have seen in Chap. 4, the key quantity, which links an antenna pattern to the E-field in its aperture, is the spectrum function. For a uniformly illuminated aperture of width a, in a two-dimensional scenario, it can be deduced using Eq. (8.2). It has the form:

$$F(\sin\theta) = \frac{E_0 a}{\lambda_0} \frac{\sin\left(\frac{\pi a \sin\theta}{\lambda_0}\right)}{\frac{\pi a \sin\theta}{\lambda_0}} \tag{8.3}$$

For the purposes of examining optical applications, it is not inappropriate to direct our attention to two-dimensional embodiments, because little loss of generality is suffered in doing so. In Eq. (8.3) as we have already seen that E_o is the E-field strength in the aperture, λ_o is the free-space wavelength, and θ is the pattern angle in space measured from the normal to the aperture face. In the context of CSP systems, the a-dimension—for example the width of a trough-shaped reflecting mirror—is of the order of 10^6 times the mean wavelength of the light impinging upon it. In terms of Eq. (8.3), this means that a finite and meaningful solution occurs only where θ is essentially equal to zero. Thus, as an antenna, a light-concentrating CSP mirror has a 'pencil-like' (no beam spreading) primary beam which can with little error be viewed as being made up of 'rays'. The ray represents an incoming plane electromagnetic wave at the frequency of light, arriving from a single direction (the sun) and normal to the aperture surface. Although λ_o has a very wide range of values for light (from 10^{-5} to 10^{-7} m), the a/λ_o ratio remains sufficiently high at all frequencies to assume that ray optics apply.

The transition to ray theory, or geometrical optics as it is more precisely termed, is developed more comprehensively and coherently in the following sections.

8.2.1 Huygen's Principle and Rays

The ray theory of optics has already been encountered briefly in Chap. 4. It is suggested there, in the discussion accompanying Fig. 4.4, that the cylindrical wave front from a small two-dimensional slit can be viewed as plane over a very restricted subtended angle, if R is not too small.

To be confident of the validity of this method, beyond the narrow slit example encountered in Chap. 4, we need to more precisely define a ray mathematically, and to link this back to Maxwell's equations. It is shown here that any arbitrary wave front can be viewed as a summation or spectrum of plane waves, and remains equivalent to the Huygen's method. In optical terms, the plane wave spectrum is considered to be a 'fan-like' distribution of light rays, with each ray as a directional vector of length $c\delta t$. In free space $c = c_0 = \frac{1}{\sqrt{\mu_0 \varepsilon_0}}$, and this means that the ray representing a plane wave component of the original wave front (see Fig. 8.2) can be depicted as straight. In a more complex medium, the wave velocity c could be a function of position in which case the rays can be curved [8]. However, this special case need not concern us in this development of solar collection principles. Furthermore, the wave front, depicted in Fig. 8.2, is viewed as an imaginary surface moving out into space in unison with a target electromagnetic wave. On it, the phase of the electric or magnetic field is everywhere the same. More specifically, this phase condition exists on the surfaces L_1 or L_2 in Fig. 8.2. In the case where the wave front is locally plane, it is possible to define the power flow density (W/m^2), at that location, by the Poynting vector [see Eq. (2.45)]. This is the essence of the ray concept which is intrinsic to geometrical optics.

The most general mathematical expressions of the relationship between electromagnetic radiation and a time-varying electric current or magnetic current source are presented in Eqs. (8.1) and (8.2), respectively. The conversion from the magnetic potential functions $\mathbf{A}$ and $\mathbf{A}_m$ to the more familiar electric and magnetic fields forms representing the radiation is achieved through Maxwell's equations and is delineated in Chap. 4, Table 4.1. On an equiphase surface S representing a propagating wave, as depicted in Fig. 8.3, the electric and magnetic fields lie in the plane

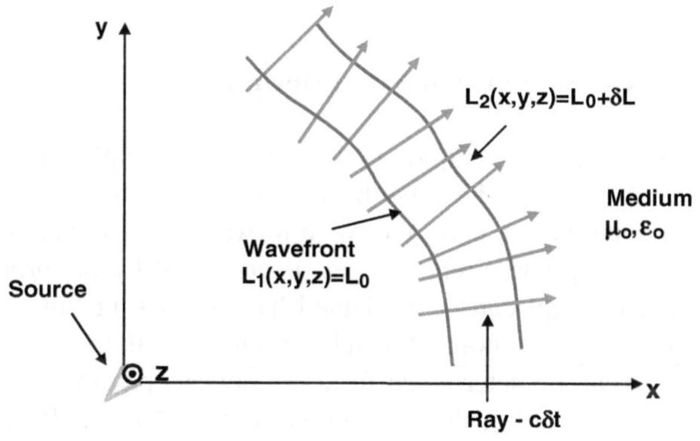

Fig. 8.2 Geometrical optics wave front

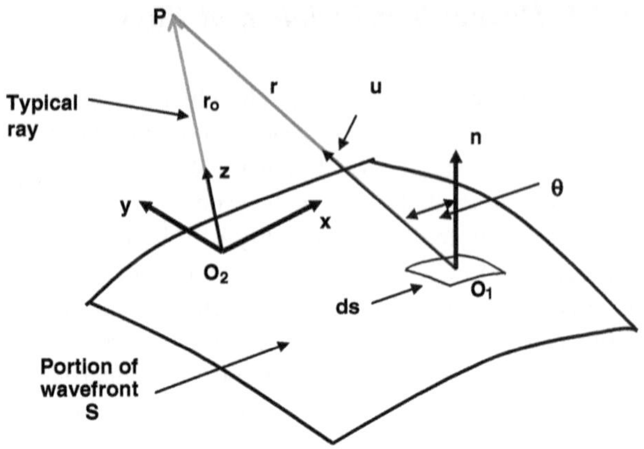

Fig. 8.3 Stationary-phase principle

of the surface. At any point on the surface O_1 (say), a magnetic current $\mathbf{J}_m = \mathbf{n} \times \mathbf{E}$ exists over the elemental area dS. Also everywhere on the wave front (S), this $\mathbf{E}$ vector is largely uni-directional and can with little error be represented by a scalar. Hence, for the entire surface S, we can use Eq. (8.2) to determine the scalar field ($E(P)$) at point P ahead of the phase front. The result is [9]

$$E(P) = \frac{j}{\lambda_0} \iint_S \gamma E(O_1) \frac{\exp\left[-\mathrm{j}\frac{2\pi r}{\lambda_0}\right]}{r} \mathrm{d}S \tag{8.4}$$

$E(O_1)$ is the tangential field at O_1 on the phase front, with O_1 representing the instantaneous coordinate values at dS. The γ multiplier is termed the obliquity factor given by:

$$\gamma = \frac{1 + \cos\theta}{2} \tag{8.5}$$

It expresses the small but finite directionality of the 'aperture antenna' of area dS (see Chap. 4).

8.2.1.1 Stationary-Phase Principle

In optical terms, Eq. (8.4) represents the discipline of physical optics. It dictates that every point on the wave front of a ray of light is a source of spherical waves which in combination carry the wave forward. The transition to geometrical or ray optics, in appropriate circumstances, is governed by the principle of stationary phase [8]. The concept can be explained by considering Fig. 8.3. In geometrical optics, the light reaching point P which lies ahead of the wave front S representing a beam of light can be determined from the ray ($\mathbf{r}_o$) passing through P and defined as normal to S at O_2. But, there are other contributions at P, such as waves in the direction

r from point O_1. How do they add to the total field? To assess this scenario, Eq. (8.4) which determines the E-field at P from magnetic currents on S can be rewritten in the form:

$$E(P) = \frac{j}{\lambda_0} \exp\left(-j\frac{2\pi r_0}{\lambda_0}\right) \iint_S \gamma E(O_1) \frac{\exp\left[-j\frac{2\pi(r-r_0)}{\lambda_0}\right]}{r} \mathrm{d}S \tag{8.6}$$

The integral in this equation can be viewed as a summation of phasors in the complex plane each with a magnitude ΔV and phase deviation $\delta\alpha$ relative to the ray, where :

$$\Delta V = \frac{\gamma E(O_1)\mathrm{d}S}{r} \tag{8.7}$$

and

$$\delta\alpha = \frac{2\pi}{\lambda_0}(r - r_0) \tag{8.8}$$

The phasor representation of the ray at O_2 in Fig. 8.3 has phase angle:

$$\varphi_0 = \frac{2\pi r_0}{\lambda_0} \tag{8.9}$$

It is clear from Eq. (8.8) that when $|r - r_0| << \lambda_0$, $\delta\alpha$ is small and slowly changing and the phasor contributions to the integral are re-enforcing—as suggested in Fig. 8.4 by the red vectors. This is termed the stationary-phase condition [8]. Otherwise, where $|r - r_0| \geq \lambda_0$, $\delta\alpha$ varies rapidly resulting in phasor cancellation (see Fig. 8.4 randomly orientated blue phasors) and hence makes a negligible contribution to the integral. Consequently, $E(P)$ comes from phasors in the stationary category. At a point P, therefore, ahead of an optical (or electromagnetic) wave front, the electric field of the developing wave at P is dependent almost wholly on contributions on the original wave front from the near vicinity of the normal at O_2, and passing through P. This is the description of a ray passing through P. It represents a very useful simplification of the electromagnetic propagation problem and forms the basis of geometrical optics which permits the modelling of large (in wavelength terms) optical or electromagnetic wave scattering structures such as solar collectors.

Equation (8.6) can therefore be simplified further to give:

$$E(P) = \frac{j}{r_0\lambda_0} E(O_2) \exp(-j\varphi_0) \iint_S \exp(-j\delta\alpha(x, y))\mathrm{d}x\mathrm{d}y \tag{8.10}$$

The γ multiplier disappears since it is essentially equal to unity in the stationary-phase range. For wave front surfaces exhibiting radii of curvature predominantly

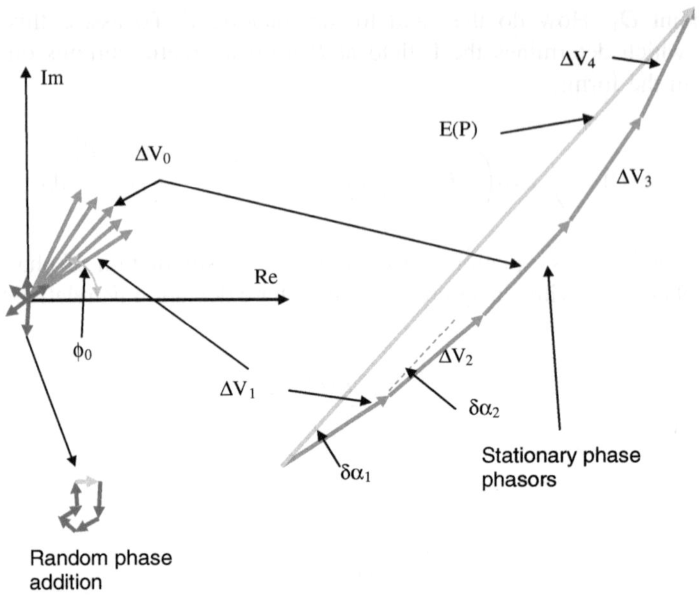

Fig. 8.4 Schematic depiction of stationary phase

equal to R_1 and R_2 (R_1 for the original wave front and R_2 for the new surface through P), this integral can be evaluated to give [9]:

$$E(P) = \sqrt{\frac{R_1 R_2}{(R_1 - r_0)(R_2 - r_0)}} E(O_2) \exp(-j\varphi_0) \quad (8.11)$$

It is important to note that the magnitude of $E(P)$ in Eq. (8.11) is frequency independent. In other words, in ray optics all frequencies in white light translate from one wave front to the next in exactly the same manner apart from an irrelevant propagation delay. This frequency independence in the practice of geometrical optics is re-enforced by the use of materials which are lossless—that is perfect dielectrics or perfectly conducting metals.

8.2.2 Geometrical Optics

Ray theory, which underpins the geometric theory of optics, satisfies, as we have seen, the requirements of electromagnetism in scenarios where stationary-phase conditions can be applied. However, without some additional refinements to it, our model will not necessarily meet the very important physical requirement of satisfying the first law of thermodynamics, namely the conservation of energy or power.

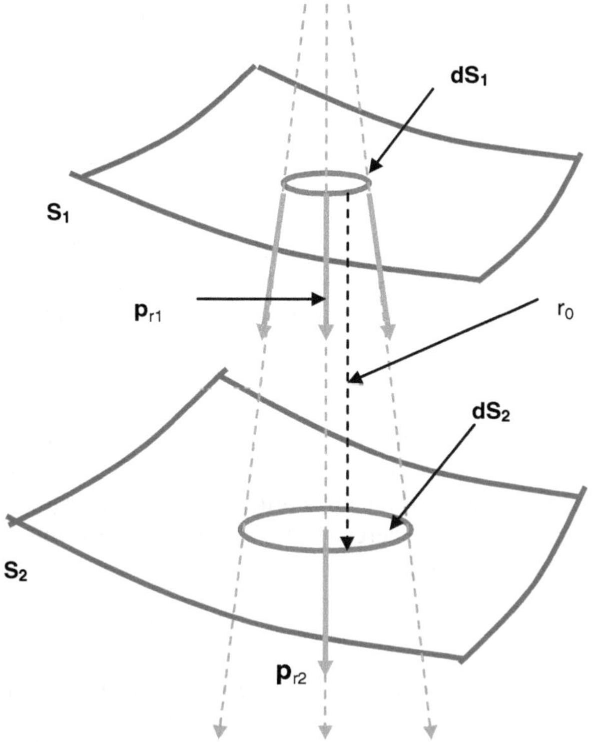

Fig. 8.5 Power flow carried by rays

In electromagnetic theory as we have seen in Chap. 2, energy flow is computed from the Poynting vector which equates to the power density at a point in space where transverse electric and magnetic fields exist (Eq. 2.45). The relationship between the Poynting vector (solid red vectors) and a conical ray (dotted arrows and blue ellipses) passing through imaginary surfaces S_1 and S_2 is suggested in Fig. 8.5. The ray interception area at S_1 is $\mathrm{d}S_1$, while at S_2, it is $\mathrm{d}S_2$. Consequently, if $\mathrm{d}S_1$ and $\mathrm{d}S_2$ are small enough, and $\mathbf{p}_{r1}$ and $\mathbf{p}_{r2}$ are the Poynting vectors at their centres, then power conservation requires that:

$$|\mathbf{p_{r1}}|dS_1 = |\mathbf{p_{r2}}|\mathrm{d}S_2 \tag{8.12}$$

In electromagnetic field terms, and in free space, where c is everywhere the same, we can rewrite this in the form:

$$E_1^2\mathrm{d}S_1 = E_2^2\mathrm{d}S_2 \tag{8.13}$$

This equation together with Eq. (8.11) ensures that ray calculations when applied properly are in accordance with both Maxwell's equations and with energy-conservation requirements.

A final element is the geometrical optics 'jig-saw' is required to accommodate reflection and refraction. This element is the concept of 'optical path', expressed mathematically as:

$$L = \int_{\Gamma} n d\ell \tag{8.14}$$

where Γ is the path taken by the light, L is the length of the path, n is the refractive index along the path and $d\ell$ is an element of path length. In a homogeneous dielectric medium (relative permittivity = ε_r) such as air or free space

$$n = \sqrt{\varepsilon_r} \tag{8.15}$$

and since $\varepsilon_r = 1$ for free space, it is thus constant along the path. However, this need not be the case in, for example, an inhomogeneous medium or where the medium changes along the path. Furthermore, it is a requirement of the principle of least action, or least energy, that a ray must expend minimum possible energy, in travelling between any two points. Hence:

$$\Delta L = 0 \tag{8.16}$$

This is also referred to as Fermat's principle and has the consequence that light travels in straight lines in a homogeneous medium such as free space. Not surprisingly perhaps, the principle also requires that the light ray adheres to Snell's laws.

8.2.2.1 Optical Path Applied to a Curved Reflector

Geometrical optics is the primary analysis tool for understanding optical systems including concentrated solar power collectors, which as Fig. 8.1 shows, are based on light manipulation using optical reflectors. It is hardly a surprise, perhaps, that the main source of applications for geometrical optics has been in imaging systems, and largely in the design of lenses and mirrors. Solar light collection is an analogous activity to imaging insofar as optical focusing is involved and so similar techniques are employed in its examination. However, these need not include the Abbe sine condition, which manifests itself where it is important for an imaging lens or mirror to produce sharp images off-axis as well as for the primary on-axis target. It is simply stated mathematically as:

$$\frac{\sin \theta'}{\sin \Theta'} = \frac{\sin \theta}{\sin \Theta} \tag{8.17}$$

where θ and Θ are the angles, measured relative to the optical axis, of any two rays emanating from the target object, while θ' and Θ' are the angles of the continuing rays after diffraction or reflection where they reach the image plane. We shall see

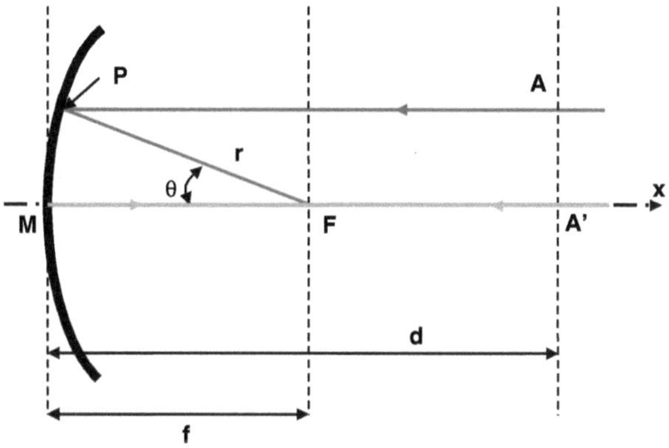

Fig. 8.6 Schematic for trough reflector shape calculation

later in this section that the sine equality in Eq. (8.17) does not necessarily apply in solar ray operations [1].

The concentrated solar power structure depicted in Fig. 8.1a is typical of an optical reflector-based system. It comprises mirrors which are long and linear in the horizontal plane and curved in the vertical plane to direct solar power onto long fluid carrying pipes running parallel to the mirrors. The question then arises as to what shape these reflectors should be in the vertical plane to optimise power collection? This can be determined quite straight-forwardly using geometrical optics as follows (Fig. 8.6).

The optical problem is one of converting parallel rays from a very distant source (the sun) into cylindrical waves which are required to converge on the heat pipe at F. If M is the apex of the mirror then the focal distance is MF = f. Also, we choose that the plane AA' at distant d from the apex M should represent a typical incoming plane wave front. So, PF and MF are converging rays due to reflection of incoming parallel rays AP and A'M. Consequently, the optical path law required for phase convergence at F can be written as:

$$\mathrm{AP} + \mathrm{PF} = \text{constant} = A'M + \mathrm{MF} \tag{8.17}$$

or

$$r + r\cos\theta + d - f = f + d$$

which leads to:

$$r = \frac{2f}{1+\cos\theta} = \frac{f}{\cos^2\frac{\theta}{2}} \tag{8.18}$$

This, as expected, is the equation of a parabola. It expresses the direct trade-off which prevails between a desirable low rate of curvature (large r-value) for the reflector but at the expense of a long focal length. The latter can be a potential source of optical inefficiency.

8.2.3 Theoretically Optimum CSP Collector

Solar power collection techniques based predominantly on optical imaging techniques have been evolving for almost seventy years with one of the earliest papers directed towards the use of static planar mirrors [10]. However, during much of this period, an important class of electromagnetic wave concentrator has been largely ignored. This novel development, first described in reference [1], was distinctive insofar as it represented a clear departure from optical imaging and in particular the Abbe sine law. Winston [1] referred to the new design as the 'ideal concentrator', but it was only ideal in a restricted optical sense. However, it does offer significantly higher levels of concentration than more conventional reflector geometries, although this is at the expense of enlarged reflector surface area—as we shall see.

In order to establish the fundamental law governing pure light concentration without the additional constraints associated with imaging, it is useful to consider a fluid flow example of concentration in a laminar flow transition from a large diameter tube to one of much smaller diameter as shown in Fig. 8.7. It shows a tube of cross-sectional area a_1 connected through a 'smooth' transition to a pipe of cross-sectional area a_2. The fluid travels at velocity v_1 in pipe 1 and at velocity v_2 in the exit pipe. In the entry aperture (1), just before the transition, the momentum of an element of fluid there is $\rho \mathrm{d}v_1 \mathrm{d}a_1$ kg/sec, where ρ is the fluid density in kg/m^3. Therefore, the total momentum in the input pipe (1), namely p_{m1} is given by:

$$p_{m1} = \rho \int_0^{v_1} \int_0^{a_1} \mathrm{d}v_1 \mathrm{d}a_1 \tag{8.19}$$

Similarly at aperture (2) in the exit pipe:

$$p_{m2} = \rho \int_0^{v_2} \int_0^{a_2} \mathrm{d}v_2 \mathrm{d}a_2 \tag{8.20}$$

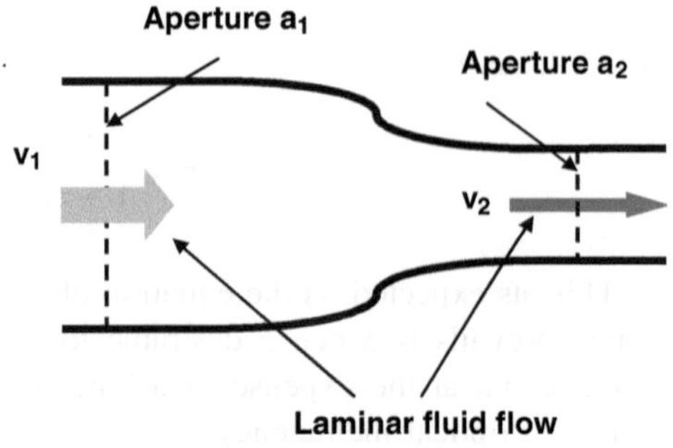

Fig. 8.7 Laminar fluid flow between pipes reducing in cross section

But conservation of linear momentum requires that:

$$p_{m1} = p_{m2} \tag{8.21}$$

in which case the following equality applies:

$$\int_0^{v_1}\int_0^{a_1} \mathrm{d}v_1 \mathrm{d}a_1 = \int_0^{v_2}\int_0^{a_2} \mathrm{d}v_2 \mathrm{d}a_2 \tag{8.22}$$

In words, this implies that velocity or momentum space is preserved in the laminar concentration process.

It may be risky to construct an analogy between fluid flow and electromagnetic ray behaviour in a concentrating device, as suggested in Fig. 8.8, but hopefully, it will illuminate the mathematical rules underpinning this enhanced optical reflector. The basic problem of light concentration has been enunciated by *R*. Winston [1] in the following manner: 'given a set of light rays with a specified angular divergence $\theta_{\max}$ distributed over the entrance aperture, how can we direct these rays efficiently into the smallest possible exit aperture?' The question is addressed by Winston by first acknowledging that light and waves possess momentum. In Eq. (6.13), it has been demonstrated that electromagnetic wave momentum is related to the propagation coefficient of the wave through:

$$\vec{p}_{m_o} = \hbar\vec{k}\,\mathrm{kg\ m/s} \tag{8.23}$$

where the constant of proportionality $\hbar$ is just Planck's constant divided by 2π. We also know that in magnitude $k = \omega/v = n\omega/c$ where n is the refractive index. Hence, the following proportionality applies:

$$k \propto n \tag{8.24}$$

On applying the conservation of momentum to the entry and exit apertures in Fig. 8.8, which for convenience have been assumed to be rectangular, we obtain an equation equivalent to (8.22), namely:

$$\int_{-\theta_{\max}}^{\theta_{\max}}\int_{-\theta_{\max}}^{\theta_{\max}}\int_0^{x_1}\int_0^{y_1} \mathrm{d}k_x \mathrm{d}k_y \mathrm{d}x_1 \mathrm{d}y_1 = \int_{-\pi/2}^{\pi/2}\int_{-\pi/2}^{\pi/2}\int_0^{x_2}\int_0^{y_2} \mathrm{d}k_x \mathrm{d}k_y \mathrm{d}x_2 \mathrm{d}y_2 \tag{8.25}$$

This equation also expresses the idea that phase space is also conserved [1] for plane electromagnetic waves. Within the equation, x_1 *and* y_1 are the height and width of the entry aperture, and x_2 *and* y_2 are the height and width of the exit aperture. The maximum acceptance angle $\theta_{\max}$ is assumed to be the same in the xz- and yz-planes. At the exit aperture, the angles θ of the rays incident on a_2 are

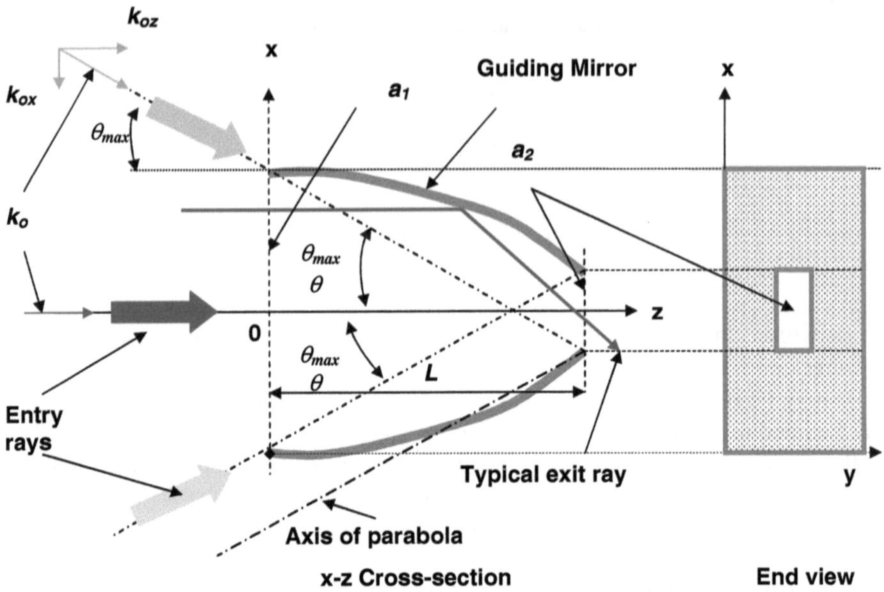

Fig. 8.8 Geometry for collector calculations on an 'ideal' compound parabolic concentrator (*CPC*)

unknown, but it can be asserted that they must lie within the range −π/2 to +π/2. Hence, on noting that $k_{x_z} = k \sin \theta : k_y = k \sin \theta$, Eq. (8.25) reduces, on using relation (8.24), to [1]:

$$n_2^2 a_2 \geq n_1^2 a_1 \sin^2 \theta_{\max} \tag{8.26}$$

This inequality provides the requisite restriction on the exit aperture size, as dictated by the desired angular divergence at the input, for an optical system that is not image forming. It essentially replaces the Abbe sine condition in image-forming optics.

The shape of the reflectors in Fig. 8.8, which has not so far been considered, is determined as described in Sect. 8.2.2, by applying Fermat's principle to typical rays. If the extremity rays (at $\theta_{\max}$) are presumed to be focused onto the opposite edge of the exit aperture, as suggested in the figure, then the result is a parabolic-shaped 'bowl' with the axis of the mirrors inclined to the z-axis by $\theta_{\max}$ (see Fig. 8.8). Of course, in practice the entry and exit apertures are more likely to be circular or elliptical than square or rectangular.

The impetus for seeking a high degree of concentration of the solar radiation in CSP systems lies in the demand for the high temperatures which are needed to maximise thermodynamic efficiency in the turbine/generator unit. Almost all CSP systems (see Fig. 8.1) use heat as the conversion medium. The exception is concentrated photovoltaic systems which are treated separately below. The heat generated at the receiver of a solar collector can reasonably be equated to the thermal absorber area (A_{th}) in which case we can define concentration C as:

$$C = \frac{A_{in}}{A_{th}} \tag{8.27}$$

In relation to Fig. 8.8, this gives:

$$C = \frac{a_1}{a_2}$$

Not all of the heat generated is transferred to the thermodynamically driven turbine system. Some is inevitably lost as radiant heat. This loss increases very rapidly with temperature T, in fact as T^4. It can be minimised only by optimising the concentration, which has become, as a result, a critical parameter in CSP design.

It is clear from Eq. (8.26) that the available level of concentration at a solar collector is strongly related to the tolerable maximum acceptance angle (θ_{max}) at the input. Equation (8.26) suggests that the ideal concentrator, for which the refractive index (n_2) for the absorber at the exit aperture is equal to the input refractive index (n_1), displays an optimum possible concentration of:

$$C_i = \frac{1}{\sin^2 \theta_{max}} \tag{8.28}$$

That $C_i \to \infty$ as the acceptance angle approaches zero simply means that if the incoming rays can be assumed to form perfectly parallel beams aligned with the z-axis then the light can in theory be focused into a pinhole (i.e. $a_2 \to 0$). Equation (8.28) represents a theoretical limit with which practical designs can be compared. Most real concentrators fall short of the ideal by at least a factor of two. For the more general case of a concentrator in air ($n_1 = 1$) with an absorber for which the refractive index is greater than unity ($n_2 = n$), then Eq. (8.28) becomes:

$$C_i = \frac{n^2}{\sin^2 \theta_{max}} \tag{8.29}$$

This equation is relevant to concentrators where the receiving medium is a photovoltaic array.

Concentrators of the compound type as shown in Fig. 8.8, while providing potentially high values of C do so at the expense of a relatively large reflecting surface area. Extensive mirror area is a source of high-fabrication cost. Some alleviation of this difficulty is possible through a technique termed 'truncation' [2].

The trade-off offered by truncation can be illustrated with reference to Fig. 8.8. Further, to keep the visualisation simple, we need to only consider rays in two dimensions, or more specifically in the xz-plane in the above figure. On imagining such two-dimensional rays entering the input aperture (a_1) on the left of the

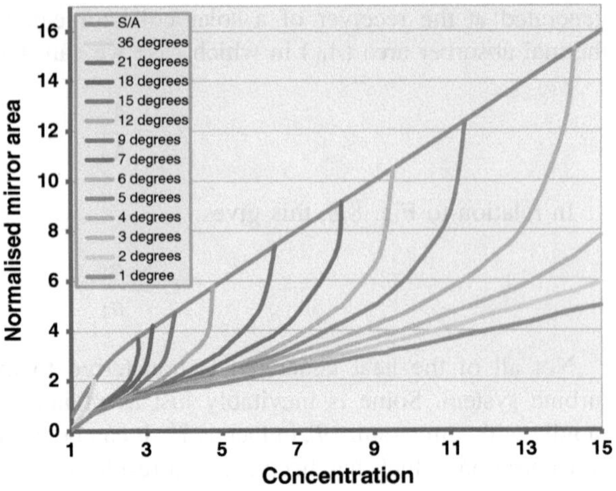

Fig. 8.9 Reflector surface area normalised to the entry aperture area as a function of solar concentration (*Source* data Ref. [2])

schematic, it is not too difficult to discern that very few rays which are guided to a_2 are intercepted by those portions of the mirrors near a_1. Consequently, some shortening of the arrangement is, in principle, possible without introducing a serious diminution of the concentrator performance. This mirror area reduction process is the essence of truncation. The effect is quantified in Fig. 8.9 where the surface area of the reflector (A_{ref}) is normalised to the entry aperture area (a_1). This ratio is plotted as a function of the ideal concentration C_i for a range of acceptance angles θ_{max}.

The dark blue curve running at ~45° across the graph represents A_{ref}/a_1 for the untruncated CPC. All other curves lie below this line with the high-acceptance angle options (e.g. 36°—yellow curve) in the low-concentration area of the graph, while the very limited acceptance angle choices (e.g. 1°—pink curve) traverses the graph at very low slope. This means, not surprisingly, that a short 36° design option is sensitive to truncation (reducing A_{ref}/a_1), while the 1° design option is not. For example, if we consider a full CPC with an acceptance half angle of 6° (say), the graph suggests that it generates a concentration level of 9.6 and requires a total mirror area of 10.6 m^2 for each square metre of entry aperture. However, when A_{ref}/a_1 is reduced to 5, a more than 50 % area drop, the concentration is still acceptable at a level of 8.2. In such a case, truncation is an attractive option.

In addition to causing some loss in collector concentration, truncation also diminishes collector performance by randomising ray arrival angles at the exit port (a_2). This change relates to the alteration in the number of ray reflections (n_r) on the mirror surfaces as truncation is implemented. At any given angle of incidence, ($\theta_i \leq \theta_{max}$) on the entry aperture a_1 the number of reflections depends both on θ_i and on the point of incidence on a_1. The average number of reflections for a given θ_i is denoted by $\langle n_r(\theta_i) \rangle$. It can vary significantly from a 'low' value to a 'high' value as shown in Fig. 8.10 which has been generated by ray tracing [2]. If the number of reflections is averaged over all incident angles within the acceptance range

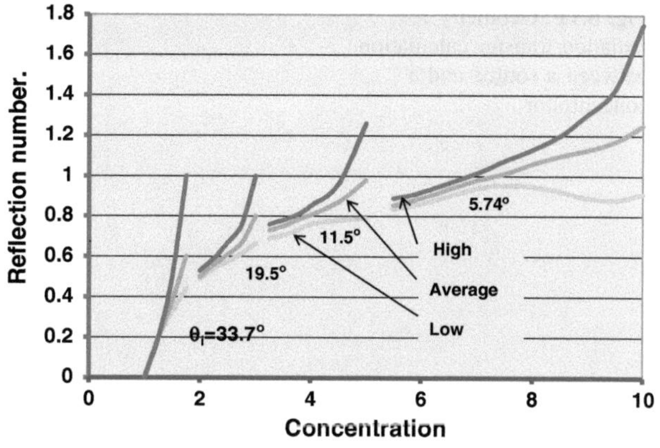

Fig. 8.10 Number of reflections per ray on concentrator mirror surfaces as a function of concentration magnitude, with 'high' (*blue*), average (*red*) and 'low' (*green*) n_r values (*Source* data Ref. [2])

$(-\theta_{\max} \leq \theta_i \leq \theta_{\max})$ for any given concentrator design, this number is denoted by $\langle n_r \rangle$. $\langle n_r(\theta_i) \rangle$ has been plotted as a function of concentration in Fig. 8.10. Since concentration decreases with truncation, the general trend shown in the figure is for $\langle n_r(\theta_i) \rangle$ to decline with truncation. At low-acceptance angles ($\theta_{\max}$) and high concentrations, both the average number of mirror reflections and the spread with entry angle are high pointing to a potential for absorber temperature variation. Large numbers of mirror reflections also have the potential to raise mirror losses in non-perfectly reflecting surfaces. This has implications for collector efficiency.

8.3 Concentrator Thermodynamics

In addition to being a ray optics problem, the solar concentrator also requires some consideration of the thermodynamics of radiative heat transfer from a source at temperature T_s (nominally the sun) to an absorber at temperature T_{th}.

For the particular case of a stationary concentrator and absorber irradiated with energy from a stationary source, as depicted in Fig. 8.11, the computation of heat transfer is relatively routine. The source in Fig. 8.11 is an isotropically radiating sphere of radius R_s with a fraction of the radiation gathered by a collector aperture of area A which is at a distance R from the source. The entry angle in this stationary case is $2\theta_a$. When the source is the sun, R is so large that we can presume that $A/R^2 \to 0$ and that $\sin \theta_a = R_s/R$. The radiation at the aperture A will be uniformly distributed over it for all angles $|\theta| \leq |\theta_a|$. In empty space, the radiation from the source S will spread out equally in all directions in accordance with the inverse square law. Hence, the radiation from S at temperature T_s is given by the blackbody relationship:

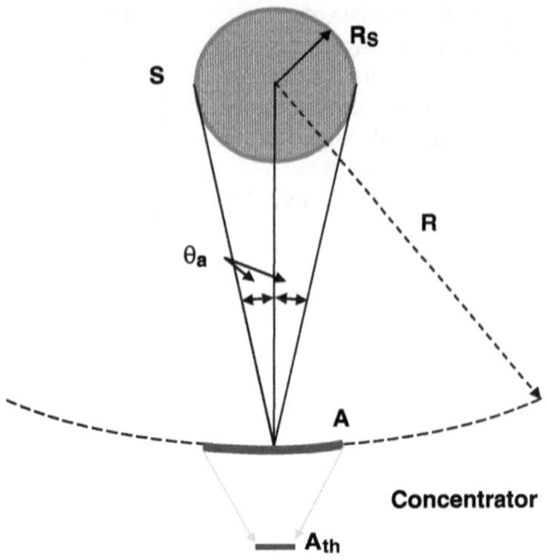

Fig. 8.11 Geometry for radiation transfer calculation between a source and a concentrator

$$Q_S = 4\pi R_S^2 \sigma T_S^4 \text{ W} \tag{8.30}$$

where σ is the Stephan–Boltzmann constant. A tiny fraction of this is gathered by the aperture A, namely:

$$q_{\text{co}} = \frac{A}{4\pi R^2} Q_S = \frac{\text{A}R_S^2}{R^2} \sigma T_S^4 \text{ W} \tag{8.31}$$

The concentrator raises the absorber to temperature T_{th} which in turn becomes a blackbody radiator emitting back into space from its surface A_{th}, an amount:

$$q_{\text{th}} = A_{\text{th}} \sigma T_{\text{th}}^4 \text{ W} \tag{8.32}$$

Evidently only a very small fraction ($f_x \ll 1$) of q_{th} returns to the source, namely q_x, which has the magnitude:

$$q_x = f_x A_{\text{th}} \sigma T_{\text{th}}^4 \text{ W} \tag{8.33}$$

In the theoretically rather limiting case where the absorber temperature is raised to that of the source, so that $T_{\text{th}} = T_S$, the second law of thermodynamics dictates that radiant heat transfer between the two bodies must be zero. In this case, Eqs. (8.31) and (8.33) yield:

$$\frac{\text{A}R_S^2}{R^2} = f_x A_{\text{th}} \tag{8.33}$$

Hence, using Eq. (8.27), we obtain:

$$C = \frac{A}{A_{\text{th}}} = \frac{R^2}{R_S^2} f_x = \frac{f_x}{\sin^2 \theta_a} = f_x C_i \tag{8.34}$$

Clearly, and not surprisingly, the thermodynamic assessment of radiant heat transfer endorses the ray optics determination of the limits to solar concentration. Given that physics dictates that $f_x \leq 1$ it follows that for practical concentrators:

$$C \leq C_i = \frac{1}{\sin^2 \theta_a} \tag{8.35}$$

Having established the importance of solar concentration and how it can be manipulated in a collector, it is useful and relevant at this stage to develop an equation relating ray concentration to the temperature to which the receiver can be raised. Equation (8.31) already gives us a relationship between the heat gathered by a collector aperture (A) from a remote source at temperature T_s namely:

$$q_{\text{co}} = A \frac{R_S^2}{R^2} \sigma T_S^4 = A \sin^2 \theta_a \sigma T_S^4$$

If the remote source is the sun, then T_S is the solar temperature (6,000 K) and $\theta_a \approx 0.25^{\circ}$. Not all of this heat is absorbed in the collector. Some is reflected or refracted away before absorption and can be represented by a loss factor τ. We also need to take account of the rate of heat absorption in the receiver material—the absorptivity α_{abs}. Hence, the receiver gathers an amount of heat given by:

$$q_{abs} = \tau \alpha_{abs} A \sin^2 \theta_a \sigma T_S^4 \text{W} \tag{8.36}$$

Some of this heat is lost at infrared frequencies in the form of blackbody radiation. It has the magnitude:

$$q_{\text{rad}} = \varepsilon_{\text{th}} A_{\text{th}} \sigma T_{\text{th}}^4 \text{ W} \tag{8.37}$$

where ε_{th} is the emissivity of the material forming the absorber. For simplicity, it is assumed here that the collector resides at an ambient temperature of zero degrees Kelvin. Hence, if η_{co} is the overall efficiency of the collector in gathering solar radiant heat, then:

$$\frac{q_{\text{rad}}}{q_{\text{abs}}} = 1 - \eta_{\text{co}} \tag{8.38}$$

Consequently:

$$(1-\eta_{\text{co}})q_{\text{abs}} = q_{\text{rad}}$$

and on inserting Eqs. (8.36) and (8.37), we obtain:

$$(1-\eta_{\text{co}})\tau\alpha_{\text{abs}}A\sin^2\theta_a T_S^4 = \varepsilon_{\text{th}}A_{\text{th}}T_{\text{th}}^4 \tag{8.39}$$

But

$$C = \frac{A}{A_{th}} \quad \text{and} \quad \sin^2\theta_a = \frac{1}{C_i}$$

resulting, on substitution, in the following relationship for receiver temperature:

$$T_{\text{th}} = T_S\left\{\frac{(1-\eta_{\text{co}})\tau\alpha_{abs}C}{\varepsilon_{\text{th}}C_i}\right\}^{1/4} \text{K} \tag{8.40}$$

Theoretically, the maximum possible absorber temperature is T_S = 6,000 K. However, it could only reach this magnitude if at high temperature $\varepsilon_{th} \to \alpha_{abs}$ and if losses were absent, resulting in $\tau \to 1$. This would also imply that $A_{\text{th}} \to 0$ and $C \to C_i$. Certainly at $T_{\text{th}} \approx T_S$, the absorber, like the sun, becomes an efficient radiator such that $\eta_{co} \to 0$ as required. However, in practice receiver temperatures in CSP systems are a long way short of the theoretical optimum. They tend to be in the 300–1,000 °C range, as we shall see below.

8.4 Architecture of CSP Systems

The differing structural characteristics of currently competing concentrated solar power systems are summarised in Fig. 8.1, while the relative performance levels of the four main categories of electrical power generators are compared in Table 8.2. These four are the most commonly discussed and debated systems at the time of writing (2013). Their physical differences are summarised in words, in the bullet points below. More extensive deliberations are provided in the next four sections of this chapter.

- Trough: uses linear parabolic-shaped mirror reflectors—usually arranged like parallel troughs in an east–west alignment with some degree of elevation adjustment to follow the seasonal movement of the sun.
- Fresnel: employs linear plane mirrors again lying in east–west alignment and segmented in the transverse direction according to Fresnel principles to focus solar power on fluid carrying receiver.
- Heliostat: comprises circular arrays of small flat adjustable mirrors which direct solar rays onto a central tower with a receiver converting the concentrated solar power to thermal energy.

Table 8.2 Relative performance of competing CSP systems

	Capacity MW	Concent-ration	Peak solar efficiency (%)	Thermal cycle efficiency (%)	Annual solar efficiency (%)	Capacity factor (%)	Land use (kW/m^2)	Relative Cost
Trough	10–200	70–80	21	30–40 ST	10–15	25–70	~0.1	Low
Fresnel	10–200	25–100	20	30–40 ST	9–11	25–70	~0.075	Very low
Heliostat	10–150	300–1000	20	30–40 ST	8–10	25–70	~0.12	High
				45–55 CC				
Dish-Sterling	0.01–0.4	1000–3000	29	30–40 SE	16–18	25	~0.12	Very high
				20–30 GT				

- Parabolic dish: the dish-shaped parabolic mirror directs the collected solar rays to a receiver, housing a Sterling engine coupled to an electrical generator, which converts the solar power directly to electricity. Each dish unit represents an independent mini power station.

The categories listed in Table 8.2 have the following meanings:

Capacity: This quantifies in megawatts the potential power capability of a single solar farm which uses concentrated collection techniques to generate heat for a conventional thermal power station. The exception is the parabolic dish/Sterling engine arrangement which generates electricity so that each unit is a mini power station in the 10–400 kW range. Of course, a farm supporting say 100 units would compare with the alternatives.

Concentration: using the basic definition provided by Eq. (8.27), it is clear from the table that concentration varies considerably between systems with the linear mirror systems at the low end and the full beam forming parabolic dish at the high end.

Peak Solar : solar efficiency is defined as the ratio of the power in the generated electricity to the solar power incident on the collector. There is little difference between systems in solar efficiency terms, although the dish/Sterling units display a marginally higher level than the others as a consequence of the much higher concentration which brings the efficiency benefits of superior-operating temperatures.

Thermal Cycle Efficiency: this parameter applies to the conversion of heat into electricity which occurs in the turbine/generator set. The level is applicable to any thermal power station however fuelled. Consequently, there is no difference between the first three systems which all feed their heat output into a conventional steam turbine (ST) with 30–40 % thermal efficiency. The Sterling engine displays a similar level. Improved efficiency is available for power stations operating under a combined cycle (CC) scheme where otherwise waste heat is used for space heating.

Annual Solar Efficiency: this is essentially solar efficiency integrated over a year. The figures presented are mainly projections or figures deduced from demonstrator systems.

Capacity factor: this is the ratio of the number of hours of useable sunshine in a year divided by the total number of hours in a year. It is a site-dependent number. The values presented are for similar sites in the USA, so not surprisingly it does not alter as the systems change.

Land Use: this figure relates to power level in megawatts which can be extracted from a square metre of land covered by a solar farm. It is assumed that all the farms are on sites subject to similar radiation levels over the period of a year. On the basis of this criterion, the Fresnel collecting systems are the least effective at 75 W/m^2, while the Dish/Sterling is most effective at 120 W/m^2.

Relative Cost: cost is a difficult parameter to establish in an area of engineering which is evolving rapidly. However, in relative terms, it is possible to state, on the basis of available evidence in 2013, heliostat systems are potential winners on cost-effectiveness, while systems of the dish/Sterling type are at the opposite end of the scale. So, as is not unusual in engineering, trade-offs between performance and cost apply, with the least costly being the least efficient and vice versa. Where the balance lies will generally be dictated by location, climate, customer base, and its extent.

8.4.1 Parabolic Trough Collector System

A schematic drawing depicting the layout and the major components of a solar farm employing long parabolic mirrors in parallel alignments is depicted in Fig. 8.12. In California, operational plants can be found with mirrors which cover areas of desert in excess of two square kilometres [7]. The parabolic troughs, designed using the optical principles outlined in the preceding sections of this chapter, provide directed solar reflections from the mirrored surfaces of the troughs onto a fluid carrying absorber tube which runs the length of each reflector. The troughs and their associated absorbers are generally arranged to lie in an east/west direction, with the receiver pipes located along the axes of the curved mirrors. However, as the sun

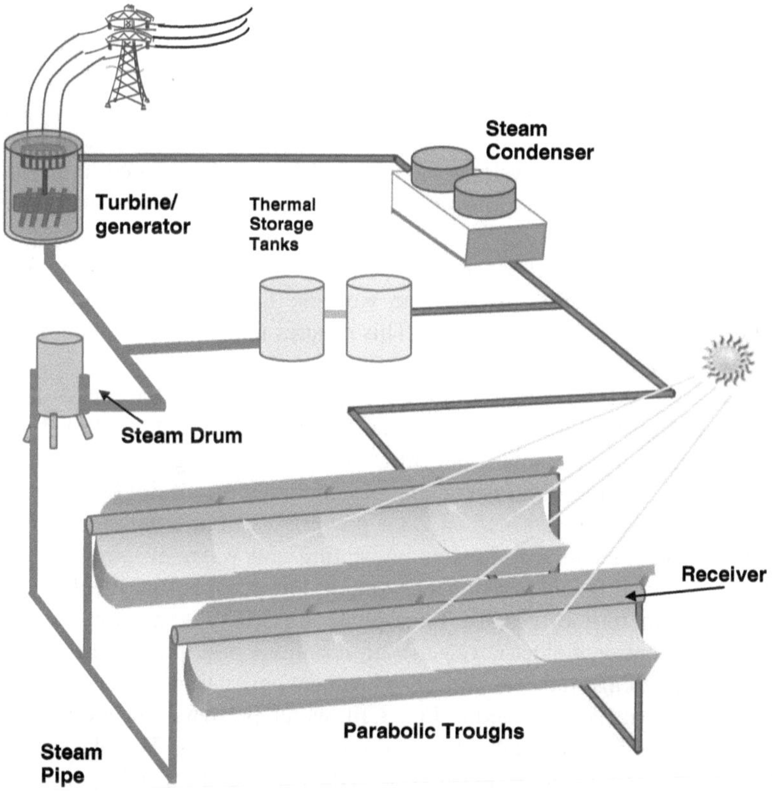

Fig. 8.12 Schematic of electricity power station driven by parabolic trough-type solar collectors

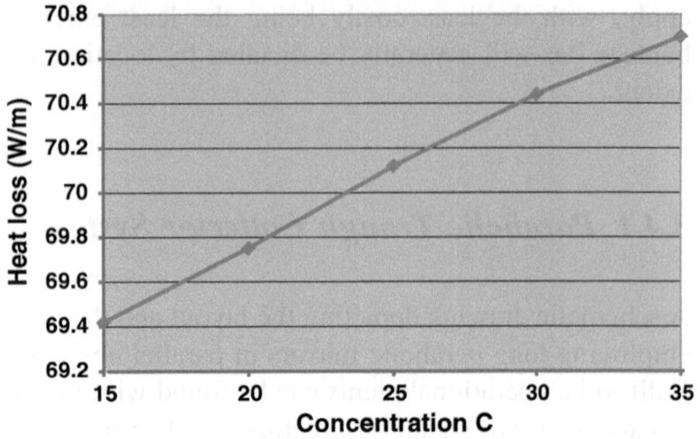

Fig. 8.13 Heat loss per unit length from a trough reflector illuminated receiver as a function of concentration (*Source* data Ref. [11])

traverses the daytime sky, the focus shifts away from the receiver, and the mirror/receiver unit as a whole has to be realigned to maintain optimum performance. The light acrylic-based mirrors are cradled on trestles which permit electronically controlled north/south rotational adjustment of each mirror/receiver unit to compensate for daily and seasonal movement of the sun.

The receiver, typically, takes the form of a black-coated metal pipe within a glass sheath which helps to ensure that the heat loss by convection is kept to a minimum. The surface coating of the metal tube is designed to maximise solar absorbance while at the same time securing a thermal emittance level which is as low as possible. It is pertinent to note here that the optics and the thermodynamic implications of parabolic trough systems have been thoroughly explored in the literature [11], where useful design relationships have been generated such as heat loss versus concentration as shown in Fig. 8.13. The working fluid in the receiver pipe can be water, but the water/steam-phase change associated with evaporation is unpredictable. In more modern systems, the working fluid is more likely to be a synthetic oil with a high-boiling point. This ensures that a more predictable thermal behaviour is available.

In practice, the oil, or water/steam, in the receiver tubes at the foci of the trough reflectors (see Fig. 8.12) can be raised to temperatures approaching 350–550 °C [6], this being achieved with solar concentration levels ranging from 70 to 100. At these levels, the receiver fluid is capable of generating electricity by being pumped to a conventional steam turbine/generator power system as suggested in Fig. 8.12. The thermal cycle usually comprises preheating and water evaporation units followed by superheating and perhaps reheating of the steam before it enters the steam turbine which in turn drives the three-phase electricity generator. However, in contrast to conventional electricity generation stations solar power farms also divert heat to thermal storage tanks at periods when heat generation is high and electricity demand is low. This is illustrated in Fig. 8.14, which presents power output over the period of one day for a typical solar power-generating plant. The black curve with diamond markers shows total power collected by the solar receivers maximising as expected in the middle of the day (8.00–18.00), while the red horizontal line

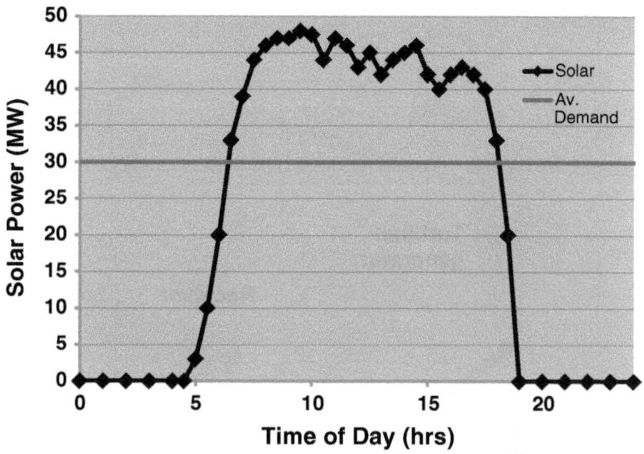

Fig. 8.14 Solar power delivery from typical power station with storage [21]

represents the average grid demand for the plant. In a farm equipped with storage capacity, in the scenario where surplus power is developed during the middle of the day (area above average demand line), the working fluid is redirected to the storage tanks. Subsequently, it is passed to the thermal cycle to compensate for the dwindling solar power after 18.00. A significant degree of levelling of the daily variations is thus possible, stretching plant operation towards 24 h.

Thermal storage is most effective with systems employing oil as the working fluid. Generally, one or more large-insulated storage tanks are used (see Fig. 8.12). These can contain molten salt, or a rock/metal aggregate, through which the hot fluid is pumped during the storing phase of operation. Subsequently, fluid can be passed through the hot tanks when the sun no longer generates the requisite power to meet demand. Incorporating storage into a renewable power plant is a costly investment, and it inevitably introduces additional heat losses into the system. However, for a given quantity of collected heat, by spreading its collection and delivery over a longer period of time, it can be collected at a lower solar flux level which means savings in optical hardware. Renewable plants with storage capability can provide a major impetus to this technology by offering a route to counteracting the endlessly expressed, yet somewhat erroneous, notion that renewables cannot operate without base load power from fossil fuel backup stations. For photographic images of typical power plants, see reference [12].

8.4.2 Linear Fresnel Reflector System

In thermal and storage terms, a solar power plant based on the linear Fresnel style reflector system is no different to the trough reflector plant described in the previous section. The contrast lies in the reflector arrangement which is depicted in Fig. 8.1 and as part of a generating plant in Fig. 8.15.

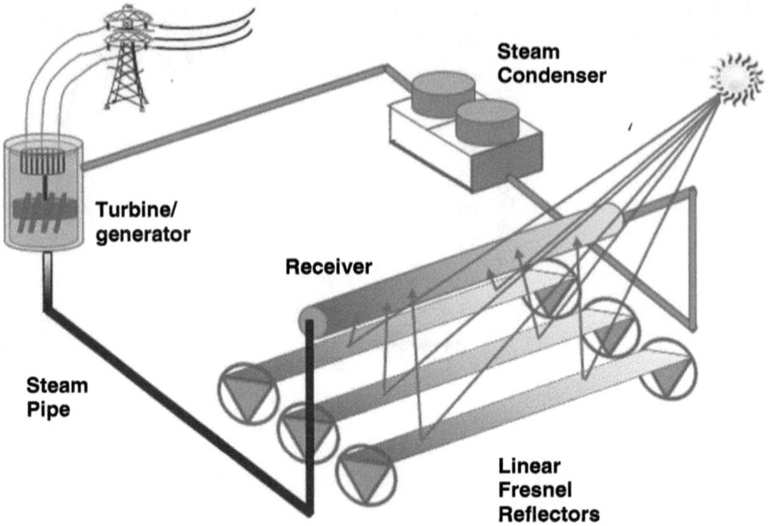

Fig. 8.15 Schematic of electricity power station driven by Fresnel-type solar collectors (for original see www.renewable-energy-info.com)

Fig. 8.16 Air cavity receiver for linear Fresnel solar collector

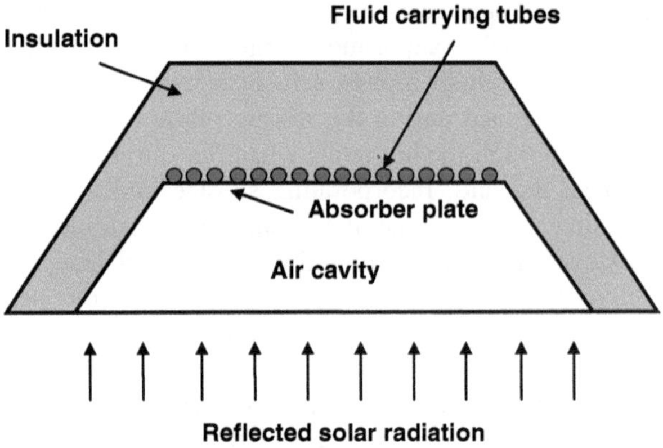

Linear Fresnel reflectors are formed from long flat mirrors arranged in parallel arrays with each mirror element angled to concentrate light on a long linear receiver mounted above it. The support tower is typically 10–15 m tall and contains fluid carrying tubes as postulated in Fig. 8.16 [13].

The flat mirrors formed from silvered acrylic material are both flexible and light, and being unshaped, as in the trough system, they are very cost-effective. Generally, each is mounted on a two-axis tracking frame which enables the mirror to follow both east–west and north–south solar movements. The flat linear mirrors can be viewed as reflecting strips of a parabolic trough each aligned to focus the reflected light towards the receiver. The layout and alignment of the individual mirrors can largely be determined using methods described in this chapter, particularly Snell's laws. Significantly, the ability with the Fresnel technique to 'reshape' the mirror surface represents a major advantage over the trough system, in that solar movement across the sky can be compensated for by simple adjustment of the mirror elements rather than requiring movement and control of the reflector/receiver unit as a whole. This simplifies the support and tracking structure leading to lower implementation costs (see Table 8.2).

Solar power stations based on linear Fresnel concentrators have opened recently in both Spain and California. To view photographic images, the reader should proceed to reference [12] on the Internet. A typical solar power generator employing Fresnel technology is provided by the German company Novatec Solar. On the basis of this successful 1.4 MW development, commissioned in 2009, the company has recently completed (in 2013) a much larger 30 MW linear Fresnel plant at Calasparra in Spain. Developments of this kind are indicative of a welcome trend towards renewable technology in the power-generation industry. However, in climate change terms, the transition away from fossil fuels remains painfully slow.

8.4.3 Heliostat Field System

In the heliostat field system, a central solar power tower gathers the focused sunlight from circular or arced arrays of flat, movable mirrors (called heliostats) individually directing reflected light from the sun towards the focus, or receiver, atop the tower.

Early designs used these focused rays to heat water, and employed the resulting steam to power a turbine as described in the preceding sections. More recent manifestations take advantage of working fluids which exhibit high-heat capacity. This means they can store the thermal energy before using it to heat water to its evaporation point and hence drive turbines with the resultant steam. Systems using liquid sodium have been demonstrated, and systems employing molten salts (40 % potassium nitrate, 60 % sodium nitrate) as the working fluids are now in operation. These storage capable designs also allow electrical power, as we have seen already, to be generated when the sun is not shining.

Heliostat field electrical power stations, as suggested in Fig. 8.17, whether under development or being planned, use flat mirrors which are much more cost-effective than shaped versions. For a photographic image, the reader should look to reference [12]. Each heliostat resides on a two-axis tracking mount and exhibits a surface area typically in the range 50–150 m^2. Optical design of the concentric mirror arrays follows classical Snell's laws and ray optics for flat mirrors [14]. The mirror alignments in the heliostat array are electronically controlled through motors to track the solar movement across the sky and also seasonal changes. The software and technology to do this is little different to that used in connection with the linear Fresnel array described in the preceding section.

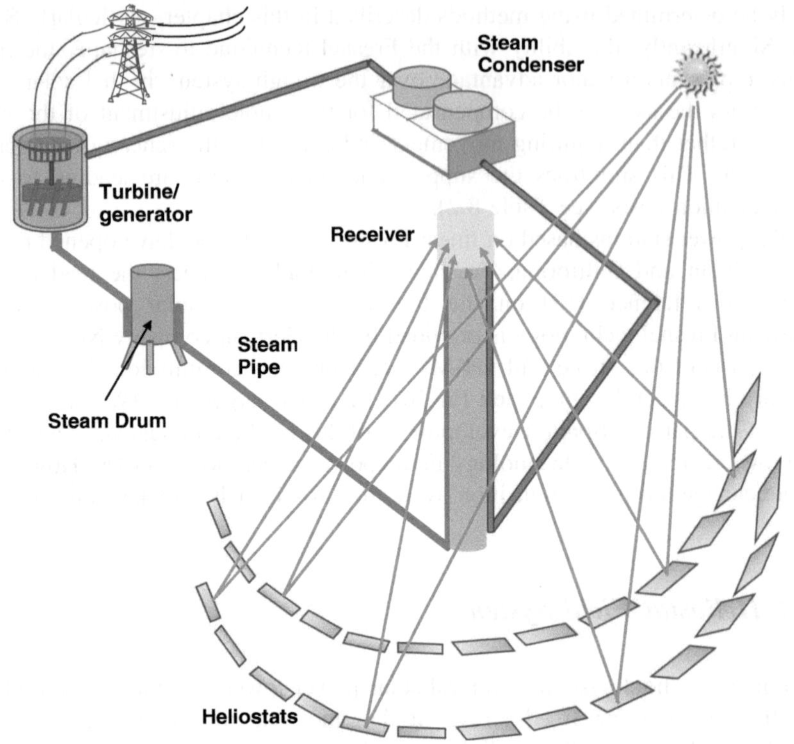

Fig. 8.17 Schematic of electricity power station driven by Heliostat-type solar collectors. In practice mirrors fully encircle the tower (see www.eere.energy.gov/basics/renewable_energy/)

Heliostats generate high levels of solar concentration (see Table 8.2), and hence, the power towers can operate at very high temperatures of the order of 1,500 °C, which has efficiency advantages. The transfer fluid as suggested above is usually molten salt, but the receivers can vary considerably in design depending on the fluid's characteristics [7]. Generally, the receiver employs tubes arranged in a circular array around the cylindrical tower top, not unlike boiler piping in a conventional power plant combustion chamber. The heated liquid is then pumped to storage tanks or heat exchangers as indicated in Fig. 8.17, thus delivering heat to the power cycle. While the thermal cycle efficiency (see Table 8.2) is potentially high for heliostat-based plants, the land area required—current installations generally use from 150 ha (1,500,000 m^2) to 320 ha (3,200,000 m^2)—tends to be more extensive per MW delivered than is common with trough and Fresnel systems.

A typical example of a heliostat field system is Gemasolar's 19.9 MW plant located near the city of Fuentes de Andalucía in the Seville province of Spain. It is mooted to be the world's first commercial-scale plant to use solar technology which incorporates a central tower receiver, a heliostat field and a molten-salt heat storage system.. It was officially commissioned in October 2011. This plant is designed to produce 110 GWh a year and supply clean energy to 25,000 homes while reducing carbon dioxide emissions by more than 30,000 tons per year.

8.4.4 Parabolic Dish System

Concentrated solar power systems employing parabolic mirrors produce high levels of light concentration (between 1,500 and 4,000). The mirrors are mounted on rigid support frames which facilitate solar tracking, thereby directing, throughout the daylight hours, solar light onto a receiver at the focal point of the parabolic dish (see Fig. 8.18). The receiver could be a cavity in which fluid is heated to high temperature then pumped to a thermal cycle comprising a turbine/generator unit as in the trough and Fresnel plants (see Figs. 8.12 and 8.15). However, it is more likely to

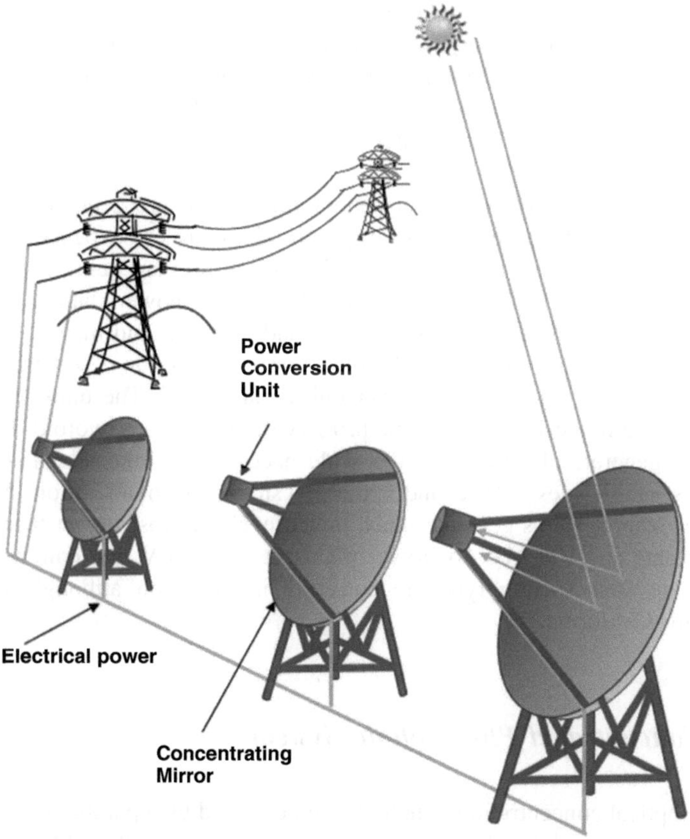

Fig. 8.18 Schematic of electricity power station driven by parabolic dish-type solar collectors (see www.eere.energy.gov/basics/renewable_energy/)

be a compact heat engine, either a Stirling engine or a micro-gas turbine, which produce electrical power directly through a coupled generator, obviating the need for fluid transfer arrangements [4].

The Stirling engine is in many ways much like the petrol or diesel engine which powers your car, except for one major difference. It is an external combustion engine rather than an internal combustion engine. While the gas in the cylinder of a petrol engine (vapourised petrol) is ignited by a spark and burnt internally, and in a diesel engine the vaporised diesel is ignited internally by pressure then burnt within the cylinder, the working gas in a Stirling engine, usually hydrogen, is sealed into the cylinder and is not burnt. Piston movement is caused by thermal expansion of the gas by the external application of heat through a heat-exchanging interface material. In the solar dish-type array, the heat is supplied by the focused sun light. At peak operation (solar irradiance greater than 250 W/m^2), the conversion efficiency from solar power collected by the parabolic dish to electrical power supplied by the generator is claimed to be 30 % on the basis of prolonged testing [15]. The system efficiency is, however, susceptible to daytime irradiance dropping below the optimum level due to clouds or haze, and this means that over time, a ~20 % conversion efficiency for solar power to electrical power to the grid, for a large farm of this type, is more representative of its real capability. This is still better than trough and heliostat systems (see Table 8.2) because of the avoidance of the losses associated with the inefficient transfer of power to the steam turbines through the agency of a hot fluid.

Parabolic mirrors for dish power-generation systems are generally constructed from one large parabola per dish/Stirling unit, although less expensive techniques such as forming the dish from an array of small mirrors are being explored. Photographic images can be found at reference [12]. Solar power stations of this type have the advantage of being highly efficient and modular, and may be constructed from dish/Stirling units on a range of scales. For installation in desert environments, they also use little water if the receiver unit is air cooled. The units are normally available in the range 15–70 kW. The primary drawbacks are twofold. Firstly, it is difficult to generate electricity on the scale necessary to take advantage of size-related cost-efficiencies, and secondly, thermal storage is not an option. As of 2013, dish-based solar systems generate much lower levels of gross power than trough or Fresnel plants at >100 MW, or tower plants at ~10 MW. No commercial solar farms of the dish/Stirling type are currently in operation although several are planned for California.

8.4.5 Concentrated Photovoltaic System

The high-optical concentration which can be generated by a parabolic dish reflector makes it an ideal candidate for concentrated photovoltaics (CPV), although the trend is towards a small concentrator per cell as part of a multi-cell PV array—rather reminiscent of car headlight technology in reverse. The approach makes it justifiable

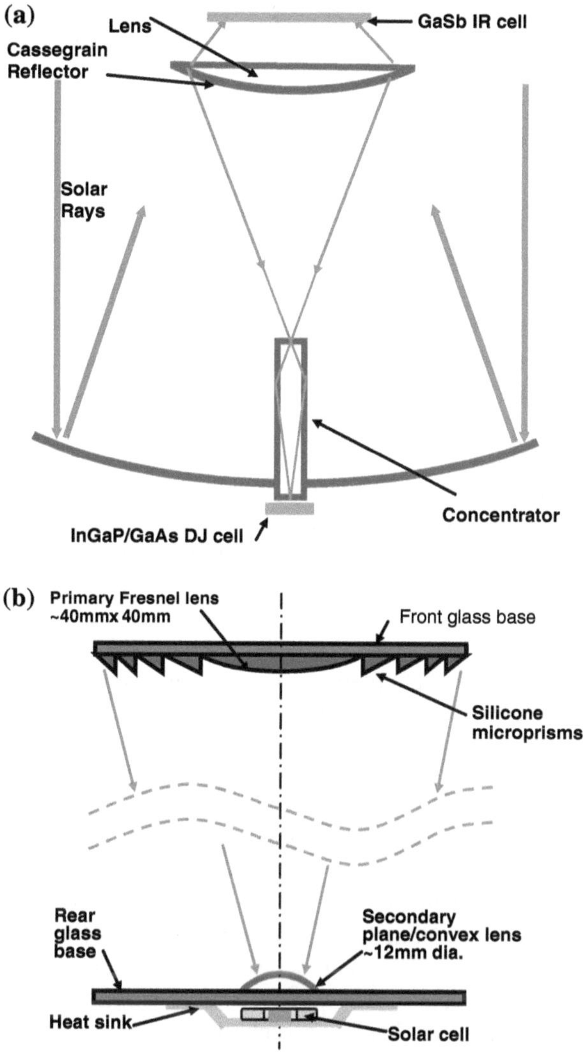

Fig. 8.19 Concentrated photovoltaic systems **a** cassegrain dual-reflector system, **b** Fresnel lens system

to adopt, at the focus of the mini-reflector, a state-of-the-art high-performance photovoltaic cell (or cells), despite the high cost of such devices. A strong solar flux focused onto a small region of cells results in relatively high levels of electrical power per unit, of sufficient magnitude to offset the large initial investment in fabrication and construction. High-performance multi-junction photovoltaic cells based on gallium arsenide, for example, are currently very costly to fabricate, but this is counterbalanced by the high-operational efficiencies ($\sim$30 %) that are possible with solar concentration. In the context of CSP, these PV technologies are remarkably heat-resistant and perform better under high-solar flux in a reflector or lens system, as suggested in Fig. 8.19. Early results on focused PV systems indicate that they can demonstrate operational characteristics that are as good as, if not better than, heat engine embodiments, and with considerably longer lifetimes.

In Fig. 8.19a, a Cassegrain reflector arrangement is employed to focus solar energy on to a solar cell(s) mounted below the main reflector. Since solar cells in a PV module may have to be repaired and replaced from time to time, this geometry gives easy access to in situ panels, thus drastically lowering maintenance costs. In the Cassegrain system, the parabolic main reflector is focused at a point behind a hyperbolic sub-reflector which directs the solar rays towards a light concentrator and hence through an aperture in the main reflector to the PV cell. The whole optical geometry can be designed [5, 14] using the ray optics discussed earlier in this chapter. The schematic also suggests that the Cassegrain dish could be reflective to light frequencies but transparent to infrared frequencies which are focused on IR sensitive cells behind it. This is not too difficult to arrange by using modern nano-technologies. Systems currently under development in California employ hybrid PV cells comprising InGaP/GaAs-GaSb multi-junctions which are claimed to be capable of achieving 40 % efficiency in converting sunlight to electricity [16, 17]. However, to date, engineering developments have generally been restricted to small, limited power, embodiments where commercial interests are currently directing their attention.

Concentrated solar power systems are almost exclusively based, as we have seen, on optical reflector geometries to procure the desired focused solar flux at a receiver. The alternative, which is focusing using lenses, is generally avoided, because in large structures mainly formed from glass, such lenses are unwieldy, heavy and expensive, even where material bulk has been reduced by employing a Fresnel design (Fig. 8.19b). A good example of such a bulky lens structure is provided by the shaped and patterned glass envelope to be seen encasing the beam producing lamp in a traditional coastal lighthouse. On the other hand, tiny lenses in conjunction with solar cells have become common to drive battery-free small electronic devices, but the approach hardly seems compatible with concentrated solar power. Nonetheless, given the history of lenses, solar developmental projects have been re-examining the use of Fresnel lenses to boost solar flux in photovoltaic power systems.

The Fresnel lens is essentially a conventional convex optical lens which is debulked by modifying and rejigging the convex surface (see Fig. 8.20a) while largely retaining the optical focusing property of the original lens. That this is possible is explicable from Snell's laws which govern the passage of light through the lens. For a lens material with a predetermined refractive index, and for a known set of incident angles for the incoming rays, the transmitted angles are dictated by the orientations of the air/glass interfaces through which they pass. So for a solid lens only the angular orientations of the outer surfaces on which light is incident and from which light is transmitted are relevant to the focusing action. The interior bulk of the lens does not contribute. For a convex lens with a flat entry surface, the portion of the lens that influences the direction of the transmitted light is its rear surface shape. Thus, the lens can be discretised as suggested in Fig. 8.20a and then adjusted into a 'flat' profile as proposed in Fig. 8.20b. This is termed the Fresnel

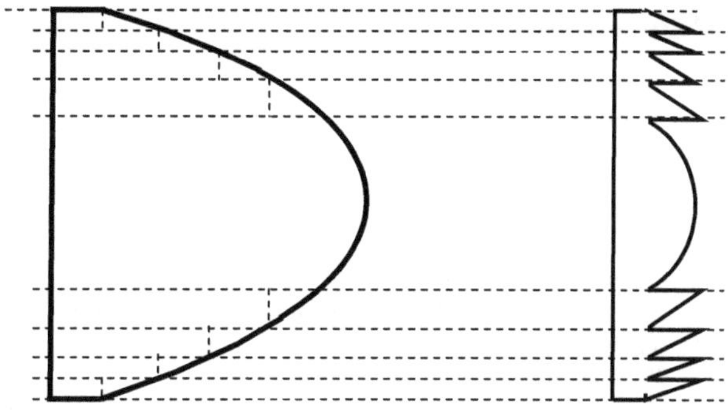

Fig. 8.20 Fresnel lens alongside an equivalent domed lens of circular cross-section: **a** conventional lens, **b** equivalent Fresnel lens

lens. Note that the discretisation, and the sharp edges it generates, which are absent in the original, is a source of unwanted diffracted rays. Consequently, the Fresnel lens is a much poorer imaging lens than the original smooth version. On the other hand, for optical concentration in the context of power collection, this disadvantage is less troublesome and is greatly outweighed by the relative lightness and compactness of the Fresnel lens over the 'smooth' original.

CPV systems deploying solar focusing by means of a parabolic reflector generally employ a single large dish to illuminate a large area PV array module (see Fig. 8.19a). A serious problem with this arrangement is that if illumination uniformity across the array is not achieved, the system is reduced to the efficiency of the least irradiated cell. This disadvantage is circumvented in lens systems by providing a lens per cell, so that a solar module takes the form of many lens-plus-cell units [4]. This solution also solves the problem of lens bulk and weight, by distributing the lenses across the PV array. Prototype systems have shown that PV cells perform more efficiently in lens concentrated light. State-of-the-art triple junction cells have evolved which demonstrate efficiencies in the range 40–41 % at light concentrations varying from a low level of 100 to an upper level of 900. At the time of writing, a state-of-the-art efficiency of 41.1 %, at a concentration level of 326 times, has been reported [18, 19].

The installation and setting up of solar concentrators in practical CSP systems have been adjudged to be beyond the scope of this book. Nevertheless, the topic has attracted considerable attention in recent years and needs to be acknowledged. Three principal methods for solar concentrator alignment exist. These include on-sun single mirror facet alignment, mechanical alignment and optical alignment using lasers and cameras. The advantages and disadvantages of these various methods have been examined in [20].

References

1. Winston R (1970) Light collection with the framework of geometrical optics. J Opt Soc Am 60 (2):245–247
2. Rabl A (1976) Comparison of solar concentrators. Sol Energy 18:93–111
3. Segal A, Epstein M (2000) The optics of the solar tower reflector. Sol Energy 69(1–6):229–241
4. Lovegrove K, Stein W (2012) Concentrating solar power: principles, developments, and applications, (Woodhead Publishing Series in Energy. Woodhead Publishing Ltd, Cambridge
5. Barlev D, Vidu R, Stroeve P (2011) Innovation in concentrated solar power. Sol Energy 95:2703–2725
6. Mehos M (2008) Concentrating solar power. AIP Conf Proc 1044:331–339
7. H. Muller-Steinhagen, F. Trieb, Concentrating solar power: a review of technology. In: Quarterly of royal academy of engineering, Ingenia 18, March 2004
8. Silver S (1965) Microwave antenna theory and design. Dover Publications Inc, New York
9. Drabowitch S et al (1998) Modern antennas. Chapman and Hall, London
10. Hottel HC, Woertz BB (1942) The performance of flat-plate solar-heat collectors. Transactions of ASME 64:91–104
11. Reddy KS, Ravi Kumar K, Satyanarayana G (2008) Numerical investigation of energy efficient receiver for solar parabolic trough concentrator. Heat Transfer Eng 29(11):961–972
12. www.bing.com/images
13. Singh PL, Sarviya RM, Bhagoria JL (2010) Thermal performance of linear Fresnel reflecting concentrator with trapezoidal cavity absorbers. Appl Energy 87(2):541–550
14. Segal A, Epstein M (2000) The optics of solar tower reflector. Solar Energy 69(1–6):229–241
15. Braun HW (1992) Solar stirling Gensets for large scale hydrogen production. Solar Energy Technol 13:21–29 SED
16. Tsutagawa MH (2011) Triple junction solar cell. US Patent No. 2011/0297213 A1, Dec 8, 2011
17. Jayarama Reddy P (2012) Solar power generation. CRC Press, Taylor & Francis Group, London
18. Geisz JF, Friedman DJ, Ward JS, Duda A, Olavarria WJ, Moriarty TE, Kiehl JT, Romero MJ, Norman AG, Jones KM (2008) 40.8 % efficient inverted triple-junction solar cell with two independently metamorphic junctions". Appl Phys Lett 89(12):123505
19. Guter W (2009) Current matched triple junction solar cell reaching 41.1 % conversion efficiency under concentrated light. Appl Phys Lett 94(22):223504
20. Xiao J, Wei X, Lu Z, Yu W, Wu H (2012) A review of available methods for surface shape measurement of solar concentrator in solar thermal power applications. Renew Sustain Energy Rev 16:2539–2544
21. Geyer M, Quaschning V (200) Solar thermal power—the seamless solar link to the conventional power world. In: Renewable Energy World, July–August, pp 184–191

Chapter 9
Solar Power Satellites (SPS)

Engineering refers to the practice of organizing the design and construction [and, I would add operation] of any artifice which transforms the physical world around us to meet some recognized need.

G.F.C. Rogers

Abstract The aim of this chapter is to review and assess the state of the art in electrical technology in 2013 insofar as it relates to any future realisation of solar power collection in space for subsequent delivery to the planetary surface by microwave beams. On the assumption that standard solar photovoltaic panels can be deployed over a sufficiently extensive area to generate gigawatts of DC power, then the relevant technology areas, which will arguably make solar power satellites (SPS) viable, lie largely in the domain of microwave power generation and microwave array antennas. Three power source candidates are addressed in Sect. 9.3, namely klystron and magnetron power tubes and solid-state microwave power amplifiers. The operational principles of each type are reviewed, while their advantages and disadvantages in the proposed SPS application are evaluated. The antenna, which will be required to beam microwave power to a receiving station on the earth's surface, is realisable only in array form, and Sect. 9.4 addresses the alternatives, ranging from slotted waveguide arrays, waveguide phased arrays to micro-strip patch arrays. Again, advantages and disadvantages of each in the SPS context are examined. In addition to power handling, gain and efficiency of these arrays, an exceedingly important requirement is that the microwave beam should remain 'captured' by the ground station. Methods for achieving this are developed in this chapter. Finally, the ground station receiving antenna is required both to capture the microwave beam and to efficiently convert the microwave power to DC power for transmission over the grid to end-users. The favoured approach is based on the rectenna (rectifying antenna) concept which is described in Sect. 9.5.

9.1 Introduction

Given the obvious difficulties of ground-based methods (Chap. 8) of garnering power from the light reaching the earth from the sun, it seems like a natural development, for a species with growing space travelling pretensions, to consider

A.J. Sangster, *Electromagnetic Foundations of Solar Radiation Collection*,
Green Energy and Technology, DOI 10.1007/978-3-319-08512-8_9

placing our photovoltaic arrays in space, rather than on the planetary surface. At the edge of the earth's atmosphere, solar insolation, as we saw in Chap. 1, is approximately 1,350 W/m^2, whereas the attenuation of light as it passes through the atmosphere, even on a clear day, reduces this level to 1,000 W/m^2 at the surface near the equator at noon. Furthermore, when we factor in diurnal variations, seasonal variations and solar station latitude, the average surface insolation is reduced even further, from the above figure, to about 250 W/m^2—six times down on the edge of space value. Other disadvantages of surface power collection range from atmospheric variability, particularly in non-equatorial latitudes, the necessity to track solar movements, restrictions in available real estate, to limited daytime necessitating storage infrastructure. However, while these terrestrial solar collection problems may disappear with the adoption of satellite methods, space technology imposes its own potentially more serious hurdles, which could possibly cripple the whole space initiative, as we shall see. Nevertheless, solar power from an orbiting platform is undoubtedly becoming technically feasible, and an assessment of the fundamental technology required to make meaningful progress will be provided in this chapter. The engineering challenges, which are unquestionably severe, will also be addressed.

The concept of solar power satellites (SPS) entered the public domain close to 45 years ago, with the appearance of a paper in 1968 outlining the idea in Science magazine. It was followed by a patent in 1973 credited to Peter Glaser. In the late 1970s, essentially in the age of space discovery, finance flowed towards the SPS concept, particularly in the USA, and there were several authoritative reports giving a generally positive appraisal of the proposal. Priorities began to change however, and by the 1980s, a rather negative report by the US National Research Council terminated major funding there. Coincidentally, interest in the topic was growing in Japan, Europe and Canada, and by the 1990s, NASA became involved in the following studies leading to a 'Fresh Look' report. As a result of research endeavours during this decade, it was possible, by the commencement of the new millennium, to read many papers [1, 2] which were emboldened to suggest that SPS was feasible and merited continuing support.

Despite the negativity of accepted 'economic' wisdom based on the 'cost-to-first-power' which has been calculated to be in the vicinity of a 'mind-blowing' 300 billion US dollars [3], proposals to fabricate, transport and assemble in space, orbiting solar power stations, are not entirely in the realms of science fiction as some suppose. This can be demonstrated by considering the economic implications of building an equivalent ground-located solar power station with a base-load electricity supply capability, a significant feature of a space-based system. A not untypical power level for such a role would be about 1 GW, and if it is to provide base-load, it needs to be available for 24 h, every day of the year. Given that in the Northern Hemisphere, daylight hours get short, but more importantly, even in favourable locations, overcast conditions in winter may last up to 5–6 days [1], it is easy to see that significant levels of storage capacity will inevitably be required on a terrestrial power station to achieve a base-load role. For currently available solar arrays with conversion efficiencies in the 20–30 % range, it is not too difficult, by employing typical insolation levels, to calculate that base-load capability requires a tracking array which extends over an area of between 80 and 120 km^2 [1]. (To generate the same power, namely 1 GW, at the planetary surface, but emanating from space, would require a solar array occupying an area of about 5 km^2.) In 2001, the basic cost of fabricating solar arrays of this description was estimated at $600–$1,200 (US) per square metre or $2–$4 per watt at the integrated system level. This equates to a cost of $40–$80 billion for the solar power system array. However, to secure the base-load performance from a ground-based station, an energy storage capacity of approximately 150 GWh is also required. This figure assumes that electrical power transmission in and out of the facility can be done at typically 80 % efficiency. Such a storage system would cost about $45 billion at 2001 prices. In total, a terrestrial solar power station, which can achieve the above aims, is projected to cost in the vicinity of $85–$125 billion. If the use of less optimistic and more realistic efficiency figures for the solar modules had to be enforced, a headline cost of nearer $250 billion could emerge, in which case the orbiting satellite power generator option becomes much less fantastic.

The SPS concept is illustrated in schematic form in Figs. 9.1 and 9.2. As the diagram shows, it is actually rather simple and elegant, although the engineering advances needed to realise it are undeniably daunting. The sunlight collecting PV array is depicted in the figure as a very large grid structure possibly occupying an area of up to 5 km × 10 km. The upper surface of the lattice is covered by solar cells which are directed towards the sun. At the lower end of the lattice, microwave power generators and a microwave array antenna are mounted. The electrical power from the PV array is there used to drive the microwave generators with the resultant microwave power being directed into the array antenna. Finally, the space craft-mounted antenna, beams the gathered power to a receiving antenna on earth. This is generally posited to be a rectenna (see Sect. 9.5) capable of converting the received microwaves directly to electricity. Other geometries, based on reflectors, are also possible, and many illustrative examples can easily be found on the Internet. In this chapter, we shall direct our focus towards the geometry shown in Fig. 9.1, which seems to be representative of the currently favoured formats [1–3].

The strategic technical and engineering issues which must be overcome to make SPS a reality have been identified in the literature [2]. They can be summarised as follows: (1) significant reduction in system mass; (2) ground/space transportation improvements; (3) much better device efficiency; (4) more effective management and distribution of satellite power; (5) thermal management solutions; (6) establishment of cost-effective space assembly of a very large systems plus economical maintenance and repair; and (7) significant further evolution of large-scale and low-cost manufacturing. Most of these are problems for the aeronautic, thermodynamic, mechanical and production engineers. However, items (3), (4) and (5) are arguably of relevance to electrical engineers for whom the issues of concern are (i) the design of the solar arrays for weight, ease of assembly, efficiency, thermal dissipation in the vacuum of space and platform control for solar and ground tracking, (ii) the design of the microwave power sources particularly in relation to efficiency, heat dissipation and delivery of microwave power to the microwave array and precise

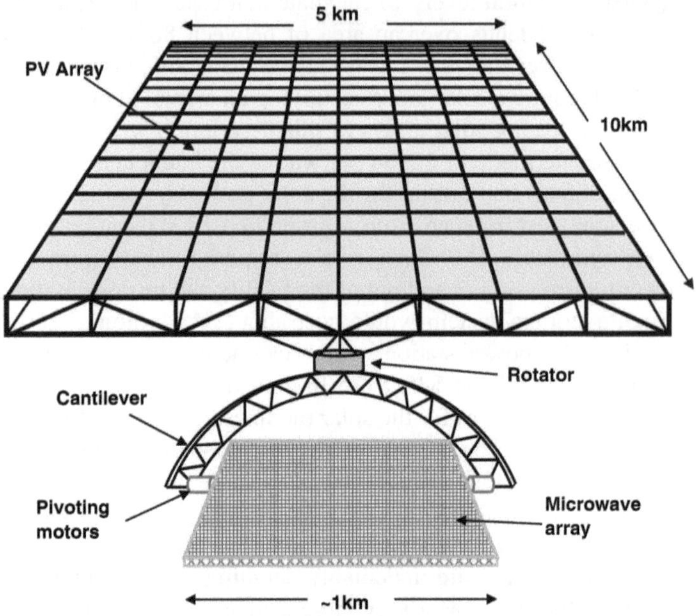

Fig. 9.1 Schematic diagram showing primary components of an ~5 GW SPS array

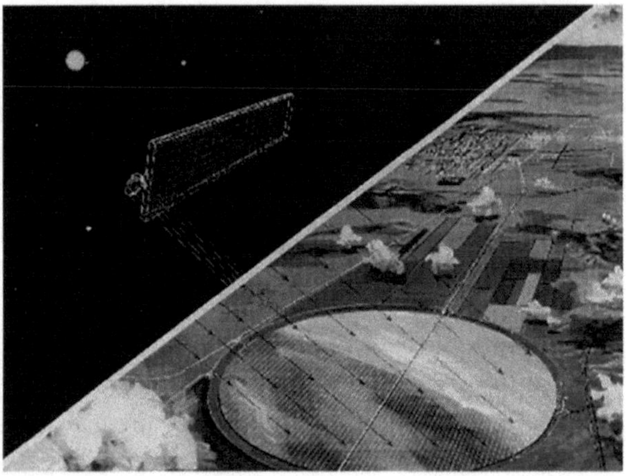

Fig. 9.2 Schematic diagram showing space (above the 45° dividing line) and ground system components (below the 45° line) of an array-based SPS system (*Courtesy* Elsevier)

beam control and (iii) the design of the ground-based microwave array in relation to safe and accurate reception of the microwave beam and efficient conversion of the received microwaves to DC power. Each of these electrically related topics will be addressed sequentially in the sections below.

It is perhaps appropriate to note here that the ultimate solar power space platform would be one which uses the earth's natural satellite, namely the moon. Such a proposal burst into the technical literature in 2013. It is termed the 'luna ring' [4] and is the brainchild of the Shimizu Corporation of Japan. The press release modestly claims 'The LUNA RING for the infinite coexistence of mankind and the Earth'. The luna ring technology uses many of the techniques reviewed in this chapter, differing from what is described herein only in that much of it will be lodged on the moon rather than on a space platform. Arguably, if 'luna ring' is remotely realisable, then surely, in technological terms, SPS can be claimed to be in the realms of the routine!

9.2 Space-Based Photovoltaic Array

The photovoltaic arrays that are likely to be installed on an orbiting platform in space will not be greatly different, electrically at least, to those which have been developed for ground-based solar power systems [1, 2]. The fundamentals of such arrays have already been described in Chaps. 7, and 8 (Sect. 8.4.5). However, cost-effective transportation into, and assembly in, space enforces a significant degree of modularity, arguably more so than for ground-based systems and perhaps, more importantly, is the incorporation of materials and construction methods, in these destined-for-orbit solar modules, which will keep launch weight to a minimum.

From an electrical perspective, the major difference between space and ground operation of solar panels is heat dispersal. On earth, heat convection and air movement around the panels, and perhaps conduction to a sink, would generally be enough to prevent overheating of solar modules. In extreme conditions, water cooling is always possible. However, in vacuum of space, convection is not an option, while fluid cooling is unlikely to be other than prohibitively expensive. This leaves radiation and the option of conduction cooling through a light metal support framework which carries excess heat to a sink. This sink can be designed to be heat tolerant so that a steady but slow heat loss by means of the mechanism of radiation is not a problem. However, the format of thermal management regimes on a future satellite solar power station is very much a live issue, and current developments seek to minimise heat generation by pressing for major improvements in solar panel efficiency, possibly by adapting the techniques alluded to in Sect. 8.4.5, but also by adopting thin-film techniques and multi-bandgap PV cells [2].

The use of superconductors such as YBCO ($YBa_2Cu_3O_{7-\delta}$) in cabling systems has also been mooted to minimise heat build-up in the orbiting structure.

9.3 Microwave Power Generation

The photovoltaic array housed on an SPS system performs the role of converting sunlight to DC power. This power can be presented (see Chap. 7) in high voltage, low current form if the array modules are series-connected, or in low voltage, high

Table 9.1 Source and array comparisons for 2 GW transmitter at 5.8 GHz [1]

Transmitter type	Klystron	Magnetron	Solid-state source
Available power (typical CW)	30,000 W	5,000 W	60 W
Operating voltage (typical)	30,000 V	6,000 V	80 V
DC-to-RF efficiency	80 %	85 %	90 %
Source weight	~14 kg	~1 kg	~0.001 kg
Operating temperature (typical)	300 °C—tube body 500 °C—collector	350 °C on radiator	300 °C at junction
No. of sources for 500-m-diameter array	~210,000	~400,000	~84,000,000
Specific weight of transmit array	~40 kg/m^2	~30 kg/m^2	~35 kg/m^2

current form if parallel connection is adopted. This difference has a considerable bearing on the method employed to generate microwave power. There are three methods of generating electromagnetic waves at microwave frequencies, which are relevant to the SPS concept. These are the klystron oscillator, the magnetron oscillator and the solid-state oscillator, which are compared in Table 9.1 within the context of a notional 2 GW SPS operating at 5.8 GHZ. The power delivered to the planetary surface by an SPS station has to be large enough to justify the cost of developing a space-based system, and 2 GW meets this criterion. However, there is an upper limit dictated by transmit array size which is also influenced by the operating frequency. The chosen frequency of 5.8 GHz is a compromise between low attenuation of the microwave beam as it penetrates the atmosphere and practical array size consistent with procuring a well-focused beam (see Sect. 9.4). The table shows that the power source alternatives offer very different electrical characteristics which have considerable bearing on the transmit array design, as we shall see.

In the following sections, the three microwave source candidates identified in the table will be examined in, hopefully, sufficient detail to help the reader appreciate their potential usefulness to SPS developments. Much more in-depth treatments can be found in Ref. [5].

9.3.1 Klystron

The predominant feature of almost all microwave tubes is interaction between an electron beam and an electromagnetic wave through the action of the wave's electric field on the motion of electrons. Generally, this is a decelerating action resulting in the conversion of kinetic energy from slowing electrons into the growth of power in the wave.

The principal elements of the klystron amplifier are depicted in Fig. 9.3. The electron beam is 'fired' from an electron gun, at voltage V_o, through two re-entrant

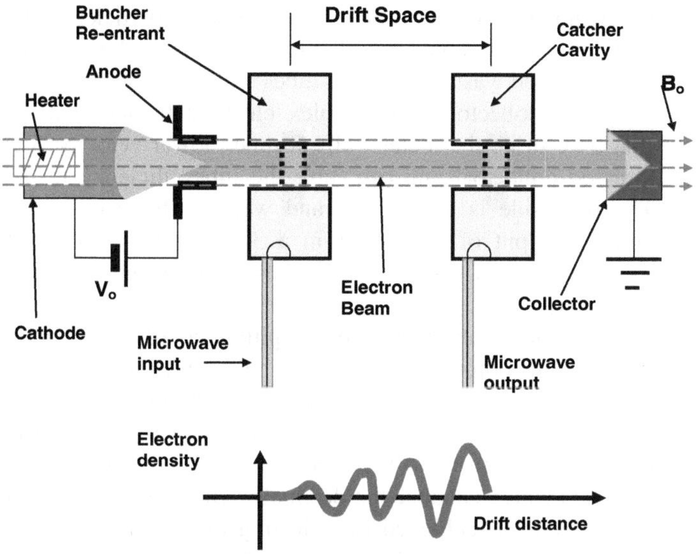

Fig. 9.3 Schematic of a two-cavity klystron amplifier

microwave cavities, separated by a distance termed the 'drift' region. Not shown is the magnetic field stack which supplies the focusing axial DC magnetic field B_o. This field maintains the beam diameter which would otherwise balloon uncontrollably due to increasing space charge forces (electron repulsion forces) as bunching proceeds. Also, the whole space occupied by the beam must be a vacuum. This is not shown for diagrammatic simplicity.

The input cavity, or 'buncher' cavity, is excited at a low-power level in a low-order cavity mode for which the electric field resides mainly in the re-entrant gap and is aligned with the beam. This oscillating field (at the resonant frequency of the cavity) periodically presents decelerating and accelerating forces to the beam electrons which are consequently slowed down or speeded up relative to the average velocity of the beam. This is referred to as 'velocity modulation'. In the 'drift' space, electron bunching occurs, as 'fast' electrons catch up with 'slow' ones. This is illustrated by an electron density curve which grows as the drifting proceeds, until saturation occurs at the point where space charge forces resist further bunching. At this point, a second (output) cavity is introduced to the beam. As it passes through the grids at the re-entrant gap, the bunched beam induces strong alternating currents in the cavity walls at the frequency of the 'buncher'. These currents induce greatly amplified modal fields in the 'catcher' cavity if it is tuned to the 'buncher'. The power that can be extracted from the output cavity, usually from a coupling loop formed with the inner wire of a coaxial line, is very large relative to the input signal. Typically, 30–40 dB of power gain is possible [5].

Beyond the output cavity, the velocity-modulated beam impinges on a 'collector' which in the simplest tubes comprises little more than a metallic cup at zero voltage. However, this means that on average, the beam particles strike the metal walls at high velocity, simply converting the remaining beam kinetic energy to heat which has to be dissipated wastefully. Such tubes are little more than 40 % efficient. In modern tubes, the collector is a complex electrostatic lens which slows the electrons so that they strike the metal walls at much lower velocity. The device is termed a 'depressed collector' and can raise the tube efficiency to ~80 % (see Table 9.1). The downside is added cost and weight. However, at this level of efficiency, a power output of 30 kW from a 30-kV tube is achievable with a relatively moderate current of 1.25 A, which is also a not insignificant consideration in the satellite context.

To create an oscillator, rather than the amplifier, as depicted in Fig. 9.3, the signal developed in the output cavity has to be coupled back to the input, so generating a feedback mechanism. Actually, this can be done more compactly by replacing the output cavity with an electrostatic reflector plate which reverses the beam direction so that it returns to the input cavity. There, the beam modulation induces wall currents which enhance the original cavity-mode fields. If the phasing is re-enforcing in character (achieved by adjusting the position of reflector plate), the oscillation process becomes self-sustaining. In this case, power can now be extracted through the 'input' coaxial port. The device is termed a reflex klystron.

9.3.2 Magnetron

In essence, the magnetron is an electronic diode of coaxial construction with the cylindrical cathode enclosed in an anode block machined to create a multi-cavity periodic structure (Fig. 9.4). The interior of the anode block is evacuated to permit free flow of electrons from the cathode, typically set at −6 kV relative to the enclosing anode which is usually connected to a zero-voltage terminal, or earth potential. In the absence of magnetic focusing, the electrons would flow radially 'down' the potential hill from the cathode to the anode. In this scenario, a high current would also flow in the external circuit generating a lot of heat at the anode. When focusing is applied by means of a pair of magnets, as suggested in Fig. 9.4, an axial magnetic field—aligned with the axis of the coaxial geometry—is formed. A simple application of the Lorentz law (Chap. 2, Eq. 2.26) dictates that the radially moving charge will experience firstly a circumferential force which sends the electrons into circumferential paths and secondly an inward radial force which pushes them back towards the cathode. With a proper balance between the applied voltage (V_o) and the axial magnetic field (B_o), the external current can be suppressed in which case the diode is said to be 'cut-off'. In this condition, the electrons trace cycloidal paths around the cathode at a frequency given by

$$\omega_c = \frac{eB_o}{m} \tag{9.1}$$

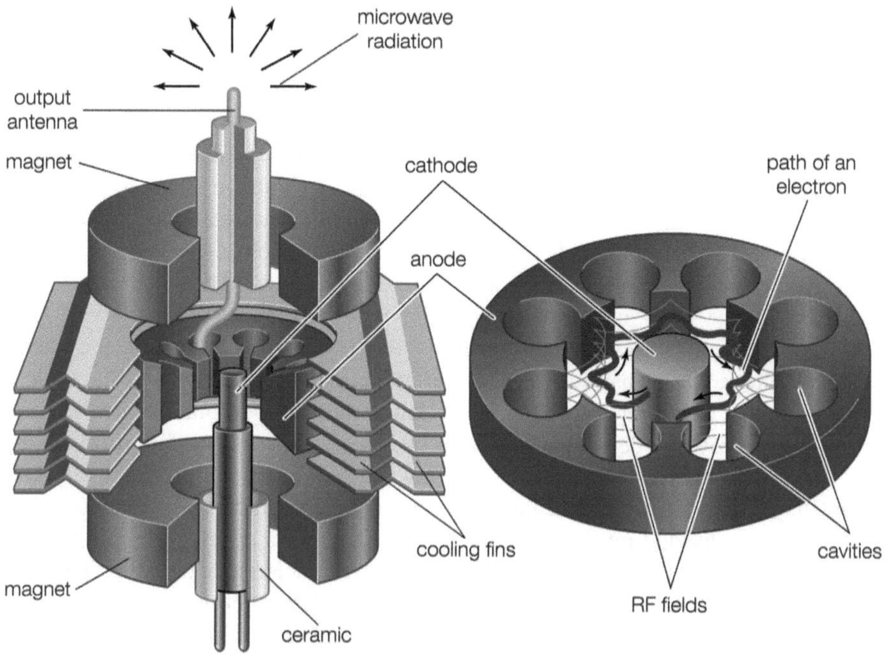

Fig. 9.4 Schematic of a multi-cavity magnetron (*Courtesy* Encyclopaedia Britannica)

where e is the electronic charge and m is its mass. The tangential velocity of the electrons is given by $v_t = \omega_c R$ m/s where R is the mean radius of the cycloidal motion, which is key to the magnetron operation [5].

In electromagnetic wave terms, the cathode together with the anode block can be viewed as a periodic waveguide, with regularly spaced metal irises, which has been moulded into a ring-shaped cavity. Such a cavity supports a resonant mode termed the π-mode (see Fig. 9.4). If the travelling wave component of this mode is synchronised to the circumferential velocity of the electrons, on average, the electrons in the beam are slowed by the π-mode electric fields. These electrons no longer maintain their cycloidal paths and drift down the potential slope towards the anode. A large current appears in the external circuit, indicating that the magnetron oscillation has been set up. The power in the π-mode fields comes from the potential energy released by the electrons as they 'drop down' the potential slope between the cathode and the anode. These magnetron oscillations generally develop quite spontaneously, with the electromagnetically noisy act of 'switch on' being enough to seed the beam/wave interaction process.

With modern design methods, and advanced materials, a tube capable of delivering $\sim$5 kW can be very efficient at $\sim$85 %, very robust and certainly light enough to be effective in an SPS role (Table 9.1).

9.3.3 Solid-State Microwave Source

While statistically the average lifetimes of electron beam tubes are placed in the vicinity of 30 years, this could actually be too short in a space application. Any given tube can fail for a number of reasons ranging from vacuum deterioration due to leakage, to cathode poisoning, to overheating, to voltage breakdown. In the context of an expensive long-term investment such as a satellite power station, such failures could be devastating. Failure rates, which are much lower than microwave tubes can sensibly deliver, are really required to make SPS truly convincing. Fortunately, such rates are provided by solid-state power sources.

Microwave solid-state sources generally employ a very high-frequency semiconductor diode or transistor as the 'active' component. The essence of the topic of semiconductors is broached in Chap. 7 in relation to the photovoltaic effect, which is not unlike microwave detection, although obviously at a very different frequency. Many textbooks exist on the subject of microwave power generation which reference [1] in Chap. 7 is typical. These usually observe that active microwave semiconductor devices can be divided into two broad classes, namely transistors providing amplification and diodes displaying negative resistance characteristics (see Fig. 9.5).

The first class encompasses conventional transistor geometries extended in operation to microwave frequencies by sophisticated fabrication processes, which are employed in the manufacture of field-effect transistors (see Fig. 9.6). Negative resistance devices include the long-established tunnel diode, plus two subgroups emanating from transferred electron diode (TED) action (e.g. Gunn diode and limited space charge accumulation (LSA) diode), and from avalanche transit-time action (e.g. read diode, IMPATT diode and TRAPATT diode).

The electrical characteristics of each of the devices alluded to in Fig. 9.5 can vary hugely, not only because of the differing physics, but also because all are

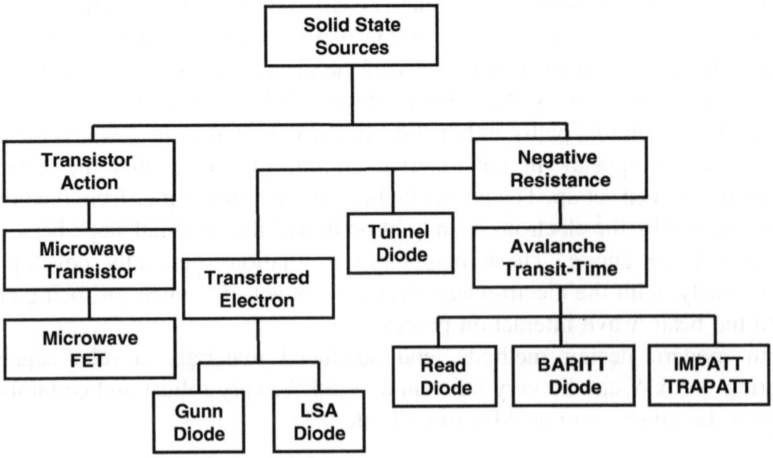

Fig. 9.5 Solid-state source semiconductor devices

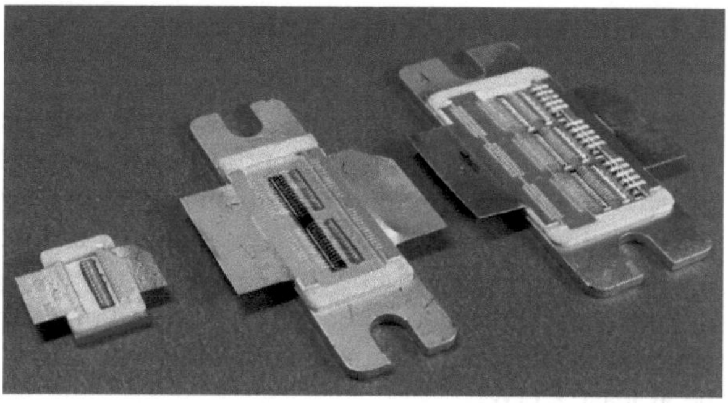

Fig. 9.6 Microwave power transistor (from the *left* 1 GHz, 45 W; 2 GHz, 60 W; 2 GHz, 200 W) (*source* IEEE Research Journal)

strongly dependent on the nature of the semiconductor material employed in their construction. There is a very wide choice, ranging from silicon- and germanium-based substrates, to those employing gallium arsenide and zinc sulphide. Transistor amplifiers and oscillators today operate over a wide range of microwave bands from L-band (0.5–1.5 GHz) to Ku-band (33–36 GHz) with CW power outputs ranging from 200 W to 10 mW generally decreasing with frequency. For oscillators, conversion efficiencies in the 70–90 % range are available, and power supply requirements are modest. Clearly, from the viewpoint of transport into and assembly in space, weight is not an issue by comparison with microwave tube alternatives as Table 9.1 illustrates.

Transferred electron diodes have also been evolved with applications ranging from 3 to 30 GHz. The most common example of this class is the Gunn diode, but like the related tunnel diode, it is restricted to low-power applications such as local oscillators in receiver systems. On the other hand, much higher power is available from the LSA diode at high microwave and millimetre-wave frequencies which are beyond the interests of SPS.

Power levels of relevance to SPS are perhaps more likely to emanate from the avalanche category of diodes. The IMPATT diode (IMPact ionization Avalanche Transit Time) uses the technique to realise active devices capable of modest power generation in high-frequency electronics and microwave devices. The BARITT (BARrier Injected Transit Time) diode has a similar profile. They can operate at frequencies as low as 3 GHz but are generally targeted at much higher frequencies reaching as high as 100 GHz or more. While high power in the range 2–10 W at these very high millimetre-wave frequencies represents a main advantage of the IMPATT diode, its characteristics are not too relevant to SPS applications. On the other hand, TRAPATT (TRApped Plasma and Avalanche Transit Time) devices, which are already being used in phased array radars, offer relatively high power in the 10–100 W range, at efficiencies in the 20–75 % range at frequencies between 0.5 and 10 GHz. Interestingly, a gallium nitride-based high-electron-mobility transistor has been announced recently which will potentially deliver 320 W at 5.8 GHz from a single device. Such a power source clearly has significant SPS potential.

9.4 Microwave Array Antennas

Arguably, the key element that makes satellite-derived solar power a tenable concept is the array antenna. The primary function of this sub-system is to beam the gathered solar power to earth, both efficiently and accurately. In the block diagram of a satellite power generating system shown in Fig. 9.7, the microwave antenna naturally follows DC-to-microwave conversion (orange box) which was examined in Sect. 9.3. Authoritative reports [1–3] based on comprehensive studies of the SPS concept dictate that for economic viability, the power transported to ground should be at least 2 GW.

Operational efficiencies for the sub-systems identified in Fig. 9.7 have been estimated in a range of reports and papers [3], and these are applied to the figure. The power conversion efficiency includes filtering requirements: the microwave array figure incorporates sub-array failures, amplitude errors, phase errors, phase quantisation errors, antenna aperture efficiency: transmission loss includes atmospheric effects and ground antenna collection efficiency: rectenna efficiency encompasses random element failures in the rectenna, filter losses and element efficiency. The overall efficiency figure suggests that to achieve 2 GW of electrical power entering the grid, 4.7 GW has to be generated by the solar array. If the solar

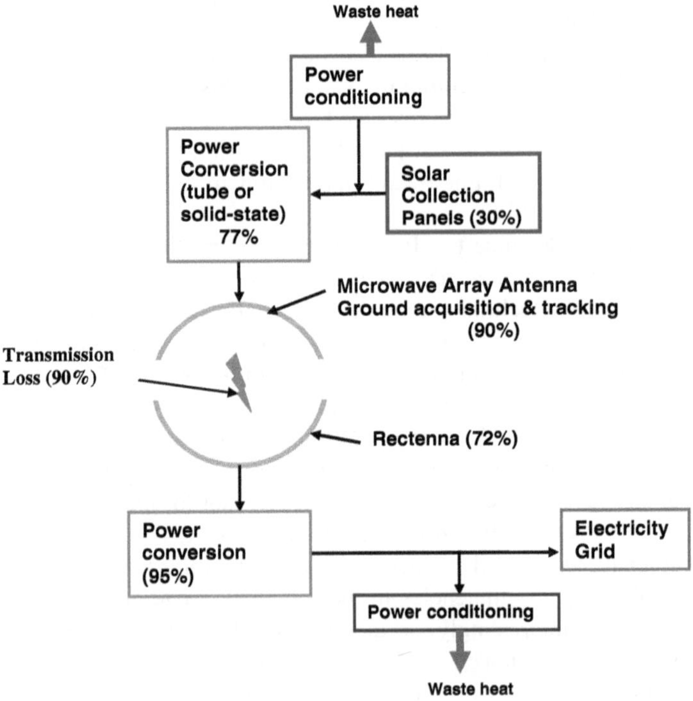

Fig. 9.7 Mandatory sub-systems required to form a satellite-based electrical power generating plant

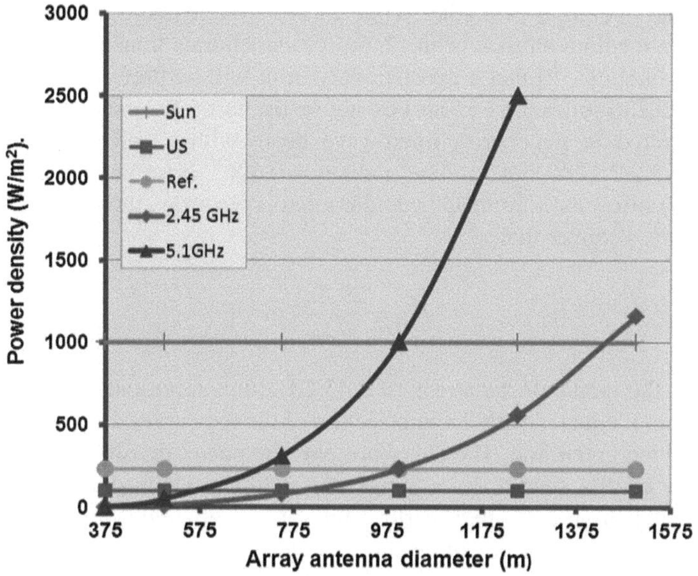

Fig. 9.8 Peak power density (W/m^2) on the ground, for 4.7 GW satellite transmitter in GEO, as a function of array diameter

panels of this array are optimistically deemed to be 30 % efficient, then the photovoltaic array (or arrays) will be required to provide a solar collection area in space of about 11,000 m^2 in order to match the power demand. An area of 11 km^2 equates in area to that of a soccer pitch, which is large in space station terms but apparently not unfeasible [1] (Note that this is not 11 square kilometres, which is 1,000 times larger).

So what are the basic constraints on the space station-mounted microwave array antenna if it is required to deliver 2 GW of EM wave power to the planetary surface safely and accurately onto a 'footprint' of tolerable dimensions? The safety criteria are summarised in Fig. 9.8. At light frequencies, the maximum power density which living creatures experience from solar radiation incident upon the planet is about 1,000 W/m^2 at noon at the equator. This is represented by the orange horizontal line (with + symbols) on the figure. Needless to say, the safe level for man-made radiation at microwave frequencies (typically 2.45 GHz) is much lower than this at 230 W/m^2 (green line with circular markers). This has been a globally agreed level until recently, when it was lowered to 100 W/m^2 in the USA (purple line—square markers). If the former global figure is used, we can make the approximate calculation, by dividing 2 GW by an averaged power density computed for a microwave beam with an approximately cosine distribution at the earth's surface with a peak of 230 W/m^2. This calculation gives the acceptable footprint of the microwave antenna and has a magnitude of the order of 12×10^6 m^2. If the beam is assumed to be approximately circular, then the footprint has a diameter of about 3.9 km.

With the footprint of the microwave beam known, the required mainlobe width of the satellite-stationed microwave array can be estimated, provided that the altitude of the satellite is available. While an SPS system could in theory be located in a low or a medium altitude orbit [2, 6, 7] to ameliorate transportation costs, it is much more likely [1, 2] that a geostationary orbit will be preferred. Consequently, located at 22,236 miles, or 35,786 km, above the earth, the satellite array antenna will be required to generate a microwave beam with a half power beamwidth (HPBW) of $3.9/35{,}786 = 9.3 \times 10^{-5}$ radians = 0.006°. If for simplicity, we choose to employ a uniformly illuminated circular microwave array, then from Chap. 5 (c.f. Eq. 5.17), we discover that

$$R_A = \frac{29.2\lambda_0}{\theta_{3\mathrm{dB}}}\ \mathrm{m} \tag{9.2}$$

where R_A is the radius of the array. At 2.45 GHz, this represents an array diameter of ~1,000 m. It equates with the intersection of the red curve (diamond symbols) in Fig. 9.8 and the green line. The curve presents the power density in the microwave beam at the earth's surface for a range of satellite antenna diameters (m), for a system located in geostationary earth orbit. Not surprisingly, as the antenna size increases, the beam becomes more focused and the power density on the ground rises. At the intersection with the 230 W/m^2 line (green line), the international safe limit for microwave radiation is breached. For a system operating at 5.1 GHz (blue curve triangular symbols), the safe limit is penetrated at a rather lower antenna size with a diameter of ~720 m.

In microwave antenna terms, these are very large arrays indeed and would normally be difficult to realise, because at these sizes, array flexing would generate intolerable pattern errors and significant real-time variations of the efficiency of power transmission to the planetary surface. Fortunately, in the zero gravity of geostationary orbit and in the vacuum of space, distorting forces become minimal, and these difficulties are potentially solvable.

In Chap. 5, it was demonstrated that to achieve a radiation pattern with a single primary lobe from an array antenna, it is essential that the radiating elements of the array are spaced by a distance of close to half-a-wavelength at the operating frequency. Consequently, for an array operating at 2.45 GHz for which the free-space wavelength is 0.12 m, a satellite-based microwave array exhibiting a diameter of ~1,000 m (see Fig. 9.1) will require about 50 million radiating elements to ensure single lobe operation. Furthermore, if such an array is required to transmit 4.7 GW into space towards the planetary surface, simple division suggests that each element of the array must be capable of handling on average about 100 W. This level of power is well within the capability of microwave components based on waveguides fed from klystron or magnetron sources and is even low enough to allow more cost-effective micro-strip or stripline technology, powered by solid-state sources, to be proposed for the microwave array. On the other hand, an array with 50 million radiating elements, together with a compact and efficient feed structure, poses a whole new technological challenge in terms of the current state of the art in array development.

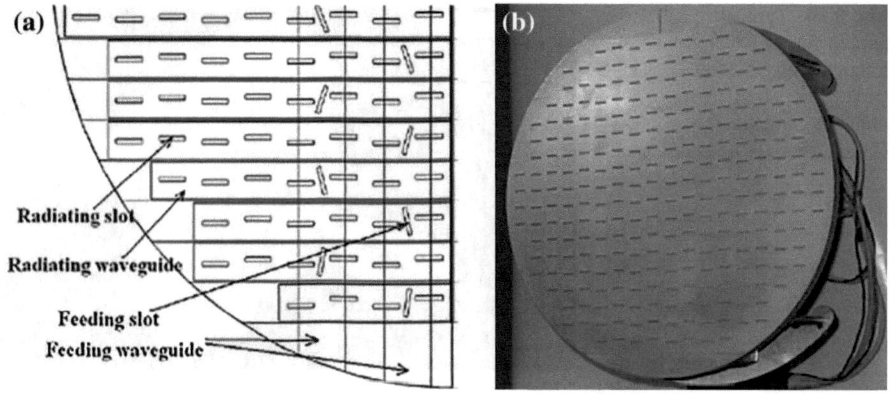

Fig. 9.9 Typical vertically polarised planar slotted waveguide microwave array antenna (source open access publisher: www.intechopen.com/books/radar-technology)

9.4.1 Waveguide Slot Arrays

Current technology points to the slotted waveguide array as arguably the most realistic solution to the procurement, for SPS, of a cost-effective space antenna comprising many millions of radiating elements. Such a flat-plate circular slotted waveguide array antenna is depicted in Fig. 9.9. Of course, the array shown, which is designed for a radar application, is miniscule by comparison with any proposed SPS antenna, but geometrically, the general concept illustrated by the figure remains valid.

Fundamentally, slot arrays are initially implemented in linear array form and most commonly in TE_{10}-mode rectangular metal waveguide (see Chaps. 2, Sect. 2.3.4 and 6, Sect. 6.5). These linear elements are then paralleled as shown in Fig. 9.9 to form a planar array antenna. Other waveguide geometries are also possible [8–10, 11–13], which offer desirable polarisation features such as circular polarisation and polarisation control.

In order for a slot to radiate effectively, it must be located in the waveguide wall in a manner which ensures that significant disruption of the local current density flow occurs. Referring back to Chap. 2 and the waveguide in Fig. 2.9, the perfect conductor boundary condition for tangential magnetic field at the walls dictates that

$$\mathbf{J}_s = \hat{\mathbf{n}} \times \mathbf{H} \tag{9.3}$$

where $\mathbf{J}_s$ is the surface current density in A/m and $\mathbf{n}$ is the unit normal to the surface. This means (again referring to Chap. 2, Fig. 2.9) that current flows in a direction orthogonal to the magnetic field, that is, radially in the top and bottom walls of the waveguide and vertically in the side walls. The slot antennas depicted in Fig. 4.1d, e, respectively, the longitudinal shunt slot and the inclined edge slot achieve this disruptive role. They are also the most commonly employed forms. As it interrupts

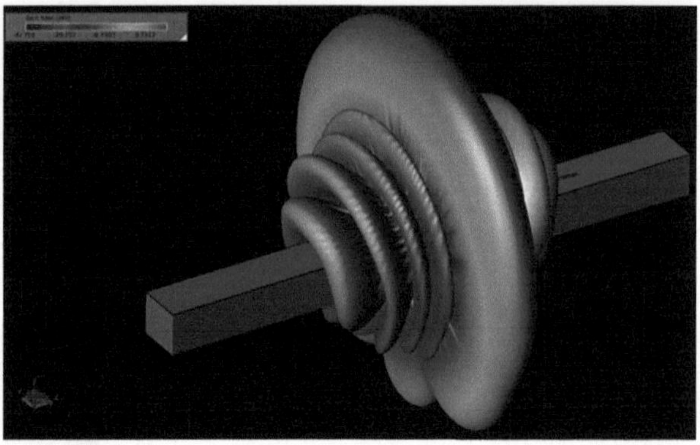

Fig. 9.10 Typical radiation pattern for an isolated longitudinally slotted waveguide, showing mainlobe and sidelobes (*source* Google images)

the current, each slot in the waveguide causes instantaneous charge build-up along its long dimension (l) and hence an electric field across its narrow dimension (w). The slot then behaves like a small aperture antenna (Chap. 4) of length l and height w, radiating into free space. When $l \approx \lambda_0/2$ the resonant length, and $w \approx l/10$, the radiation strength is about optimum.

The array in Fig. 9.9 employs offset longitudinal slots of resonant length to form a planar array. The radiated field pattern for a single longitudinal slot is a simple 'doughring' shape. When several such slots are combined (see Chap. 5) as in Fig. 9.10, where nine slots are represented, the radiation pattern takes on a directional character which is the feature primarily desired by any practical antenna. The pattern has been created by a commercial EM solver and essentially depicts an isopower density surface representative of the radiation (see Chap. 4). The added colour merely emphasises the pattern shape. It is linearly polarised with the E-field orthogonal to the waveguide axis.

For a space array antenna operating at a frequency of 2.45 GHz, the appropriate TE_{10}-mode waveguide is WG9 [14] which cuts off, for this mode, at 1.7 GHz and operates optimally in the frequency range 2.2–3.3 GHz. The cross-sectional dimensions for this waveguide are 3.5 in. × 1.75 in. (8.9 cm × 4.45 cm), which give it a power handling capability of 21 MW, determined by electric field breakdown in air. In cold empty space, power handling is much higher than this being dictated by breakdown in vacuum which occurs due to ionisation and free-electron emission at internal metallic surfaces. It is termed multi-pactor effect [15]. It is also relevant to note that the very extensive waveguide runs inherent to an SPS array are susceptible to wall heating generated by resistive losses, and these can become a source of unwelcome inefficiency. For all practical purposes, however, the power handling of waveguides employed in space arrays is unlikely to be a significant issue for SPS.

Mere common sense dictates that a square or circular slotted waveguide array antenna formed from WG9, in a light metal such as aluminium, and occupying $\sim$800,000 m^2, will have to be constructed from panels or modules, which can readily be fabricated on earth, be transported into space and assembled there as subarrays to form the transmitting antenna as implied in Fig. 9.1. Such an array with a

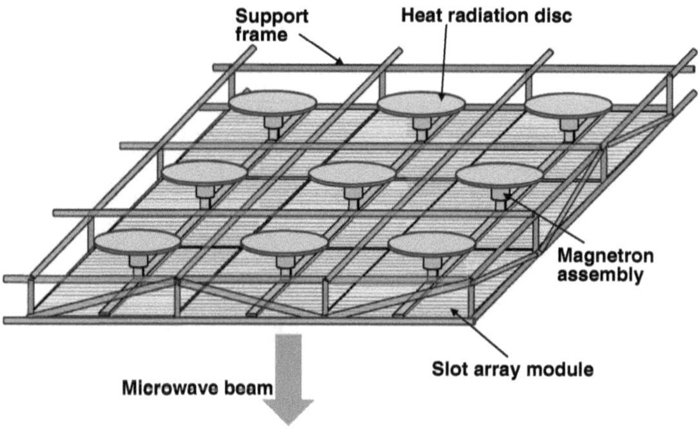

Fig. 9.11 Section (typically 4 m × 4 m) of a slotted waveguide SPS array antenna (not to scale) driven from injection-locked magnetrons

modular format presents a synergistic relationship with a power delivery system based on magnetrons [3, 16]. For example, in a square transmitting array, the square sub-array will be, typically, 4 m × 4 m in area comprising typically 16 panels with 40 linear slotted waveguide arrays each with 52 radiating slots. A simple calculation suggests that there will be 50,000 of these sub-arrays in a ~5 GW transmitting antenna. Assuming that 5 kW magnetrons are employed to power the array, then central 4 m × 4 m sub-arrays will require a microwave feed arrangement involving (say) 5 × 5 = 25 injection-locked magnetrons, and if the aperture distribution is tapered in order to lower the sidelobe levels (see Chap. 5), edge array magnetrons will be of lower power and/or the number could be tapered down to, perhaps, 3 × 3 = 9 magnetrons at the edge sub-arrays. The concept is illustrated in Fig. 9.11 which presents an elementary representation of a possible 3 × 3 corner section of an SPS array antenna (not to scale). A magnetron delivering ~5 kW is presumed to source each slotted waveguide array module, in this edge array case. Furthermore, the illustration suggests that the heat generated in each vacuum tube can be dissipated into space by infrared radiation [16] from high-emissivity pyrolytic graphite discs. These are shown in the orange colour in the figure. The slotted waveguide array panels are attached below a section of the support frame which is presumed to be approximately 4 m × 4 m in size. They are excited by the feed waveguides, running from the front to the back of the image in Fig. 9.11, and subsequently direct microwave power downwards towards earth through the multiplicity of radiating slots on the lower face.

From the array theory presented in Chap. 5, it is evident that if a planar array is to form a highly directional beam normal to its face, the individual elements of the array structure must radiate in phase. For the proposed SPS array antenna powered by a multiplicity of magnetron sources, it follows that these microwave generators cannot be free running. They must be phase-locked to each other by injecting into

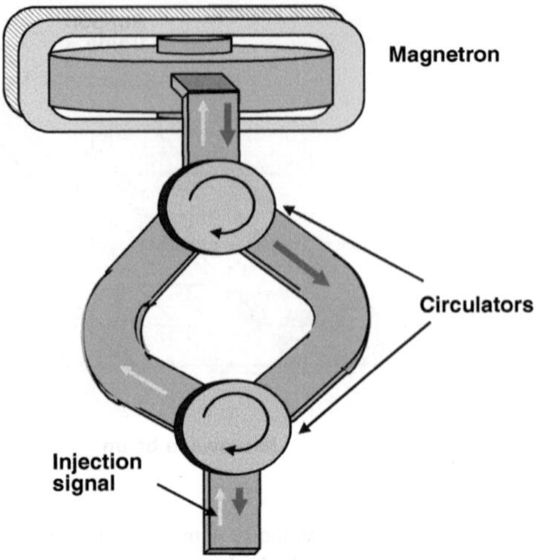

Fig. 9.12 Injection-locked magnetron assembly

each magnetron a controlling signal from a single stable-frequency source. This source could be part of the SPS hardware, but it is actually more convenient and economical to beam a stable signal up to the SPS antenna from a generator on the ground. This signal will be detected at every panel forming the SPS array and directed to its associated magnetron by means of a directional feed arrangement, much as suggested in Fig. 9.12.

Rather than the microwave output from the magnetron being coupled directly to the T-junction of the adjacent module (see Fig. 9.13), it is passed, first, through a pair of microwave circulators (a directional microwave device based on the use of a magnetised ferrite material [17, 18]) before entering the slotted waveguide panel by way of the T-junction. The combination permits minimal interference or interaction between the high-power-transmitted signal from the magnetron, now locked in frequency, and the incoming stabilising signal emanating from the ground. The concept depicted in the figure is rudimentary and is arguably not the most practical or efficient way of achieving duplex operation, but it very effectively illustrates the principle.

However, it is not sufficient for the SPS array (see Fig. 9.1) to be simply directing microwave energy towards the ground in an uncontrolled way, for both efficiency and safety reasons, as already alluded to above. The primary beam of the array must be accurately directed towards the ground collecting station and be held in this position throughout the period of power transfer. With a system based on slotted waveguide array modules, this control is most easily achieved by adopting a 'mono-pulse' or 'split-beam' technique most commonly associated with radar [11]. The basic concept is illustrated in Fig. 9.14. In principle, it could be applied to a single array module which would be quartered into four sub-array panels, as depicted in Fig. 9.14. This permits two modes of operation. The injection signal from the ground, when it illuminates this module, either can act conventionally when the sub-panel received signals are added in phase, or can operate as a sensor

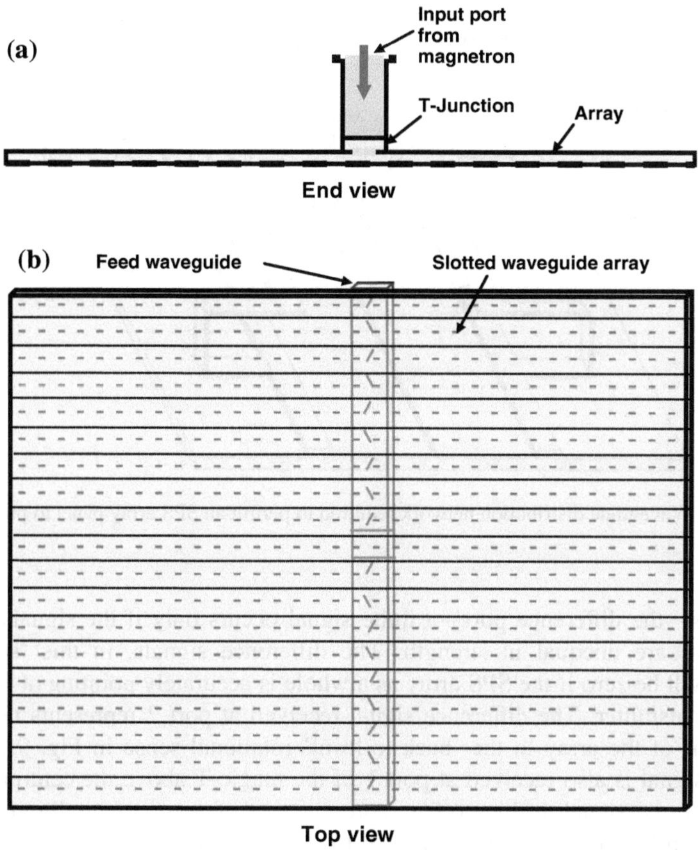

Fig. 9.13 Slotted waveguide antenna module for SSP array

with the sub-panel signals combined in anti-phase, thus determining whether or not the satellite array antenna as a whole is directed towards the ground station antenna. The secret to this behaviour is a now ubiquitous microwave device termed the magic-T [17, 18]. A possible waveguide feed arrangement, permitting signal reception and array control by incorporating three magic-Ts, is suggested in Fig. 9.14.

The guidance microwave signal from the ground station, with superimposed coding for identification, is received equally by the four quarters of the array panel and is added into the waveguide distribution system shown at the first H-plane T-junctions behind each sub-array. From there, the signals enter magic-Ts on the left and right of the centreline of the array, where two events occur. Firstly, the signals are added into the lower transverse waveguide bridge where they enter the third magic-T. Here again, there are two options, and in the adding port or sum port, to the front of the transverse waveguide (see Fig. 9.14), the received signal is equivalent to that of an unquartered standard array. Into the vertical arm of the

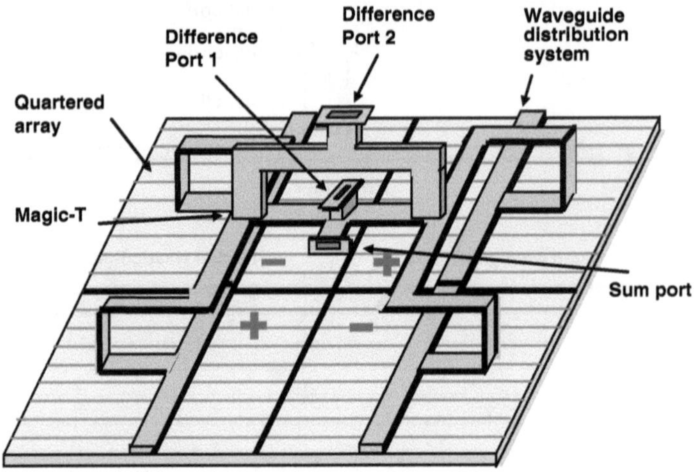

Fig. 9.14 Waveguide distribution network required to permit an SPS array panel to provide signal tracking

magic-Ts—the difference port—a finite signal occurs only if the signals from the sub-arrays are unequal in strength. The difference signals, in this 'split-beam' option, will be zero if the SPS array as a whole is accurately pointing at the ground station transmitter. The difference signal received at port 2 represents non-perfect alignment of the array in the 'back-to-front' rotational sense in Fig. 9.14. A difference signal is also received at port 1 of the magic-T above the centreline of the array. A finite signal at this port is representative of non-perfect alignment in the 'left-to-right' rotational sense in Fig. 9.14. The difference signals from this split-beam array will be fed to electronic control circuits which evaluate the deviations of the array from the ideal and generate control voltages to drive motors on the array pivoting system (see Fig. 9.1) such that the required pointing direction is restored. The degree of pointing accuracy, which will be demanded of an SPS array, in order to reassure populations living in close proximity to a ground station remains unsettled. It is possible that a 'split-beam' control system based on a single slotted waveguide array module may provide insufficient accuracy, and its application to modular groups such as 4 × 4 or 6 × 6 may be necessary. The trade-off for this increased directional control is more complex and more extensive waveguide 'plumbing' and an undesirable additional weight load to be transported into space.

With the above control system in operation, the SPS array antenna will point to the ground station as long as a guidance signal from the ground station to the satellite exists. Failure of the ground station signal would not result in the microwave beam damagingly irradiating neighbouring territory, because the guidance signal is also the magnetron injection signal. In the absence of the injection signal, the unlocked, free-running magnetrons cannot accomplish the phase coherence required to form a powerful ground-directed beam from the array. A more serious setback would be a failure in the control electronics. Secondary electronic circuits will be essential, together with robust monitoring arrangements to anticipate any control deterioration. By contrast, catastrophic 'end of life' for the SPS platform would occur if any components in the bearings, gear systems or drive systems of the suspended and rotatable slotted waveguide array antenna (see Fig. 9.1) were to fail. Duplication in this case is not really possible.

It is also appropriate to note here that the slotted waveguide array antenna, as described above, is a format which is strictly linearly polarised, and this means that the transmission efficiency of the SPS system is susceptible to rotational movements or deviations of the space platform as a whole. In this case, the resultant system losses cannot be counteracted by the split-beam control scheme described above.

9.4.2 Waveguide Phased Array

A major disadvantage of the adoption of a slotted waveguide array antenna to transmit microwave power from the SPS system to a ground antenna lies with the mechanical nature of the beam guidance hardware (see Fig. 9.1), simply because it has the potential to fail catastrophically. There is an obvious answer, which is to dispense with the pivoting microwave array antenna and to embrace waveguide phased array technology for this sub-system. With this change, beam pointing or guidance towards the ground target can now be achieved purely electronically (see Chap. 5). Needless to say, there are a very wide range of phased array configurations which could be considered for an SPS antenna, but since the resultant space-based system would have to be competitive economically with the slotted array alternative, with regard to transport to, and assembly in, orbit round the earth, the possible options are inevitably limited.

Of the widely encountered phased array configurations, which can be considered to be suitably compatible with implementation in robust waveguide hardware, as demanded by space-based operations, those shown schematically in Fig. 9.15 are the most commonly favoured. In both cases, it is assumed that a substantial power source is available and this power can be repeatedly divided through a waveguide 'tree', either in a parallel fashion by means of T-junctions (Fig. 9.15a) or in a series mechanism by means of power dividers (Fig. 9.15b). In both cases, phase shifters, generally of the ferrite latching type [17, 18], are key components and are usually located immediately behind the open-ended waveguide or horn radiators. In practice, the linear arrays depicted in the sketches will be combined with banks of similar arrays to form the desired planar array structure. Note that the process by which the beam direction of the array as a whole can be controlled simply by adjusting the signal phase at each radiator is fully described in Chap. 5.

In the literature, the favoured power source for arrays of this description is the klystron amplifier [1–3, 5], each generating in excess of 50 kW. We have already seen that the delivery of 2 GW of microwave power to a ground station rectenna

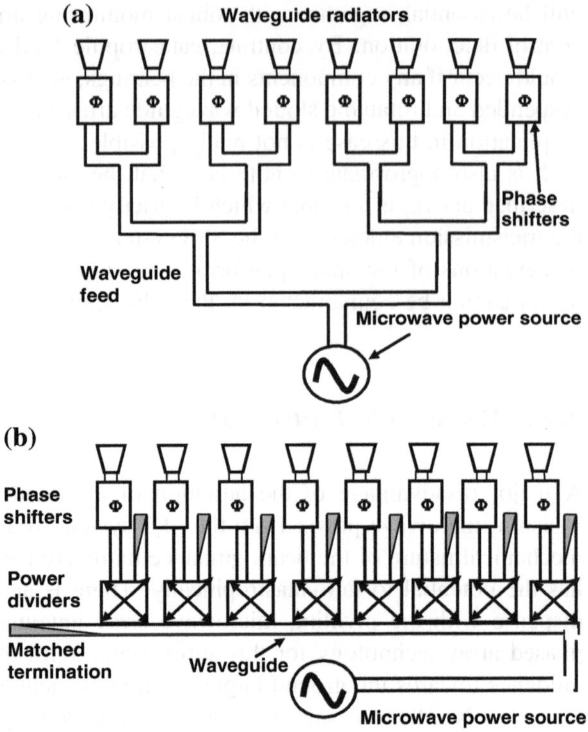

Fig. 9.15 Linear phased array antenna configurations. For SPS applications, the linear arrays will be combined to form a planar array antenna. **a** Parallel feed. **b** A series feed arrangement

requires a space-located antenna system with 50 million array elements each radiating 100 W. Therefore, a simple division calculation dictates that each klystron is required to feed 500 radiators and 500,000 klystrons are needed to generate the gross power. Beam guidance and locking into a ground-based receiving antenna can be secured, in principle, by employing the split-beam method advocated for slot array systems. The technique can be applied to the array as a whole or to a sub-array section within it. For a genuine phased array, with a phase shifter at every element, switching from the normal single-beam mode to split-beam operation is entirely routine. Consequently, the nulling of the pilot signal from the ground (see Sect. 9.4.1), within a split-beam sub-array, involves monitoring and control electronics, not too different to that described for the slot array system. However, in this case, the control is applied to the adjustment of the element phase shifters distributed throughout the array, rather than to servos and motors. Once again, if properly implemented, optimum beam pointing at the ground target is ensured during full power transfer scenarios. Obviously, a major advantage of this arrangement, over the slot array alternative, is that catastrophic failure is highly unlikely since there are no mechanical servos, gears and bearings. Any random failure among the many phase shifters will merely result in a gentle deterioration in the efficiency of the system. Nevertheless, the relatively complex monitoring and control electronic circuits continue to remain a potential source of systemic weakness.

9.4.3 Retro-directive Array Techniques

Needless to say, given the potential danger of catastrophic mechanical failure with the slotted waveguide antenna system, and the complexity and cost of the true phased array alternative, much effort, which is reflected in the literature, has been directed at securing a less 'fragile' beam locking formula for the SPS microwave array. The solution currently seems to lie with the concept of retro-directive transmission [19–29].

The concept is not new. Retro-direction, whereby an incoming radar signal on striking a target is returned back in the direction from which it came, is achieved by a 'corner reflector', often employed on the masts of small yachts [21]. The device increases the radar cross section of the target, thereby enhancing its detectability against the background of the sea. The two-dimensional schematic shown in Fig. 9.16a gives a good idea of the mechanism. The incoming ray or received wave (E_r say) from a distant source (red inward-pointing vector) on encountering the perfectly conducting surface of the 90° corner reflector reflects in accordance with Snell's laws. Consequently, the ray is directed horizontally to the opposing face of the corner reflector whereupon it is directed back towards the source at the second reflection. If we opt to take the point where the ray crosses the z-axis as reference, the received and transmitted rays can be expressed mathematically as follows:

$$E_r = A \exp j\omega t \exp j[k_0 z + k_0 x] \tag{9.4}$$

and

$$E_T = A \exp j\omega t \exp -j[k_0 z + k_0 x] \tag{9.5}$$

where A is an arbitrary magnitude and k_0 is the free-space phase coefficient. The simplistic but interesting observation is that the phasor E_T(x, z) is the conjugate of E_r(x, z). That is,

$$E_T(x, z) = E_r^*(x, z) \tag{9.6}$$

This conjugated phase relationship between the incoming and returning waves, or echo, is key to retro-directive action. In the case of the corner reflector, retro-directivity occurs for all signals within ±45° of the z-axis as the green ray suggests. At all angles, conjugation occurs. In Fig. 9.16b, an array equivalent [21] of the corner reflector is presented. It comprises a linear array of open-ended waveguides (with or without horn flanges) which are joined together in pairs through equal-length empty waveguide inserts, as shown. The arrangement ensures that a plane electromagnetic wave received by the array from an angle θ will be returned towards the source at the same angle by dictating that the transmitted wave phase is the conjugate of the received signal phase. For example, if we consider the outermost elements on the right and left of the array, at the instant when the incoming wave reaches point O (set to zero phase for convenience), the signal at the left

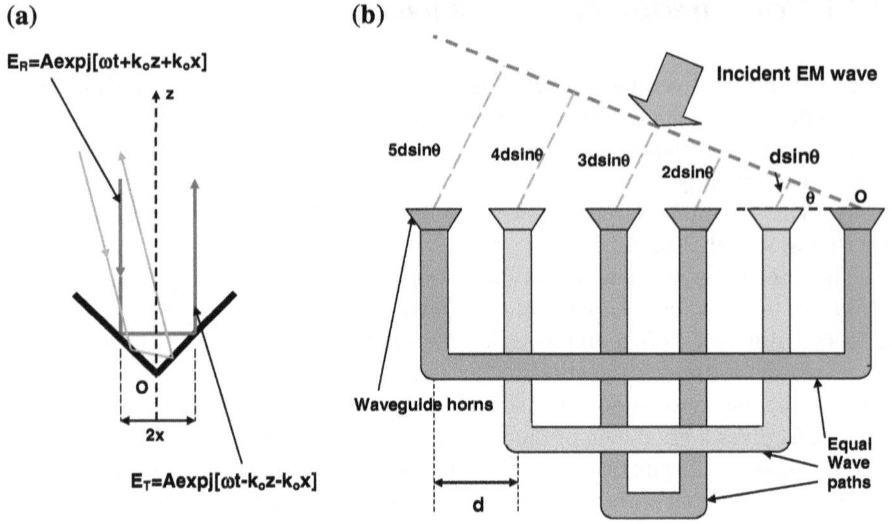

Fig. 9.16 Fundamental passive retro-directive structures. **a** Corner reflector. **b** Equivalent linear array

element is phase advanced relative to the wave front through O by $k_0 5d \sin\theta$. That is the wave at the left port can be expressed as follows:

$$E_{r\text{left}} = A\exp(j\omega t)\cdot\exp(jk_oR)\cdot\exp(jk_o 5d\sin\theta) \tag{9.7}$$

Given that any phase delays within the waveguides cancel out for the left and right travelling waves, the received signal at O on reaching the left port is delayed on transmission by—$k_0 5d \sin\theta$ relative to the simultaneous transmission from the port at O. Hence, we can express the transmission from the left port as follows:

$$E_{T\text{left}} = A\exp(j\omega t)\cdot\exp(-jk_0R)\cdot\exp(-jk_0 5d\sin\theta) \tag{9.8}$$

which, in phasor terms, is the phase conjugate of the received wave, assuming that A is real. The same result can be developed for all ports, and the transmitted, or reflected, wave is conditioned to travel back towards the source of the received wave.

The retro-directive arrangements depicted in Fig. 9.16, which are passive in character, are actually of little relevance to active arrays of the SPS type. Such arrays are also required to actively radiate microwave power, which has been generated within it, towards a target. To achieve retro-directivity in an active array, phase conjugation has to be procured in some other way.

Interestingly, it is a property of signal mixing that when a received signal is mixed with the output of a reference local oscillator at exactly twice the frequency of the incoming signal (or the carrier frequency if modulated), the resultant

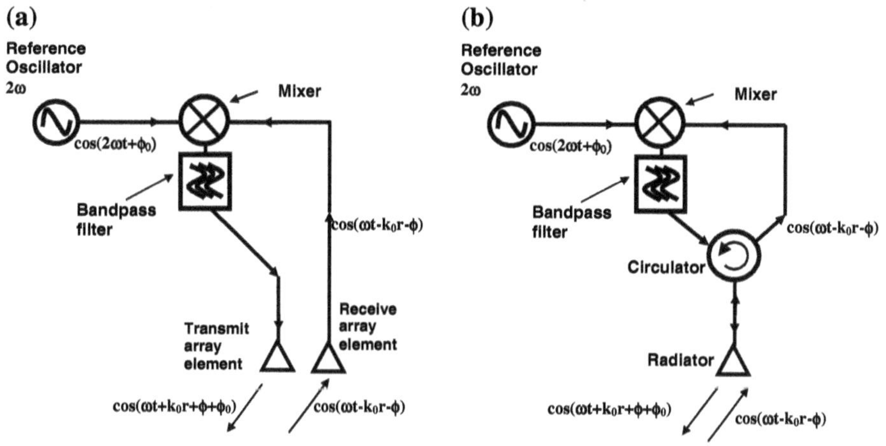

Fig. 9.17 Mixer-based phase conjugation circuit (PCC). **a** Separate Tx/Rx array elements. **b** common Tx/Rx array element

difference signal is the complex conjugate of the original [21–26]. This is illustrated in Fig. 9.17a, b where a conjugating mixer circuit is attached to one element of a supposed array. All other elements of the array would be backed by exactly similar circuits, all fed from the common local oscillator. It is not difficult to see that the transmitted (difference) signal is the conjugate of the incoming wave except for the fixed phase reference ϕ_0 introduced by the local oscillator. This phase is common to all elements, so is irrelevant to the operation of the system. The phase ϕ occurs when the incoming wave direction is at an angle θ to the array normal. It is basically the phase delay of the received signal for two adjacent elements separated by a distance d and is given by

$$\phi = k_0 d \sin\theta$$

At the mixer, therefore, the reference signal and the received signal, when multiplied, yield

$$\cos(2\omega t + \phi_0)\cos(\omega t - k_0 r - \phi) = \cos(\omega t + k_0 r + \phi - \phi_0) + \cos(3\omega t - k_0 r - \phi + \phi_0) \tag{9.9}$$

A band-pass filter between the mixer and the Tx array element suppresses the sum component, and only the difference signal (first term on right of equals sign) gets transmitted to the antenna. Note that it is phase-conjugated relative to the received wave ($-\phi$ is replaced by ϕ). So phase conjugation occurs automatically, for any angle of arrival of the received wave, at the retro-directive array. The circulator is introduced to enable transmission and reception to appear on the same antenna element.

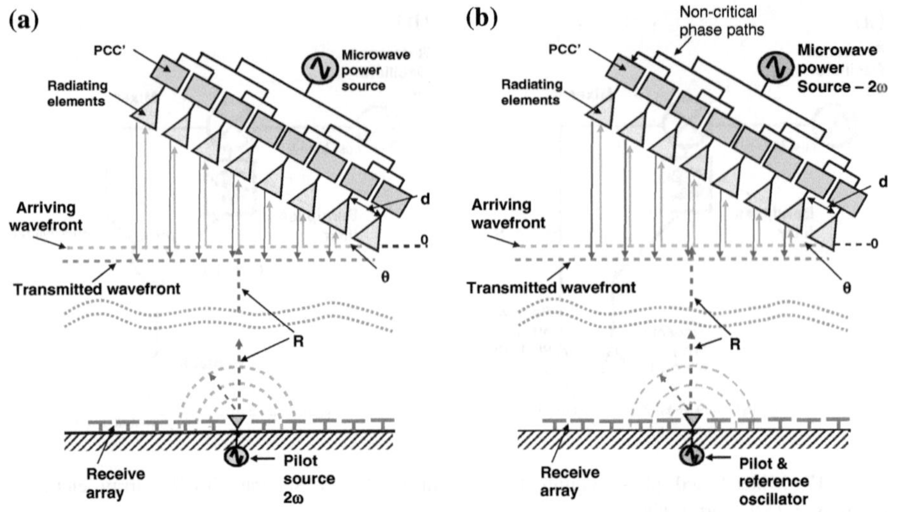

Fig. 9.18 Space arrays with retro-directive guidance **a** with space-based reference oscillator and **b** with ground-based reference oscillator

Unfortunately, a potential disadvantage of this circuit arrangement, in addition to the awkward requirement for a local oscillator at double the operating frequency, is either the need for separate transmit and receive arrays or the presence of the circulator. The latter has to be a high-quality device, to minimise tracking errors, and at microwave frequencies, such a device is certainly not inexpensive. Furthermore, this cost is self-evidently multiplied by the number of elements in the array. Eliminating the circulator by employing separate transmit and receive antennas (as in Fig. 9.17a) is hardly practical in the SPS context where the arrays are vast. Other alternatives can be found throughout the literature [22–29].

The two-dimensional schematics depicted in Fig. 9.18 display possible array antenna configurations which achieve tracking through retro-directive circuits. The precise form of the circuits themselves is not specified since it is not important to the mechanism described. Note that the phase conjugation circuits (PCC) replace the phase shifters in the arrays shown in Fig. 9.15. This represents a potentially significant advantage of the adoption of retro-directive array technology in SPS, since it introduces simpler, cheaper, yet reliable electronics into the space array. In the absence of the PCCs, the arrays in Fig. 9.18 would each form a mainbeam and sidelobes, as described in Chaps. 4 and 5, and these would be formed in a direction normal to the face of the array. Consequently, each array would be optimally sensitive to a signal coming from that direction. Unfortunately, if the array normal does not coincide with the direction to the ground-located microwave power collecting antenna, the SPS will be ineffective. The retro-directive mechanism provides the required correction to any physical misalignment of the array antenna by automatically steering the mainbeam towards the source of the pilot signal from the ground. This is achieved by enforcing on the transmitted signals at each element, through the agency of the PCCs, a phase which is the conjugate of the received pilot signal at that element.

The principle of retro-directivity is demonstrated for two representative cases in Fig. 9.18. They show, again in two dimensions for simplicity, a 'squinted' (at angle θ) space array above a ground-based, microwave power collecting system. In case (a), retro-directivity is secured by incorporating a stable oscillator on the ground which radiates omnidirectionally a 'pilot' signal at 2ω into space. The phase-locked microwave power sources at the array operate at frequency ω with the transmitted power being guided through a transmission line branching system, via PCCs, to the antenna elements. To represent the system mathematically, it is convenient to identify the right-hand element at O as $n_0 = 0$ and the left-hand, or eighth, element at $n_7 = 7$ with sequential values in between. Therefore, at element n_i, ($i = 0, 1 \ldots 7$), we can express the pilot E-field wave front at n_i as follows:

$$E_{ri} = A_{ri} \cos[2\omega t - 2k_0 R - 2k_0 n_i \sin\theta] \tag{9.10}$$

Since the phase delay $2k_0R$ is common to all elements, and setting $k_0 n_i \sin\theta = \phi_i$, then the voltage at the PCC can be expressed as follows:

$$V_{ki} = A \cos[2\omega t - 2\phi_i] \tag{9.11}$$

At the same PCC, the voltage of the signal from the microwave power source can be written in the form:

$$V'_{ki} = A' \cos[\omega t + \Phi_0] \tag{9.12}$$

Hence, following the mixing and filtering, the transmitted radiation from element k_i of the space array displays the mathematical form:

$$E_{Ti} = A_{Ti} \cos[\omega t + \phi_i + \Phi_0] \tag{9.13}$$

which is phase-conjugated relative to a wave arriving at the array from the direction of the pilot source and will be returned in that direction, provided that Φ_0 applies equally to all elements. This latter condition dictates that the transmission paths (L) from the microwave source to the PCC must be equal in length, typically $\Delta L < \lambda/12$. This is a very stringent requirement in lines which could be up to a kilometre long. A technique termed central phasing [22] has been advocated to circumvent this difficulty. An added disadvantage of the arrangement suggested in Fig. 9.18a is that the array elements have to encompass a very wide frequency range with the receive wave at twice the frequency of the transmit wave (an octave bandwidth). This rules out waveguide arrays and horn antenna elements unless the transmit and receive arrays are separated, with the elements arranged as depicted in Fig. 9.17a, but as we have already observed, this is a rather impractical solution in the SPS context.

Another option is shown in Fig. 9.18b, where the reference oscillator, at twice the nominal transmission frequency, also provides the microwave power. In this case, the pilot source frequency coincides with the transmission frequency. The retro-directive Eqs. (9.10)–(9.13) are unaltered by this change and the bandwidths of the transmit/receive array elements are now narrow and therefore easy to implement in waveguide hardware. For an SPS array delivering microwave power to the ground at 2.45 GHz (the generally preferred frequency), the scheme in Fig. 9.18b implies satellite-based microwave oscillators operating at 4.9 GHz. At frequencies of this magnitude, or higher, the line length phasing problem alluded to above is exacerbated and this tends to mitigate against using a waveguide array with its long electrical path lengths within the power distribution system. A space array with a multiplicity of low-power sources located close to the radiating elements is one possible circumvention route (see Sect. 9.4.4).

A potentially more damaging feature of the Fig. 9.18b array configuration is mutual coupling. Mutual coupling occurs in an array antenna when the elements are close enough together for a measurable proportion of the radiated power on one element to 'leak' into neighbouring elements through the agency of sidelobes or near fields [11]. In the case of an SPS array, the primary effect is to introduce errors into the retro-directive alignment process [22] which are undesirable. Many solutions are advocated in the literature [22–30]. These are mainly directed at wireless sensors and satellite communications and tend to revolve around abandoning 'inexact' conjugation through mixing and moving to 'exact' conjugation based on phase-locked loops or on undersampling techniques. This allows the adoption of differing pilot and array transmission frequencies, so negating mutual coupling. The frequencies are usually chosen to be close enough together to obviate the need to employ costly array elements exhibiting wide band characteristics.

9.4.4 Micro-strip Patch Array

The introduction of retro-directivity into postulated designs for SPS antenna arrays [1–3], with a requirement to provide low-power mixing devices behind each antenna element, is tending to encourage a drift towards systems operating at increasingly high microwave frequencies. This in turn is directing a trend towards the adoption of micro-strip technology (see Chap. 2) into SPS embodiments. This technology greatly reduces potential fabrication difficulties associated with the need to embed, for retro-directive purposes, small specialist devices or components into an array. This is particularly the case if it is served by a microwave transmission medium which is 'closed' (i.e. physically inaccessible) such as waveguide or coaxial line. Micro-strip line on the other hand is 'open' and accessible for component insertion. Microwave power distribution networks are also easier to realise in micro-strip, which is especially important in arrays with potentially millions of radiating elements.

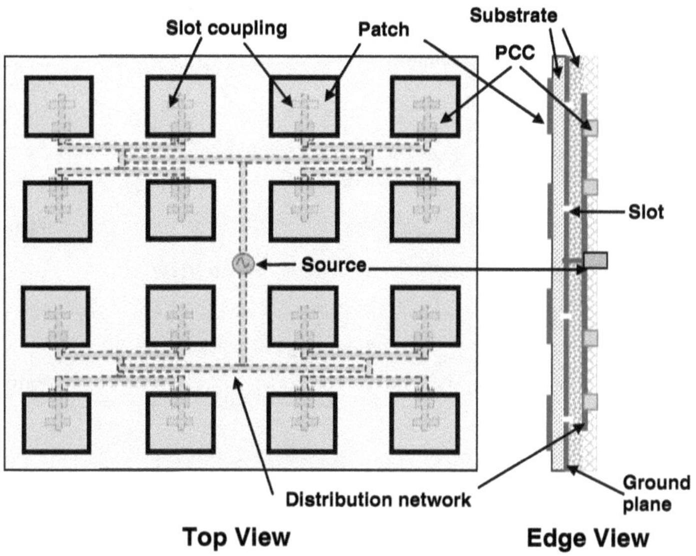

Fig. 9.19 4 × 4 sub-array for micro-strip patch antenna for SPS

Figure 9.19 depicts a 4 × 4 sub-array of square patch radiators on a dielectric substrate. At a frequency of typically 5 GHz, each patch is ~approximately 4 cm square and the sub-array occupies an area of about 22 cm × 22 cm. The patches (see Fig. 9.20) are excited by slots in the ground plane of the upper substrate, and in turn, the slots are excited in phase by micro-strip lines on the back surface of the second substrate. Each patch antenna itself is formed from a short section of 'open-circuited' micro-strip line which is $\lambda_d/2$ long to trap the first standing wave pattern. The modified TEM-mode fields on the transmission line for this terminated geometry are shown in Fig. 9.20. The schematic demonstrates that while the E- and H-fields are essentially uniform in the direction normal to the page, there is a pronounced variation in the longitudinal direction. The E-field maximises at the left- and right-hand edges of the slot dropping to zero on the centreline where the coupling slot is located. The magnetic field falls to zero at the open circuits, as the boundary conditions dictate, and it maximises on the centreline. The current flow (J A/m) on the surface of the patch is orthogonal to the magnetic field, again as required by the boundary condition at a conducting surface. It also maximises on the centreline and drops to zero at the left and right edges of the patch. The non-resonant length coupling slot is located at the mid-point of the patch with its major dimension aligned with the magnetic field. This ensures optimum disruption of the current flow and hence strong coupling.

For the patch to be an effective radiator, the terminations cannot be true open circuits (o.c.), and the schematic (Edge View) illustrates this by showing the evolution of leakage fields. In fact, the o.c. edges can be viewed as slot radiators (see Chap. 4), which emit strongly if the o.c. edges are about $\lambda_o/2$ long. These edge

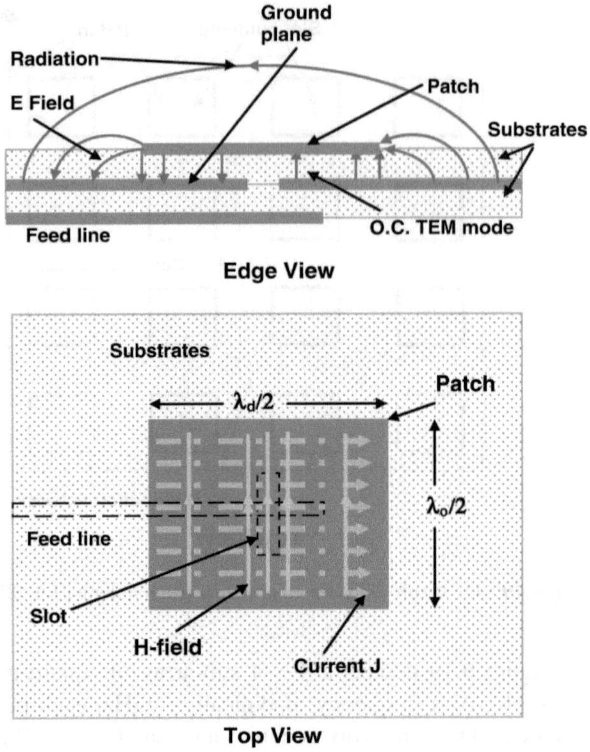

Fig. 9.20 Schematic depicting the radiation mechanism for a slot fed patch antenna

slots then form a two element array (see Chap. 5) with the emitted fields summing to create a 'beam' in a direction normal to the patch face. The radiated power is of course supplied by the coupling slot, which is in turn excited by a source on the power distribution or 'feed' substrate. It should also be noted that the antenna substrate is usually formed from a lower permittivity material, relative to the feed substrate, to encourage radiation.

In Fig. 9.19, the power to each patch is presumed to emanate from a solid-state source (see Sect. 9.3.3) located at the centre of the board and embedded in the micro-strip feed line. The line lengths from the source to each slot are designed to deliver equal phase. The phase conjugation circuits (PCCs) are also embedded into the micro-strip substrate, with each located a short distance in front of the relevant slot coupler. The sub-array is presumed to be attached to a supporting and stiffening structure as suggested in the diagram. The SPS array as a whole will then be formed from a mosaic of these sub-arrays, with the sources phase-locked to a stable reference oscillator.

The specifications for a micro-strip array antenna of this description are to be found in the literature [1–3]. A solid-state array designed to operate at 5.8 GHz has been described there. Currently, sources that are compatible with micro-strip technology can deliver mean power levels in the 50 W range. In a 4 × 4 patch module, this implies a radiated power per patch of the order of 3 W, which is well within the capability of this technology. To deliver in excess of 1 GW to the ground at 5.8 GHz, it is estimated that a 500-m-diameter array would be required, and this equates to 4 million array modules and 4 million solid-state sources. The numbers seem daunting, but to advocates of SPS, they are obviously not excessively so.

9.5 Rectenna-Based Receiver Arrays

The microwave beam directed towards earth from a future SPS system will deliver an essentially unmodulated constant wave (CW) signal to the ground array antenna. This array (see Fig. 9.2) can therefore be relatively unsophisticated with a simple diode rectifier behind the port of every receiving element. The close integration of the antenna and rectifying diode has resulted in the arrangement being termed a rectenna. It has the purpose of extracting the microwave power from the beam by converting it to a DC voltage.

Rectennas have a range of applications and have been under development since the early 1960s, but introducing the idea into SPS technology is generally attributed to Brown [31]. A basic rectenna circuit is depicted in Fig. 9.21. In the early developments of the SPS concept, the preferred operating frequency was 2.45 GHz simply because microwave components and devices were well established at frequencies in this range. The dipole, as the figure shows, was the antenna of choice largely because at 2.45 GHz, it is compact, yet robust, and furthermore, its receiver characteristics were totally predictable. Also, linearly polarised transmission from waveguide-based space arrays was envisioned; thus, dipoles aligned to the incoming E-field represented a natural mode of reception. The dipole (Fig. 9.21) is followed by a low-pass filter to reject higher-frequency electromagnetic waves associated with, for example, radar and communication systems. This is necessary since a dipole is strongly receptive to signals at multiples of its design frequency, while it is insensitive to frequencies below it. The filter is also necessary to suppress re-radiated harmonics developed in the nonlinear diode. In the figure, the rectifier is depicted as a simple half-wave circuit comprising a single diode and a single capacitor in parallel with it. In practice, more complex full-wave circuits [7, 31] are employed to maximise DC conversion efficiency defined as follows:

$$\eta = \frac{\text{Harvested DC power}}{\text{RF power input}} = \frac{P_{\text{DC}}}{P_{\text{RF}}} \tag{9.14}$$

It is assumed that the rectifying circuit is matched to the transmission line which is in turn matched to the antenna. With competent design, even the early rectenna developments [31] were capable of efficiencies in excess of 90 %. Table 9.2 presents some measured conversion efficiencies for a range of practical rectenna types from dipole to circular patch and from linearly polarised to circularly polarised options. The best efficiencies are evidently available at the lower frequencies.

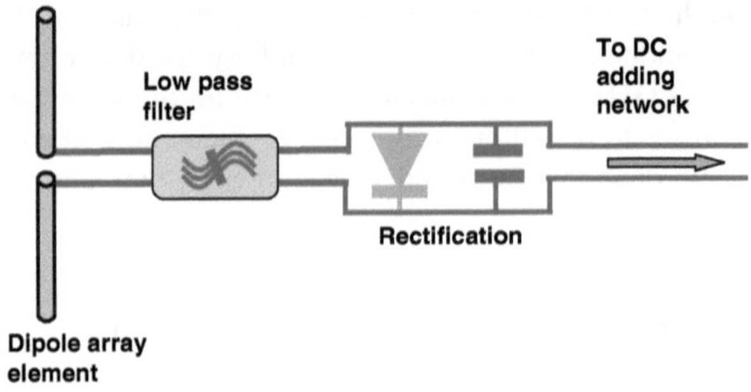

Fig. 9.21 Basic rectenna circuit for half-wave rectification

Table 9.2 Conversion efficiencies for typical rectenna circuits [3]

Rectenna type	Frequency GHz	Polarisation	Peak power (W dc)	Efficiency %
Printed dipole	2.45	Linear	5	85
Printed dipole	2.45	Linear	0.094	84.4
Printed dipole	2.45	Dual	1	70
Printed dipole	5.8	Linear	0.052	82.7
Printed rhombic	5.61	Circular	0.084	78
Square patch	8.51	Dual	0.065	66
Circular patch	2.45	Dual	5	81
Circular patch	5.8	Linear	3	76

In rectenna arrays, where care has been taken to minimise mismatch errors, the diode is the critical component in securing high efficiency, because it is the primary source of power loss. Good rectification efficiencies in excess of 80 % have been demonstrated using barrier diodes based on silicon or gallium arsenide semiconductors [3]. Note that a genuine rectenna array formed from circuits, such as that shown in Fig. 9.21, is an array only in a physical sense of being a lattice comprising a grid of receivers spread over a collecting area. It is not an array in the electrical sense since no phase addition is involved. It actually has more in common with a photovoltaic array as described in Chap. 7 than an antenna array. Each rectenna element is basically an independent microwave collector/rectifier generating a DC output. The 'array' requirement in this case is essentially to ensure that the rectenna elements are positioned so that their combined, or summed, effective areas match the 'footprint' on the ground of the SPS beam (see Fig. 9.2).

Ideally, each rectenna element should be omnidirectional to accommodate the inevitable and unavoidable north and south satellite drifts with the seasons. This is arguably best achieved with dipoles aligned with the earth's lines of latitude. Less ideal arrangements, offering some form of compromise, may be adopted, as suggested in Table 9.2, to accommodate in addition satellite rotation. Examples are the dual polarised or circular polarised antenna designs.

It has been suggested that rectification could perhaps be applied after some phase addition has been performed, in which case antenna array phased addition techniques become available and could be introduced to improve collection performance. If this were the case, then retro-directivity [7] would become relevant to the ground-based rectenna array, for example, to more effectively accommodate satellite seasonal drift relative to the planetary surface.

References

1. Mankins JC (2001) Space solar power: a major new energy option. J Aerosp Eng 14(2):38–45
2. Mankins JC (2009) New directions for space solar power. Acta Astronaut 65(1–2):146–156
3. McSpadden JO, Mankins JC (2002) Space solar power programs and microwave wireless power transmission technology. IEEE Microwave Mag 3(4):46–57
4. www.shimizu.co.jp/english/theme/dream/lunaring.html
5. Tsumring SE (2006) Electron beams and microwave vacuum electronics. Wiley, New Jersey
6. Ren YJ, Chang K (2006) New 5.8 GHz circularly polarized retrodirective rectenna arrays for wireless power transmission. IEEE Trans MTT 54(7):2970–2976
7. Olgun U, Chen CC, Volakis JL (2011) Investigation of rectenna array configurations for enhanced RF power harvesting. IEEE Antennas Wirel Propag Lett 10:262–265
8. Sangster AJ, Lyon RW (1982) A moment method analysis of a T-shaped slot radiator in bifurcated waveguide. IEE Proc-H 129(6):299–306
9. Sangster AJ, Lyon RW (1990) An efficient moment method for radiating slots in thick-walled waveguide. In: Hansen RC (ed) Moment methods in antennas and scattering. Artech House Inc, Boston, pp 368–376
10. Sangster AJ, McCormick AHI (1990) Theoretical design/synthesis of slotted waveguide arrays. In: Hansen RC (ed) Moment methods in antennas and scattering. Artech House Inc, Boston, pp 385–392
11. Johnson RC (1993) Antenna engineering handbook. McGraw-Hill Inc, New York
12. Brady MM (1971) Single slotted-waveguide linear arrays. In: Young Leo (ed) Advances in microwaves. Academic Press, New York
13. Wilkinson EJ (1961) A circularly polarised slot antenna. Microwave J 4:97–100
14. The Microwave Engineers Handbook and Buyers Guide (1963) Microwave J
15. http://multipactor.esa.int/whatis.html
16. Dickinson RM (2000) Magnetron directional amplifier space solar power beamer concept design. In: 35th IECEC proceedings. Las Vegas, NV, pp 1469–1479
17. Liao SY (1980) Microwave devices and circuits. Prentice-Hall Inc, New Jersey
18. Laverghetta TS (1996) Practical microwaves. Prentice-Hall Inc, New Jersey
19. Pon CY (1966) Retrodirective array using heterodyne technique. IEEE Trans Antennas Propag 12:176–180
20. Ghose RN (1964) Electronically adaptive antenna systems. IEEE Trans Antennas Propag 12 (2):161–169
21. Van Atta LG (1959) Electromagnetic reflector. US Patent No. 2,908,002. 6 Oct 1959
22. Chernoff RC (1979) Large active retrodirective arrays for space applications. IEEE Trans Antennas Propag 27(4):489–496
23. Leong K, Wang Y, Itoh T (2004) A full duplex capable retrodirective array system for high speed beam tracking and pointing applications. IEEE Trans Microwave Theor Tech 52(5):1479–1489
24. Shiroma GS, Miyamoto RY, Shiroma WA (2006) A full-duplex dual-frequency self-steering array using phase detection and phase shifting. IEEE Trans Microwave Theor Tech 54(1): 128–134

25. Myamoto RY, Qian Y, Itoh T (2000) An active integrated retrodirective transponder for remote information retrieval. IEEE MTT-S Internat Microwave Symp 3:1431–1434
26. Myamoto RY et al (2003) Digital wireless sensor server using an adaptive smart-antenna/retrodirective array. IEEE Trans Veh Technol 52(5):1181–1189
27. Karode SL, Fusco VF (1997) Frequency offset retrodirective antenna array. Electron Lett 33(16):1350–1351
28. Karode SL, Fusco VF (1999) Multiple target tracking using retrodirective antenna arrays. In: National conference on antennas and propagation. Publication No. 461, pp 178–181
29. Fusco VF, Soo CB, Buchanan N (2005) Analysis and characterisation of PLL-based retrodirective array. IEEE Trans Microwave Theor Tech 53(2):730–738
30. Sun J, Zeng X, Chen Z (2008) A direct RF undersampling retrodirective array system. In: IEEE radio and wireless symposium. pp 631–634. ISBN:1-4244-1463-6/08
31. Brown WC (1984) Performance characteristics of the thin film etch-circuit rectenna. In: IEEE MTT-S international microwave symposium digest. San Francisco, California, pp 365–367

Chapter 10
Optical Antennas (Nantennas)

If there is a silver bullet in energy, I think it's solar power.
M.A. Geyer

Abstract In response to inherently low levels of efficiency in the collection of light from photovoltaic cells, the nantenna has recently become a feature of the solar power gathering landscape. In simple terms, as this chapter illustrates, it is a conventional wire-type antenna, for transmitting or receiving electromagnetic waves, but expanded in its operational capability from the microwave and millimetre wavebands, up into the infrared and optical ranges. Unfortunately, as the chapter also emphasises, frequency scaling laws introduce significant implementation difficulties. In nantennas, the current carrying wires shrink in their cross-sectional dimensions to sizes in the nanometre range (radii less than 100 nm). In addition to the obvious fabrication problems which are encountered, even when sophisticated lithographic methods are adopted, these nanoscale dimensions impose additional limitations. The current flows in such fragile wires, enforced by the laws of physics are restricted in unexpected ways. By focusing on the dipole antenna at the nanoscale, the chapter demonstrates the negative effects, on its radiation efficiency, of enhanced field penetration into filamentary conductors and of electron kinetic effects in such wires, both of which become significant at radii of less than 100 nm. Given that in space, temperatures close to absolute zero are difficult to avoid, then for orbiting solar platforms at least, it seems possible that rectennas employing supercooled and superconducting materials could offer a route towards high efficiency light gathering systems. This new technology avenue is briefly addressed towards the end of the chapter.

10.1 Introduction

In order to efficiently collect microwave power beamed down to earth from a solar power satellite, it has been proposed, as outlined in Chap. 9, Sect. 9.5, that this could perhaps best be done by employing arrays of ground based rectennas, which directly convert the microwave power received to a DC voltage in a rectifying

A.J. Sangster, *Electromagnetic Foundations of Solar Radiation Collection*,
Green Energy and Technology, DOI 10.1007/978-3-319-08512-8_10

circuit located at the antenna terminals. The question that arises, particularly in connection with solar power gathering, is why this concept cannot be applied to much higher light frequencies in order to engineer light collectors which are more efficient than those provided by currently available photovoltaic arrays? An answer framed in the electromagnetic engineering terms outlined in Chap. 6 will be presented in this chapter.

Any aerial engineer contemplating the notion of collecting the sun's rays on a conventional antenna structure would be immediately daunted by the vast spectrum of frequencies to be accommodated. However, unlike signal reception in communication systems, the optical waves from the sun, as we have already been made aware, are incoherent. Phase is irrelevant if only power collection is of interest. Consequently, the essence of the nantenna, which is evolved from the rectenna, lies in the fact that the currents induced in the antenna are immediately rectified in a semiconducting diode mounted in the antenna terminals. The DC currents are then accumulated in a power building process. This basic concept is suggested at in Fig. 10.1.

The literature on optical rectennas (where they are increasingly referred to as nantennas) is very recent but wide ranging [1–7]. It suggests that the notion of adapting the rectenna concept to the gathering of light frequencies was first raised in 1972 [1] and was further reinforced by several patent applications in the 1980s [2–4]. At microwave frequencies of the order of 2–10 GHz, rectennas offer conversion efficiencies in the 70–80 % range. On the other hand, achieving better than 10 %, even in proof-of-principle rectenna prototypes at infrared and visible frequencies, remains an immense task [5]. The reason for this is not difficult to identify. At these frequencies, the wavelength criteria (Chap. 4), which are intrinsic to antenna operation, drive circuit and array elements into the nanometre scale of dimensions where reliable fabrication is still formidable. This is exacerbated by the requirement for ultrafast rectifying diodes capable of functioning effectively at

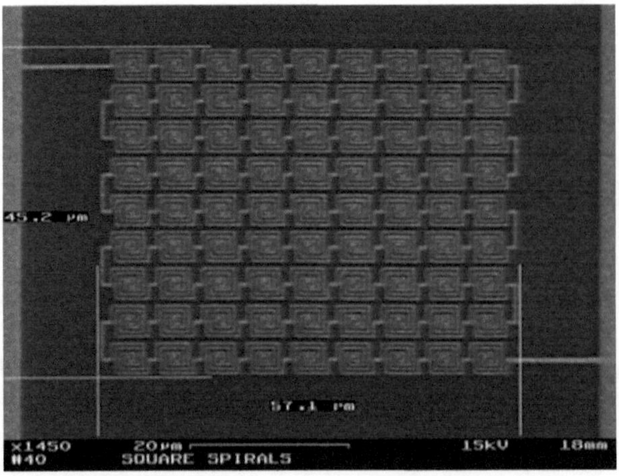

Fig. 10.1 Optical nantenna array with nanoscale conducting wires (*Courtesy* of ASME)

optical frequencies. Furthermore, even as the fabrication difficulties are being overcome, recently unknown challenges, which are inherent in subjecting electrical materials to EM fields at optical frequencies, are difficult to circumvent. For example, at optical frequencies, in marked contrast to lower frequencies, radiation penetrates deeply into metals as atomic separation distances become comparable to the wavelengths of the incident waves. The electromagnetic effect is for the metal electrons to behave rather like a space charge cloud, or strongly coupled plasma, which supports oscillations termed plasmons. At this juncture, antenna frequency scaling fails, unless exotic materials are employed to counteract field penetration into the antenna structure [7, 8].

Interest in nantennas for light collection applications has been partly stimulated by the inefficiency of photovoltaic junctions (see Chap. 7). At optical frequencies, the crystal-like lattices of the doped materials in a PN junction present periodic structure or filter-like behaviours, in electrical engineering terms, which means that only over a limited frequency band (at a bandgap in quantum mechanical terms) of light frequencies is photovoltaic operation effective. Unfortunately, the solar spectrum (see Fig. 1.15) has a broad dumb-bell shape which means that a very wide spectrum of frequencies, both above and below the PN junction conversion frequency, exists in the incident radiation. The energy in these frequencies tends to be absorbed by the junction material causing lattice vibrations (phonons) which are a source of power loss. For a single junction cell, this process leads to an upper limit for optical conversion efficiency of about 20 %.

10.2 Antenna Efficiency at Nanoscale

Antennas at the optical scale of dimensions remain noteworthy for their continuing absence from advanced technological developments, and needless to say, this is largely attributable to the issue of extreme smallness. As we have seen in Chaps. 3–5, antennas depend on dimensional characteristics which are wavelength related, and at light frequencies, this implies that practical devices require fabrication capabilities which can achieve dimensional accuracies of better than 10 nm. However, the newly burgeoning fields of nanoscience and nanotechnology are beginning to provide this level of fabrication capability, by the adoption of novel techniques such as ion beam milling and electron beam lithography. The emergence of these techniques, which are now extending antenna developments into the optical range, offers extensive opportunities ranging from spectroscopy to photodetectors. An optical antenna array formed from logarithmic spirals is shown in Fig. 10.1. The 9 × 9 array is 57.1 μm wide and 45.2 μm high, so that each spiral is approximately 6 μm × 5 μm, thus placing the width of each filamentary conductor in the 100 nm range.

Generally, research into antennas at the nanoscale of dimensions has been directed at the conventional dipole structure. This is evidently because, at below infrared frequencies, the device is well understood theoretically, it has been comprehensively examined in the literature, and since it is geometrically simple, it is realistic to consider that it is realisable at optical frequencies. The primary characteristics of the dipole antenna which are germane to the discussion in this chapter are presented below. It is particularly important to note that in all modelling of the conventional dipole, the radius (a) of the wire forming it is very small in wavelength terms (i.e. $a \ll \lambda$) but is by no means small in relation to the atomic scale of dimensions. For example, in practical microwave and millimetre-wave embodiments, a is never less than a few microns and could be as large as several millimetres. Furthermore, the dipole is constructed from a metal such as copper or aluminium, which can be assumed to be highly conducting at sub-infra-red frequencies.

10.2.1 Conventional Dipole

The nature of the standard low-frequency dipole antenna can probably be illustrated most effectively by examining its evolution from a two-wire transmission line supporting a TEM mode which exhibits a characteristic impedance of Z_o (say), as shown in Fig. 10.2. On a line, open circuited at the right-hand end and excited by an AC signal of voltage V and current I, entering from the left, a standing wave is set up with peaks $2V$ and $2I$ displaced by a distance equal to $\lambda/4$ on the line. The voltage (or E-field) nulls and the current (or H-field) nulls of the standing waves are separated by a line distance equal to $\lambda/2$.

A half-wave dipole is formed on the line in Fig. 10.2 by re-routing end sections of each wire so that they are at right angles to the original wires. This should be applied at a position where the standing wave voltage is zero and the current is a maximum, as shown. At this instant, current flows strongly into the dipole wires, carrying with it a magnetic field. This time-varying field becomes a source of radiation as described in Chap. 3. A quarter cycle later, the current disappears as electrical charge builds up on the wire extremities which in turn develops localised electric fields thus maintaining the radiation mechanism. The process continues, cycling between predominantly electric field and predominantly magnetic field, as long as power is fed into the left end of the feed line. In addition to sourcing radiation, these fields also store energy, so that, electrically, the dipole behaves not unlike a resonant circuit as the lower diagram suggests. Here, the capacitance C represents near-field energy stored in the electric fields, and the inductance L represents near-field magnetic energy, while R represents the radiation resistance, usually denoted by R_{rad}.

To compute values for these circuit parameters for a given dipole, it is first necessary to determine the fields generated by the dipole. This can be done by employing the Hertzian dipole, described and defined in Chap. 3, as a building block. It is shown there that the electric field and magnetic field surrounding a current element $I_0 dl$ are given by:

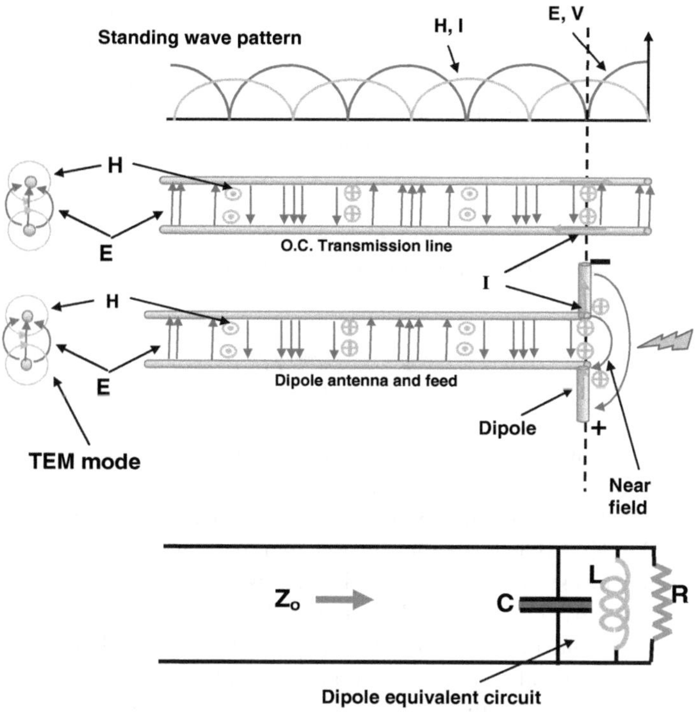

Fig. 10.2 Electrical fundamentals of conventional dipole antenna

$$H_\phi = \frac{I_0 dl}{4\pi}\left[\frac{jk_0}{r} + \frac{1}{r^2}\right]\sin\theta\exp(-jk_0 r) \tag{3.52}$$

$$H_\theta = H_r = 0 \tag{3.53}$$

$$E_\theta = \sqrt{\frac{\mu_0}{\varepsilon_0}}\frac{I_0 dl}{4\pi}\left[\frac{jk_0}{r} + \frac{1}{r^2} - \frac{j}{k_0 r^3}\right]\sin\theta\exp(-jk_0 r) \tag{3.54}$$

$$E_r = \sqrt{\frac{\mu_0}{\varepsilon_0}}\frac{I_0 dl}{2\pi}\left[\frac{1}{r^2} - \frac{j}{k_0 r^3}\right]\cos\theta\exp(-jk_0 r) \tag{3.55}$$

$$E_\phi = 0 \tag{3.56}$$

Consequently, the fields surrounding a finite length dipole (l) can be generated by dicing up the wire filament into a ladder of elements $I_0(z)\mathrm{d}z$ where z is the direction of the dipole axis. For a half-wave dipole, $I_0(z)$ is known, being essentially a half-cosine function. Hence, it becomes possible to sum up all the contributions $I_0(z)\mathrm{d}z$, essentially by performing, in the limit of $\mathrm{d}z \to 0$, an integration from $z = -l/2$ to $z = +l/2$, and the dipole field can be computed [9]. Furthermore, with the fields now known, as a function of current I_0, in the space enveloping the dipole, the stored energy in this space can be determined and the power radiated through any enclosing spherical surface is calculable. It is then a simple step to determine values for L, C and R_{rad}. These parameters are shown as a function of length for a typical dipole in Fig. 10.3, where $X_{\mathrm{in}} = X_L = j\omega L$ for

$$l > \lambda/2 \quad \text{and} \quad X_{\mathrm{in}} = X_C = -j/\omega C \quad \text{for} \quad l < \lambda/2.$$

For a finite length dipole supporting a peak current $|I_0|$, the radiation resistance is defined as [9]:

$$R_{\mathrm{rad}} = \frac{2P_{\mathrm{rad}}}{|I_0|^2}\,\Omega \tag{10.1}$$

where P_{rad} is the power in watts passing through an imaginary sphere enclosing the dipole. Radiation resistance is plotted as a function of normalised dipole length (diamond annotated curve) in Fig. 10.3. It rises slowly from zero at $l/\lambda = 0$ to a value of 73 Ω at $l/\lambda = 0.5$. Thereafter, it rises to a maximum of ~200 Ω at $l/\lambda = 0.8$ and then diminishes steadily. If the line characteristic impedance Z_o is also 73 Ω (in practice, it is usually 75 Ω), then the line will be perfectly matched, the standing wave disappears, and the radiation efficiency is 100 %. The half-wave dipole is viewed as a very efficient radiator but with typical values of nearer 95 % because of small mismatch effects and material losses. Also, at $l/\lambda = 0.5$, $X_L = X_C$ (see dashed line in Fig. 10.3), while $X_C \to -\infty$ as the frequency tends to zero to the left, and $X_L \to +\infty$ at a dipole length near λ where the current at the input terminals of the dipole $I_{\mathrm{in}} \to 0$. At this length, the dipole behaves electrically like an open circuit.

The input resistance for a dipole antenna is usually defined as [9]:

$$R_{\mathrm{in}} = \left|\frac{I_0}{I_{\mathrm{in}}}\right|^2 R_{\mathrm{rad}} \tag{10.2}$$

and so at $l/\lambda = 1$, the magnitude of the dipole input resistance R_{in} tends towards infinity. On the other hand, for the resonant half-wave dipole, $|I_{\mathrm{in}}| = |I_0|$ and hence $R_{\mathrm{in}} = R_{\mathrm{rad}}$. This is also approximately true for short dipoles as the solid red curve in Fig. 10.3 demonstrates. Note that *in theory* at $l/\lambda = 1$, I_0 is assumed to remain finite, even when $I_{\mathrm{in}} = 0$, and is hence a continuing source of finite power radiated and finite energy stored. The infinities depicted in Fig. 10.3 are the result.

10.2.2 *Efficiency Anomaly*

At communication frequencies, from the low megahertz to the high gigahertz, it is evident from the above summary that the dipole antenna is viewed as a very efficient radiator. In order to extend the operating range of the dipole into infrared

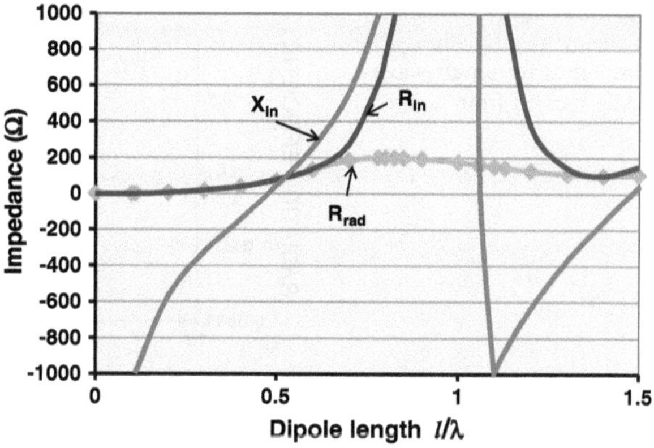

Fig. 10.3 Impedance conditions for dipole antenna

and light frequencies, its dimension has to be scaled down with the diminishing wavelength, so that the dipole wire becomes increasingly filamentary as the radius reduces towards the nanometre scale of dimensions. This has a significant effect on radiation efficiency, as Fig. 10.4 shows [10]. It provides a plot of dipole radiation efficiency at three frequencies (10, 100 GHz and 1 THz) as a function of increasing wire radius from 10 nm to 10 μ. The high radiation efficiency of greater than 90 % persists for all three frequencies down to wire diameters equal to or greater than 1 μ. Below this value, the 10-GHz efficiency level dives drastically, while the critical radii are 0.5 μ for 100 GHz and 0.1 μ for 1 THz. So what causes this behaviour?

Given that for DC and low frequencies, the resistance of a straight filamentary wire of length l, cross-sectional area *A* and conductivity *σ* in *S/m* is given by:

$$R = \frac{\ell}{\sigma A}\Omega \tag{10.3}$$

where $A = \pi a^2$, and the source of the dipole efficiency loss in Fig. 10.4 for very small values of *a* is perhaps not surprising—the wire resistance becomes extremely large. However, at very high frequencies, Eq. (10.3) can no longer be applied in the simple way indicated above, because skin effect, whereby current begins to flow through a thin surface layer, complicates the issue.

To better understand the loss, or attenuation, mechanisms on a highly conducting filamentary wire, we need to revisit the Maxwell equations which apply to all field problems of this description. To be more precise, we need to solve the problem of propagation on a long straight filament of radius *a* and carrying an oscillatory current I_z as suggested in Fig. 10.5. The wire is assumed to be located in air with permittivity ε_o and to have a complex permittivity ε_m given by:

$$\varepsilon_m = \varepsilon_o - \frac{j\sigma_m}{\omega} \tag{10.4}$$

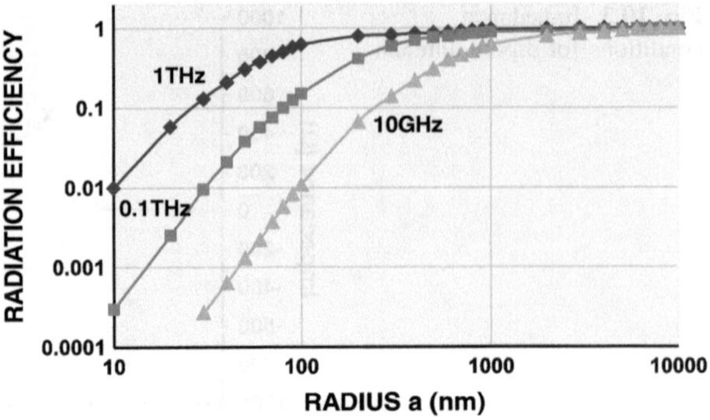

Fig. 10.4 Dipole radiation efficiency as a function of wire radius for a half-wave dipole formed from copper

where σ_m is the conductivity of the wire in *S/m*. In Chap. 3, it has been demonstrated that the fields generated by an arbitrary current source of density **J** A/m^3 are solutions of a second-order differential equation for the magnetic potential **A**. The Eq. (3.31) is repeated below for convenience.

$$\nabla^2 \mathbf{A} + k_0^2 \mathbf{A} = -\mu_0 \mathbf{J} \tag{10.5}$$

Not unexpectedly, this equation can form the basis of a differential equation governing the behaviour of the fields at the surface of the wire filament. If the wire is infinitely long, there is no radiation, and the solution represents the primary propagation mode on the wire. For a wire with conductivity σ_m and permeability μ_o, carrying a current at frequency ω, the skin depth δ is given by:

$$\delta = \sqrt{\frac{2}{\omega \mu_0 \sigma_m}}\ \mathrm{m} \tag{10.6}$$

and so for the current I_z on the straight filament, we can form the equation:

$$J_z = \frac{I_z}{2\pi a \delta}\ \mathrm{A/m^3} \tag{10.7}$$

For this z-directed current density, only the z component of the vector **A** has significance. Hence, for a long straight conducting filament, the fields are the solution to [11]:

$$\frac{\partial^2}{\partial z^2} A_z + k_0^2 A_z = -\frac{j\mu_o I_z}{2\pi a \delta} \tag{10.8}$$

Note that the slight difference between Eq. (10.8) and the equations in Ref. [11] can be attributed to the switch from the cgs system to the MKS system. Also, for completeness, King introduces a delta function voltage to represent the conditions at the dipole input terminal. However, this is not necessary when only the propagating nature of the basic mode on the filament is being sought.

Equations of the form of (10.8) are usually solved by converting it to an integro-differential equation for the unknown parameter—in this case $I_z(z)$ [9, 11]. The conversion, for the straight conducting wire of cylindrical cross section and radius a, requires the formulation of a boundary condition relating the surface fields on the outside to the surface current on the inside of the wire. Such a boundary condition is:

$$I_z(z)\varsigma_s = E_{zs}(z) \tag{10.9}$$

where $E_{zs}(z)$ is the longitudinal electric field at the curved boundary, while ζ_s is the surface impedance per unit length. It is given by [10, 11]

$$\varsigma_z = \frac{\gamma J_0(\gamma a)}{2\pi a \sigma_m J_1(\gamma a)} \approx \frac{1+j}{2\pi a \sigma_m \delta} = \frac{Z_s}{2\pi a} \tag{10.10}$$

where

$$\gamma = (1-j)\sqrt{\frac{\omega \mu_0 \sigma_m}{2}} \tag{10.11}$$

γ is the TEM mode propagation coefficient for a lossy conductor of infinite extent. Also for the exterior surface fields, Maxwell's equations yield $\frac{\partial}{\partial r} H_{\theta s} = j\omega\varepsilon_0 E_{zs}$ while the fundamental relationship relating magnetic flux density B to magnetic vector potential A (Chap. 3) provides the definition $B_{\theta s} = \frac{\partial}{\partial z} A_{zs}$. These relationships lead to [11]:

$$\left(\frac{\partial^2}{\partial z^2} + k_0^2\right)\frac{k_0}{c}\int_{-\infty}^{\infty} I_z(z')\frac{\exp(-jk_0 r)}{r}dz' = -\frac{j\mu_o I_z(z)}{2\pi a \delta} \tag{10.12}$$

where

$$r = \sqrt{(z-z')^2 + a^2} \tag{10.13}$$

with z representing the axial distance to the field point and z' the axial distance to the source point as indicated in Fig. 10.5.

10.2.3 Modal Attenuation

Equations of the nature of (10.12) can be solved efficiently by employing well-established mathematical techniques such as transforms or moment methods [9].

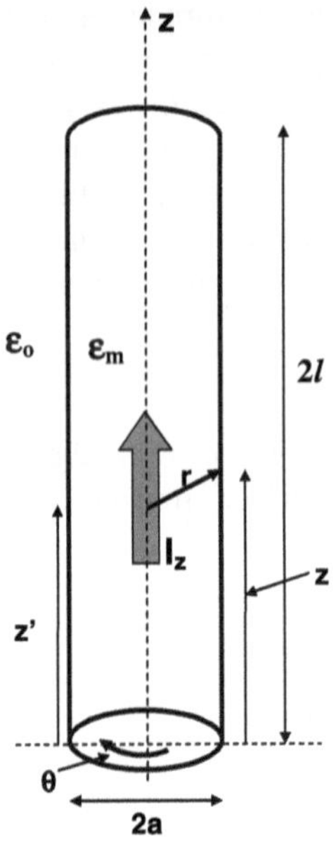

Fig. 10.5 Geometry of a long filamentary conductor carrying a current I_z

The details of the solution need not be rehearsed here. Exact and approximate solutions for both infinite and finite conducting filaments are presented in the literature [10–12]. These show that the current displays wave characteristics which are modal having the basic field structure of circularly symmetric TM modes. The predominant mode that is of interest here is the TM_{10} mode [10, 12], defined by its complex propagation coefficient γ_{m10} where

$$\gamma_{m10} = \beta_{m10} - j\alpha_{m10} \tag{10.14}$$

β_{m10} is the phase coefficient for the TM_{10} mode in radians/m while α_{m10} is its attenuation coefficient in nepers/m. These parameters, normalised to the free space propagation coefficient k_0, are presented in Fig. 10.6, for two different frequencies, as a function of wire radius a in nanometres. For relatively substantial filaments of >100 nm in radius, the phase coefficient tends towards that of free space for the principal mode which implies that the energy of the mode exists largely in the exterior surface fields. This is confirmed by the trend in α_{m10} which decays towards zero as the filament radius increases. For a finite length wire at these radii, the implication is that, at the terminations of the finite wire where the fields must diffract, the reflected component of the TM mode is similar in magnitude to the incident component forming a standing wave on the wire, in endorsement of the half-cosine shape which is known to be set up on the wire current in the case of a half-wave dipole.

As a is reduced below 100 nm, both α_{m10} and β_{m10} become very large, and furthermore, $|\alpha_{m10}| \rightarrow |\beta_{m10}|$. The latter result is suggestive of plane wave

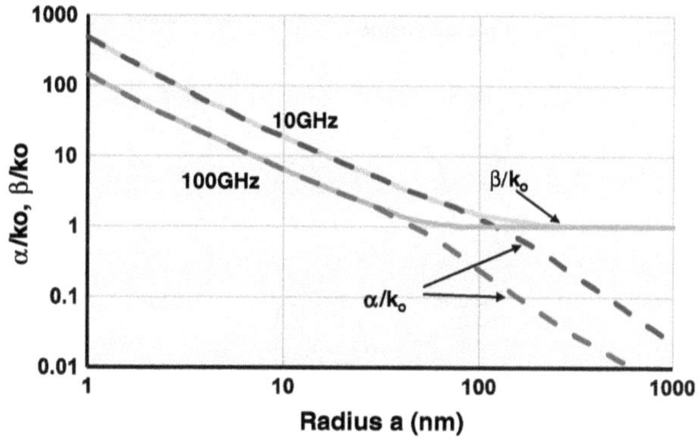

Fig. 10.6 Mode propagation coefficients β_{m10} and α_{m10} normalised to k_o for a highly conducting filament

behaviour within a lossy medium of infinite extent (see Eq. (10.11)), which in turn implies that as a decreases the field increasingly penetrates the metal of the filament. Additionally, given that the TM mode phase velocity is inversely proportional to β_{m10},

$$v_{m10} = \frac{\omega}{\beta_{m10}} \tag{10.15}$$

it becomes increasingly slow as the wire radius reduces to filamentary size. This again is suggestive of field penetration into the metal. Note that since $k_o(=\omega/c)$ is much larger at 100 GHz, than at 10 GHz, the TM mode attenuation value in nepers/m at the upper frequency is actually higher than at the lower frequency, although the graph confusingly suggests otherwise. Evidently, the loss of radiation efficiency in filamentary dipoles is associated with increasing power loss in the wire due to field penetration into it.

10.3 Impedance and Conductivity Issues

Rising levels of modal attenuation along a filamentary wire of high conductivity as its radius diminishes can evidently, as we have seen in the previous section, be ascribed to increased field penetration into the wire. However, this still does not fully explain the drastic rise in the attenuation level in a metal wire of seemingly high conductivity. To do so, we need to examine conductivity at the microscopic level in nanoscale filaments [13, 14].

At the atomic level within a conducting material, current flow due to an applied voltage or field is associated with electron drift through the atomic lattice (see Fig. 10.7). The model depicted in the figure represents an application of kinetic theory, in so far as it assumes that the microscopic behaviour of electrons in a solid may be treated classically. A good analogy is an inclined pinball machine. Like the pinballs rolling down the table under the force of gravity, rattling against fixed pins

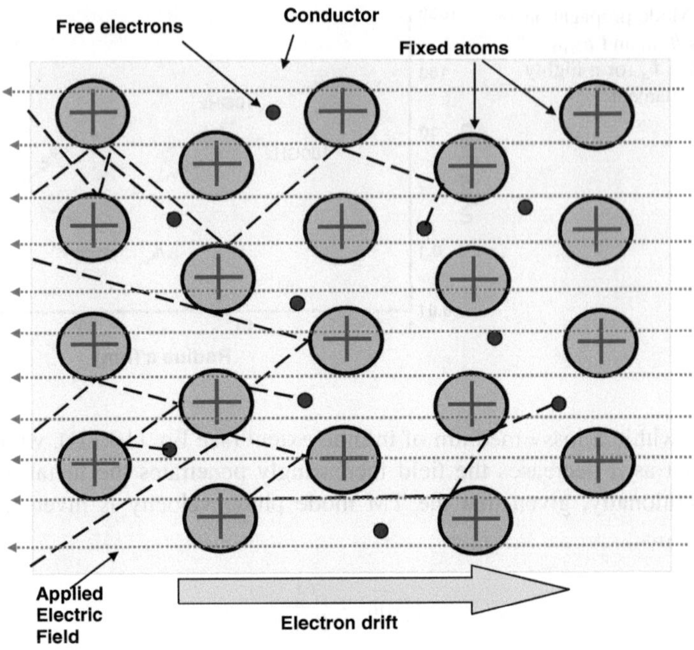

Fig. 10.7 Electron drift in a conducting material

as they go, electrons under the influence of the electric field bounce and re-bounce off heavier, relatively immobile positive ions.

The simplest analysis of this kinetic model (termed the Drude model) assumes that electric field **E** is both uniform and constant. We distinguish the relatively infinitesimal amount of momentum change dp between collisions, which are presumed to occur, on average, every τ seconds, where τ is the mean free time between ionic collisions. Note that the thermal velocity of electrons is much higher, but the motions are random and can be ignored in this context.

Then, an electron isolated at an instant in time t will on average have been travelling for time τ since its last collision and consequently will have accumulated momentum on average of:

$$\langle d\mathbf{p} \rangle = q_e \mathbf{E} \tau \tag{10.16}$$

Throughout its travels through the lattice the momentum accumulations, by an electron between collisions, could be equally either additive or subtractive (see Fig. 10.7), so that at a particular instant in time the current momentum level, $\langle \mathbf{p} \rangle = \langle d\mathbf{p} \rangle$ on average. Hence

$$\langle \mathbf{p} \rangle = q_e \mathbf{E} \tau \tag{10.17}$$

Furthermore, if the average drift velocity is represented by $\langle v_d \rangle$, then we can write:

$$\langle \mathbf{p} \rangle = m_e \langle \mathbf{v}_d \rangle \tag{10.18}$$

Also

$$\mathbf{J} = n q_e \langle \mathbf{v}_d \rangle \tag{10.19}$$

where m_e, q_e and n are, respectively, the electron mass, charge magnitude and number of electrons per unit volume. Hence,

$$\mathbf{J} = \left[\frac{n q_e^2 \tau}{m_e} \right] \mathbf{E} \tag{10.20}$$

Equation (10.20) is essentially the basis of Ohm's law. But $\mathbf{J} = \sigma_0 \mathbf{E}$ (see Chap. 3) where σ_0 is the DC conductivity of the bulk material. Hence,

$$\sigma_0 = \frac{n q_e^2 \tau}{m_e} \ \mathrm{S/m} \tag{10.21}$$

This relatively simple equation contains the 'seeds' of high inefficiency in filamentary dipoles, basically through the transit time component. As the radius a of the current carrying wire is gradually reduced towards atomic dimensions, the fixed atoms in the wire loom larger and larger in electron mobility terms. In the pinball machine analogy, it is as if the 'pins' are gradually increased in size as the table gets narrower resulting in more collisions and a slowing of the drift of balls through the machine. In electron mobility terms, when the effective radius is comparable to, or smaller than, the mean free path (e.g. 40 nm for copper at room temperature) of the electrons, then wire size has a significant bearing on drift velocity and hence conductivity [15]. The radius-dependent conductivity at these wire sizes [10] can be accounted for by the following relation:

$$\sigma(a) = w(a) \sigma_0 \tag{10.22}$$

where σ_0 is the bulk conductor conductivity value (e.g. 5.7×10^7 S/m for copper and 4.6×10^7 S/m for gold). The multiplier $w(a)$ is the nano-size reduction factor. From reference [10], we also have:

$$w(a) = \frac{1}{\left[1 + (1 - p) \frac{L_m}{2a} \right]} \tag{10.23}$$

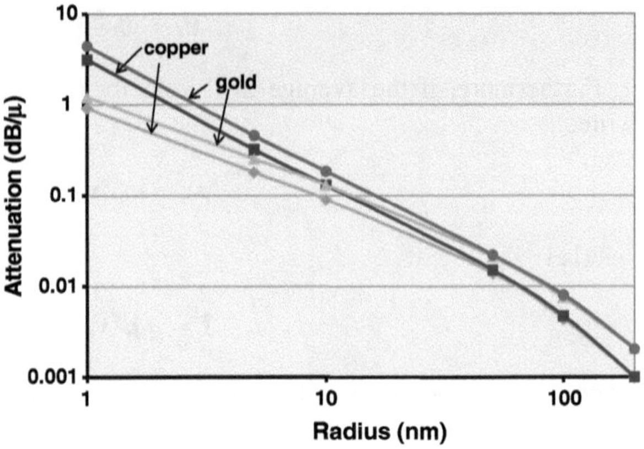

Fig. 10.8 TM mode attenuation α_{m10} (dB/µ) as a function of wire radius a at a frequency of 100 GHz. *Blue/green (diamond/triangle)* = bulk conductivity, *red/mauve (square/circle)* = reduced conductivity

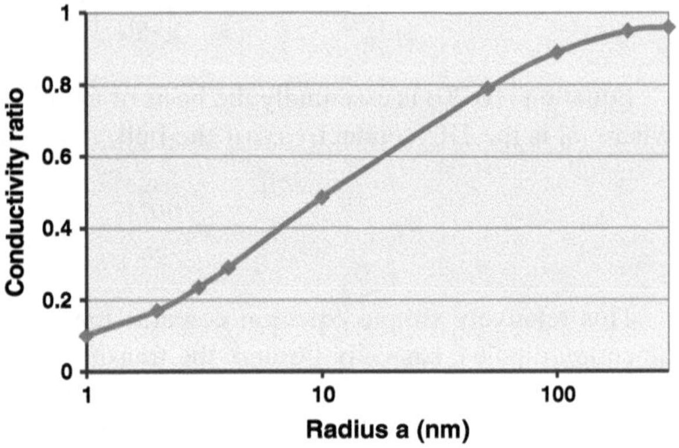

Fig. 10.9 Conductivity reduction ratio $w(a)$ as a function of wire radius for a copper wire at 100 GHz

In this equation, L_m is the mean free path for the metal, and p (typically = 0.5) is a specularity factor for electron reflection at its surface.

The influence of the Drude model of electron drift on the TM mode attenuation in a filamentary wire of radius a is depicted in Fig. 10.8. Employing the relationship,

$$\text{Attenuation} = 8.686\alpha_{m10} \text{ dB/m}$$

the figure presents attenuation trends for copper and gold as a function of radius a in nanometres. Two cases are highlighted, a computation for attenuation using the bulk conductivity in calculations and an evaluation when the Eq. (10.22) form is used. It is clear that for conventional metals, the reduced conductivity associated with kinetic effects cannot be ignored at wire radii below 100 nm. This is emphasised in Fig. 10.9 where $w(a)$ is plotted as a function of radius a at 100 GHz.

10.4 Radiation Efficiency of a Filamentary Dipole

Severe conductor attenuation in filamentary dipoles due to field penetration and enhanced kinetic effects seems to undermine any potential aims at achieving efficient dipole radiation and reception of electromagnetic waves at infrared and optical frequencies. However, this need not necessarily be the case as is explored in the following section.

To assess the implications for antenna radiation efficiency (η_{rad}) of operating with filamentary wires, we revisit the basic definition of this parameter, namely

$$\eta_{\text{rad}} = \frac{R_{\text{rad}}}{R_{\text{rad}} + R_{\text{cond}}} \tag{10.24}$$

Here, R_{rad} is the radiation resistance (=73 Ω for a half-wave dipole) and R_{cond} is a resistance representing metallic or ohmic losses. Given the difficulties alluded to above that in filamentary dipoles, conductivity becomes dependent on wire diameter, we need to know at what conductivity levels, if any, acceptable radiation efficiency is obtainable. This can be approached by re-expressing Eq. (10.24) in the form:

$$R_{\text{cond}} = R_{\text{rad}}\left[\frac{1}{\eta_{\text{rad}}} - 1\right] \tag{10.25}$$

So, for a desired level of radiation efficiency from a filamentary dipole antenna, with known R_{rad}, R_{cond} becomes the wire conductivity which is required to achieve it. To identify this desired wire resistance, R_{cond} is replaced by desired radius R_{dd}. Hence,

$$R_{dd} = R_{\text{rad}}\left[\frac{1}{\eta_{\text{rad}}} - 1\right] \tag{10.26}$$

For a wire of effective length L_{eff}, which is dependent on the current distribution along the wire, we can express R_{dd} in terms of a per unit length resistance R'_{dd} such that:

$$R_{dd} = R'_{dd}L_{\text{eff}} \tag{10.27}$$

R'_{dd} is just the real part of the surface impedance resistance per unit length ζ_s. This definition emanates from the basic definition that the resistance R_{cond} which represents ohmic loss in the dipole must be proportional to the power loss (P_o) in the wire due to heat. That is

$$R_{\text{cond}} = \frac{2P_o}{|I_{\text{in}}|^2} \tag{10.28}$$

where I_{in} is the peak AC current at the dipole terminals, while

$$P_o = \text{Re}(\varsigma_s) \int_{-L/2}^{+L/2} \frac{1}{2}|I(z)|^2 \text{d}z \tag{10.29}$$

The z coordinate is directed along the length of the wire forming the dipole of physical length L. Collecting together, these relationships yields:

$$L_{\text{eff}} = \frac{1}{|I_{\text{in}}|^2} \int_{-L/2}^{+L/2} |I(z)|^2 \text{d}z \tag{10.30}$$

For a half-wave dipole ($L = \lambda/2$), as we have seen in Chap. 3, the current $I(z)$ is essentially cosinusoidal in shape between the wire extremities. Hence, Eq. (10.30) gives, for this case:

$$L_{\text{eff}} = \frac{\lambda}{4} \tag{10.31}$$

On the other hand, for a very short dipole carrying an almost uniform current (see Chap. 3), we get $L_{\text{eff}} = L$.

By combining Eqs. (10.26) and (10.27), the desired resistance per unit length to achieve a given radiation efficiency η_{rad} can be expressed as:

$$R'_{dd} = \frac{R_{\text{rad}}}{L_{\text{eff}}} \left[\frac{1}{\eta_{\text{rad}}} - 1 \right] \tag{10.32}$$

The nature of this equation is summarised in Fig. 10.10 for three frequencies ranging from microwaves (10 GHz) to infrared (1 THz).

It is clear from Fig. 10.10 that for a conventional dipole antenna, ‘good’ radiation efficiency is synonymous with very low conductor resistance. The trend towards higher resistance as the operating frequency rises is simply due to the reducing magnitude of L_{eff} which is proportional to wavelength. Previous observations on filamentary dipoles required for operation at very high frequencies, namely that conductivity deteriorates both with higher field penetration into the filamentary wire and due to enhanced electron kinetic effects, are in conflict with the trend shown in Fig. 10.9 linking high radiation efficiency to low dipole resistance. This difficulty can be better illustrated by looking directly at dipole conductivity. An explicit equation for desired conductivity (σ_{dd}) can be formulated for

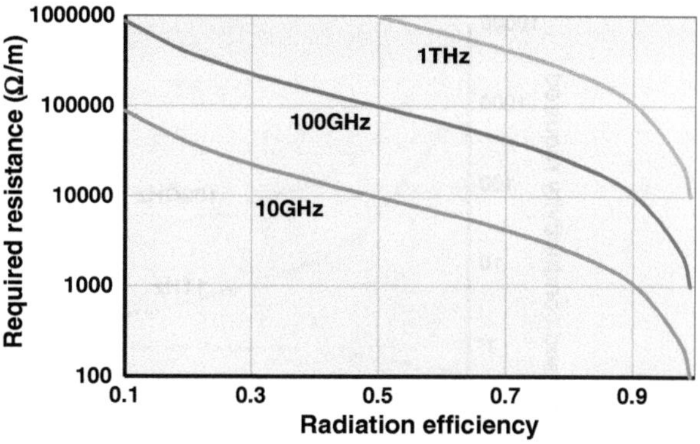

Fig. 10.10 Distributed resistance requirements for filamentary half-wave dipole to secure good radiation efficiency

the case of filamentary wire radius $a \ll \delta$, by and $\omega\tau < 1$, using Eq. (10.32) and noting that $R'_{dd} = \frac{1}{\pi a^2 \sigma_{dd}(a)}$. Hence,

$$\sigma_{dd} = \left\{ \frac{\pi a^2 R_{\text{rad}}}{L_{\text{eff}}} \left(\frac{1}{\eta_{\text{rad}}} - 1 \right) \right\}^{-1} \tag{10.33}$$

σ_{dd}, normalised to σ_m for copper (5.9×10^7 S/m), is plotted in Figs. 10.11 and 10.12, with radiation efficiency as parameter in the former case and frequency as parameter in the latter.

Figure 10.11 clearly highlights the fact that high radiation efficiency, as predicted by a standard model of a dipole antenna, dictates the adoption of metal wires of optimum conductivity and that this demand is much easier to fulfil with wires of large radius. Furthermore, given that the horizontal dotted line at the unity level on

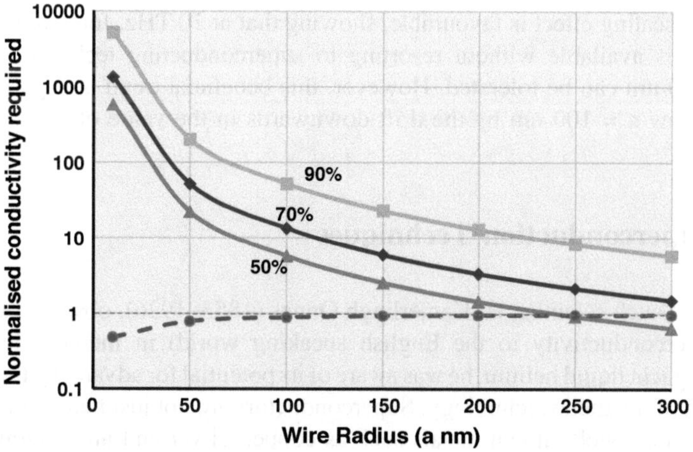

Fig. 10.11 Normalised dipole conductivity as a function of wire radius at 100 GHz (*solid curves*), with radiation efficiency as parameter. Conductivity reduction factor w (*dashed curve*)

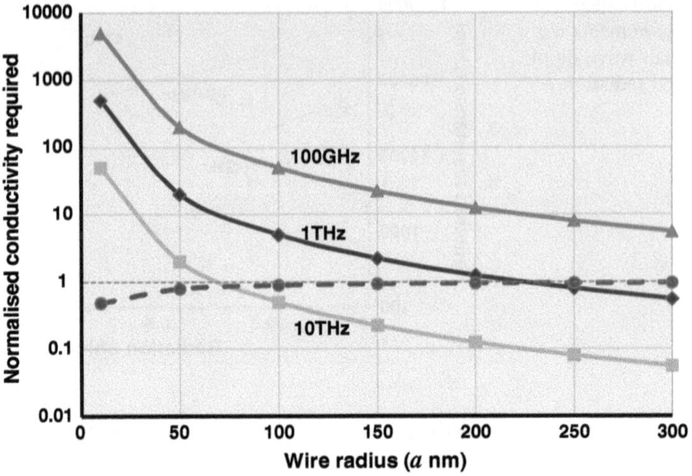

Fig. 10.12 Normalised dipole conductivity as a function of wire radius at 70 % radiation efficiency with frequency as parameter. Conductivity reduction factor w (*dashed curve*)

the left-hand scale represents a wire with the conductivity of copper, the figure suggests that dipole radii of 250 nm or more are required to ensure a radiation efficiency of greater than 50 % [10]. The region of the graph above the unity value of normalised conductivity is accessible only with superconducting materials, which we shall consider in the next section. The dashed (mauve) curve, which represents the conductivity reduction factor $w(a)$, is included to show that the field penetration and kinetic effects examined in the preceding section are insignificant above $a = 100$ nm.

The influence of operating frequency on the conductivity levels necessary to secure 'good' radiation efficiency is presented in Fig. 10.12. What is clear is that the frequency scaling effect is favourable, showing that at 10 THz, for example, a 70 % efficiency is available without resorting to superconducting techniques provided that $a > 70$ nm can be tolerated. However, this beneficial trend is negated to some extent below $a = 100$ nm by the drift downwards in the value of $w(a)$.

10.5 Superconduction Techniques

When the Dutch scientist, H. Kamerlingh Onnes (1853–1926), observed *suprageleider* (superconductivity to the English speaking world) in mercury in 1911 by immersing it in liquid helium, he was aware of its potential for advancing the influence and reach of magnetic technology. Superconductors are not just better conductors of electricity than well-known metals such as copper, silver and aluminium, but they represent a completely different phenomenon in electrical science, as is suggested by the resistance versus temperature graphs in Fig. 10.13, originally due to Onnes.

Fig. 10.13 Comparison of the temperature behaviour of a conventional and a superconductor: gold = *triangular symbols*: mercury = *square symbols*

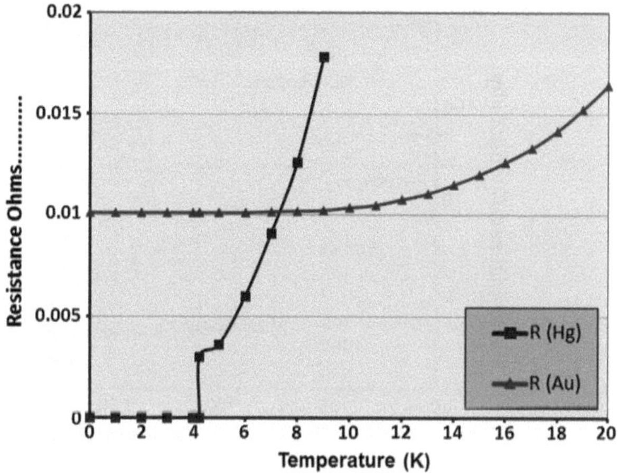

In general, most conventionally conducting materials display an increasing resistance with temperature, because fixed ions within the material become more agitated as it becomes hotter. When viewed as particles, as is normally the case in electrical engineering, 'free electrons' in the hot solid experience a much more disruptive and tortuous path through it, because of the interfering vibrating ions, and hence, the material is deemed to exhibit high resistance. But even at absolute zero (0 K), such materials possess some residual resistance because the electrons as particles are still impeded by the lattice of atoms (Fig. 10.13—red curve and triangles). For a superconductor, the behaviour at low temperature is quite different, with zero resistance or perfect conductivity being possible (blue curve/squares). There is a critical temperature (T_c) above which the superconductor acts normally, exhibiting high resistivity ($\sim$4 K in Fig. 10.13). Below this temperature, it switches to a zero resistance state. The electrons are unimpeded as they move through the superconductor even although it still possesses a lattice of fixed ions. So how do the electrons behave so differently, in superconductors as compared to conventional conductors, below T_c?

The answer to the above questions, although still incomplete, lies in quantum mechanics, and there are many general science and specialist books available which attempt an exploration of superconductivity from this viewpoint [16–18].

The observation that superconductivity is a distinctly odd phenomenon in electrical science manifests itself by the fact that classical electromagnetism and classical electrodynamics are no longer reliable analysis tools when applied to 'simple' electrical circuits containing a superconducting component. However that need not concern us here.

Highly conducting metals such as gold, silver and copper are not superconducting, but much in the same way as for gold, as shown in Fig. 10.13, their resistivities diminish (increasing conductivity), as the temperature of the material reduces towards absolute zero. The improvement arises purely from reduced thermal agitation of the metal atoms (In pinball machine terms, the balls drift more

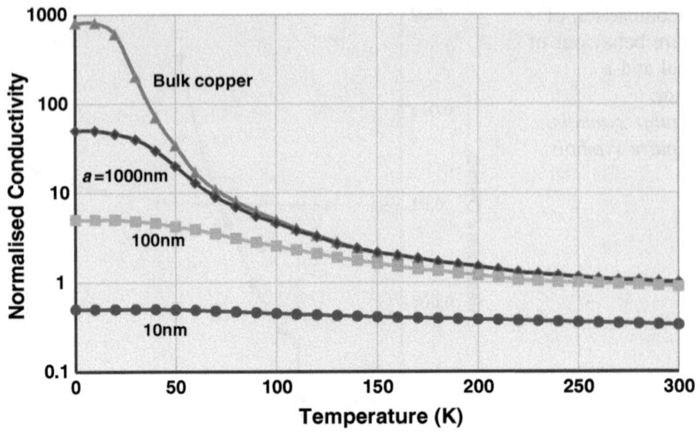

Fig. 10.14 Normalised conductivity for copper as a function of absolute temperature, with wire radius as parameter

freely down a table with fixed pins than down one with oscillating pins). This trend is depicted more clearly, for copper, in Fig. 10.14 where the bulk copper conductivity (blue curve with triangular symbols), normalised to the room temperature value of 5.9×10^7 S/m, increases from unity at 300 K to ~1,000 at 0 K. For a solar collector in space where temperatures can reputedly drop to 3 K, copper nantennas should potentially be capable of very effective radiation efficiency levels, even for filamentary dipoles exhibiting radii of less than 20 nm (see Fig. 10.12). Unfortunately for optical antennas, this apparent low temperature improvement in conductivity is negated by the nanoscale radius effect, since the $w(a)$ curve in Fig. 10.12 still applies. This is because even at absolute zero, the transit time and kinetic behaviours of electrons do not disappear in these conventional non-superconducting metals. The effect of these balancing influences is shown in Fig. 10.14 for three cases of dipoles with 1,000 nm (red + diamond symbols), 100 nm (green + square symbols) and 10 nm (mauve + disc symbols) wire radii. Evidently as the radius a diminishes, the conductivity improvement with lowering temperature begins to disappear by the time a reaches 10 nm, because of the nanoscale effect.

While conductivity enhancement in nano-wires formed from conventional metals may have limited application, superconductors which transition sharply at a characteristic temperature, as shown in Fig. 10.13, from non-superconducting to superconducting regimes, are more promising candidates for nanoscale antennas displaying high conductivity. For nano-wire applications, compounds such as yttrium barium copper oxide (YBCO), niobium tin and magnesium diboride are good examples of where developments are being directed. Today, superconduction is readily obtained in a wide spectrum of compounds ranging in transition temperatures from 5 to 100 K, those at the high end of this range being referred to as 'high-temperature' superconductors. At very low temperature where the ionic lattice in a material such as niobium tin becomes devoid of vibrational energy, the lattice becomes a filter whose passband is aligned to frequencies associated with electron waves (the mechanism is not unlike that of an electromagnetic waveguide —particularly periodic waveguide). Importantly, for infrared and optical frequency dipoles, this means that the nanoscale phenomenon associated with discrete electron motions is no longer active [16]. That superconductivity persists in nanoscale filamentary wires has been reported in the literature [19] where it has been demonstrated in filaments with radii in the <10 nm range. For a niobium tin half wavelength 3 cm long dipole with a wire radius of 20 nm, radiation has been secured at 10 GHz with an efficiency of 75 %, when the operating temperature has been reduced to the superconduction level of 4.2 K—approximately the temperature of outer space.

References

1. Bailey RL (1972) A proposed new concept for a solar-energy convertor. J Eng Power 94:73–77
2. Marks AM (1984) Device for conversion of light to electric power. US Patent 4,445,050, 24 Apr 1984
3. Marks AM (1986) Ordered dipolar light-electric power convertor. US Patent 4,574,161, 4 Mar 1986
4. Marks AM (1988) Femto diode and applications. US Patent 4,720,642, 19 Jan 1988
5. Corkish R, Green MA, Puzzer T (2002) Solar energy collection by antennas. Sol Energy 73:395–401
6. Bharadwaj P, Deutsch B, Novotny L (2009) Optical antennas. Adv Opt Photonics 1:438–483
7. Novotny L, van Hulst Niek (2011) Antennas for light. Nat Photonics 5:83–90
8. Novotny L (2007) Effective wavelength scaling for optical antennas. Phys Rev Lett 98(1–4):266802
9. Balanis CA (2005) Antenna theory. Wiley, New Jersey
10. Hanson GW (2008) Radiation efficiency of nano-radius dipole antennas in microwave and far-infra-red regimes. IEEE Trans Antennas Propag Mag 50(3):66–77
11. King RWP, Wu TT (1966) Imperfectly conducting cylindrical transmitting antenna. IEEE Trans Antennas Propag AP-14:524–534
12. Hanson GW (2006) On the applicability of the surface impedance integral equation for optical and near infra-red copper dipole antennas. IEEE Trans Antennas Propag AP-54:3677–3686
13. Kotter DK, Novack SD, Slafer WD, Pinhero P (2008) Solar nantenna electromagnetic collectors. In: Proceedings of energy sustainability conference, Jacksonville, Florida
14. Alda J, Rico-Garcia J, Lopez-Alonso J, Boreman G (2005) Optical antennas for nano-photonic applications. Nanotechnology 16:230–234
15. Steinhogl W, Schindler G, Steinlesberger G, Travin M, Engelhardt M (2005) Comprehensive study of the resistivity of copper wires with lateral dimension of 100 nm and smaller. J Appl Phys 79(023706):1–7
16. Tinkham Michael (1996) Introduction to superconductivity. Dover Publications Inc., New York
17. Lehndorff BR (2001) High-T_c superconductors for magnet and energy technology. Springer, Berlin
18. Moon FC (1994) Superconducting levitation. Wiley, New York
19. Zgirski M, Riikonen K-P, Touboltsev V, Arutyunov K (2005) Size dependent breakdown of superconductivity in ultranarrow nano-wires. Nano Lett 5:1029–1033

Index

A
Abbc, 182
Absorptivity, 191
Action at a distance, 122, 130
Aerosols, 9
Air breakdown, 222
Alignment
  errors, 234
  latitude lines, 238
Aluminium, 244, 258
Amber, 123
Ampere law, 52, 63
Analogy
  pin ball, 259
Analysis
  z-transform, 98
Anode
  block, 214
Antenna
  aperture, 73, 80, 84, 98
  aperture field, 79, 84
  array, 97
  array phase, 97
  beamwidth, 86
  dielectric rod, 73
  dipole, 54
  directivity, 86, 87
  element, 98
  far-field pattern, 84
  feed line, 71
  gain, 87
  half power beamwidth, 87
  half-wave dipole, 244
  Hertzian dipole, 65
  horn, 73
  illumination function, 86
  isotropic, 71, 87, 94
  isotropic sink, 72
  leaf, 2
  lens, 73
  linear aperture, 79
  logarithmic spiral, 243
  lossless, 87
  nano-scale, 259
  omni-directional, 233
  optical, 243
  patch, 234
  principal plane, 84
  receive, 232
  rectangular aperture, 81
  reflector, 54, 73
  slit, 81
  transmit, 232
  uniform illumination, 90
  wavelength, 243
Antenna array
  linear, 98
  pattern multiplication, 98
  periodic, 98
  planar, 98
Antenna pattern
  cuts, 84, 92
  directivity, 94
  iso-density surface, 92
  mainbeam, 94
  mainlobe, 86
  major planes, 84
  nulls, 86, 93
  omni-directional, 87
  polar form, 84
  rectangular form, 84
  sidelobe, 86
  three dimensional, 84
  two dimensional, 84
Approximation
  successive, 157
Array
  active, 230
  alignment, 225
  analysis, 99

A.J. Sangster, *Electromagnetic Foundations of Solar Radiation Collection*,
Green Energy and Technology, DOI 10.1007/978-3-319-08512-8

Array (*cont.*)
beam guidance, 228
beam locking, 228, 229
beam steering, 232
beamwidth, 102
binomial, 114
broadside, 101
conjugate phase, 230
deterioration, 228
directional control, 226
directivity, 103
efficiency, 115
element phase shift, 101
element spacing, 112
end-fire, 101
end-fire proximity, 105
exact conjugation, 234
flat plate, 221
Fourier series, 118, 119
grating lobe, 101, 105
ground based, 227
guidance signal, 226
half-power beamwidth, 103, 104
ideal pattern, 118
inexact conjugation, 234
linear, 98, 227
low cost, 221
microstrip, 234
monitoring, 226
mosaic, 236
no sidelobes, 114
null beamwidth, 102, 104
open-ended waveguide, 229
optimal spacing, 111
optimum gain, 105
pattern, 100
pattern null, 101
pattern synthesis, 115
phase coherence, 226
phase locked, 223
phase scanning, 105
phased, 228
planar, 98, 223, 227
pointing accuracy, 226
polarisation, 221
polynomial, 106
primary lobe, 101
principal lobe, 102
principal maximum, 101
rectenna, 237
retrodirective, 229, 231
root diagram, 109
root location, 109, 114
scan limit, 105
scanned, 105
secondary maximum, 101
sidelobes, 105
slotted waveguide, 221
slotted waveguide module, 223
split beam, 228
statistics, 227
sub-array, 223
supergain, 114
synthesis, 99, 106
tracking, 105, 232
tracking errors, 232
true synthesis, 118
uniform end-fire, 103
uniform linear, 100
unit circle diagram, 107
waveguide phased, 227
Astronomy
Almanac, 11, 12
Atmosphere
CO_2 concentration, 6
pressure, 12
refraction, 5, 12
Atomic lattice, 251, 258
Atomic scale, 244
Atoms
acceptor, 146
donor, 34, 124, 146
Automatic tracking, 105
Auxiliary functions, 58
Avalanche diode, 216

B
Bandpass filter, 231
BARITT diode, 217
Barium, 260
Beam pointing, 227
Beamwidth, 86
Biosphere, 3
Biot-Savart law, 63
Boltzmann, 152
Boron, 161
Boundary
granular, 141
Boundary condition, 141, 142
Boundary value problem, 175
Brewster angle, 46
Brown, W.C., 237

C
Carbon dioxide, 6
Carriers

majority, 146
minority, 146
Cartesian frame, 149
Cathode
cylindrical, 214
Cell
concentrator, 202
Characteristic impedance, 244, 246
Charge
conservation, 124
distribution, 61
electric, 56
electric field, 52
isolated, 52
magnetic, 56
magnetic field, 52
movement of, 52
Circuit
full wave, 237
half-wave, 237
phase conjugating, 232
rectenna, 237
retro-directive, 232
Civilisation, 7
Climate
science, 3
Coaxial cavity, 214
Coaxial line, 46, 50
Coherence
sources, 132
waves, 132
Collector
topologies, 17
Complex permittivity, 247
Compton, 126
Computer
modelling, 88
software, 92
Conductance
incremental, 167
Conduction current, 56
Conductivity
brine, 6
bulk, 254
copper, 257
normalised, 258
water, 6
Conductor
filamentary, 243
perfect, 142
Conjugate phase, 229, 230
Conservation
energy, 3
Continuity equation, 56
Control circuit, 226
Copper, 258
Copper oxide, 260
Corner reflector, 229
Covalent bond, 146
Curl, 35
Current
electric, 56
magnetic, 56, 77

D

De Broglie, 136
Del operator, 35
Delta function, 248
Detection
probability, 131
Detector
event, 131, 132
Differential equation
second order, 78
Diffraction
interference pattern, 129
power density, 80, 129
Diode
negative resistance, 216
nonlinearity, 237
semiconducting, 242
transferred electron, 217
Diode rectifier, 237
Dipole, 237
capacitance, 244
complex power, 68
conductivity, 256
conventional, 244
dough-ring, 67
efficiency degradation, 247
energy flow, 69
filamentary, 247
finite length, 246
half-wave, 244
harmonics, 237
Hertzian, 244
impedance conditions, 246
inductance, 244
input resistance, 246
nanoscale, 247
near field, 67

Dipole (*cont.*)
  power flow, 69
  radiation efficiency, 246
  radiation fields, 67
  radiation pattern, 67
  radiation resistance, 70
  stored energy, 67
Dipole antenna
Directivity
  optimum, 103, 105
Dirichlet condition, 36
Divergence, 35
Doppler shift, 126
Drude, 252, 254
Duality
  wave/particle, 122, 123

E
Earth, 207, 223
  antenna, 209
  Orbit data, 11
Efficiency
  collection, 14
  heat loss, 15
Einstein, 124, 125, 136
Electric charge, 124
Electric current, 30
Electric current source, 55
Electric field, 141
Electrical
  circuits, 123
  devices, 123
  engineering, 134
  generator, 17
  inductor, 134
  machines, 123
  path length, 234
  real estate, 123
  transformer, 17
Electrical Engineering, 28
Electrical power, 37
Electricity
  base-load, 208
  generation, 174
  power transmission, 209
Electrodynamics
  classical, 134
  collective, 122
  cumulative form, 137
  Feynman, 124
  probabilistic, 124
  quantum, 124, 128, 132
  quantum theory, 123
Electromagnetic wave
  absorption, 4
  antenna, 174
  broadcast, 28
  concentrator, 184
  cylindrical, 80
  cylindrical front, 81
  diffraction, 41, 80
  free space, 37
  guiding, 46
  incident, 41
  infrared, 4, 28, 242
  microwave, 28
  millimetre wave, 28
  phase velocity, 48
  plane solution, 37
  PN junction, 153
  polarisation, 40
  propagation, 125
  radio, 28
  reflected, 41
  source free, 54
  source present, 54
  spectrum, 28
  stationary pattern, 47
  transmission, 125
  transmitted, 41
  trapped pattern, 48
  trapping, 46
  ultraviolet, 28, 34
  visible, 242
Electromagnetics
  boundary conditions, 36, 43
  classical, 122
  classical theory, 138
  curl, 35
  del operator, 35
  deterministic, 125
  divergence, 35
  duality, 78
  equivalence theorem, 74
  evanescent wave, 34
  fields, 28
  finite element, 140
  half-space, 77
  impressed current, 55
  moment method, 140
  phasor notation, 38
  plane wave, 34
  potential functions, 58

power transmission, 46
radiating aperture, 76
radiating wave, 34
radiation, 2, 53
reflection coefficient, 44
scattering, 16
simulation, 140
TEM wave, 34, 37, 39, 52
theory, 28
transmission coefficient, 45
wave equation, 36, 139
wave-front, 34
wave impedance, 39
wave power density, 38
wave reflection, 16
wave theory, 123
waveguide ring, 138

Electromagnetism
atom, 124
boundary conditions, 33
charge, 30
circulation, 30
classical, 136
electric current, 30
field dimensions, 34
flux, 30
homogeneous medium, 36
linear medium, 36
microscopic level, 124
photons, 52
surface current, 34
vector field, 30, 143

Electron
bunching, 212
coherent wave, 136
collisions, 252
coupled, 136
cumulative wave, 137
drift, 212
emission of, 125
energy, 126
free, 258
free emission, 222
kinetic effects, 256
kinetic energy, 125, 212
mobility, 253
momentum change, 252
particle, 258
scattering, 126
wave, 28, 143, 260

Electron beam, 212, 243
Electron drift, 251
Electron gun, 212
Electron wave, 30
Electronic
beam control, 227
Electronic guidance, 227
Electronics
consumer, 18, 145
Electrostatic lens, 213
Electrostatics
Gauss law, 60
potential, 61
Emissivity, 223
Energy
conservation, 127
convertor, 17
field storage, 244
quantum of, 136
resources, 7
storage, 197
transitions, 3, 124
Engine
external combustion, 202
Equator, 12, 207
Equinox, 12
Equivalence principle
Love's, 74–76
Equivalence theorem, 74

F

Faraday, 136
Faraday's law, 68
Far-field pattern, 98
Feedback, 214
Fermat, 182
Ferrite material, 63
Ferrite phase shifter, 227
Feynman
diagrams, 123, 128, 130, 141
Filamentary dipole
attenuation, 255
conductivity, 253
efficiency loss, 249
field penetration, 255
radiation efficiency, 255
resistance/unit length, 256
Filamentary wire
attenuation, 254
attenuation coefficient, 249
boundary conditions, 249

Filamentary wire (*cont.*)
current flow, 249
field penetration, 251, 258
phase coefficient, 249
phase velocity, 249
radius effect, 259
resistance, 247, 255
skin depth, 248
standing wave, 249
surface fields, 248
TEM mode, 249
TM mode, 249
Filamentary wire attenuation, 247
Filter
passband, 260
Finite element method, 140
Fluid
receiver tube, 17
steam, 17
Fluid transporter, 17
Force
electric, 123
gravitational, 123
magnetic, 123
strong nuclear, 123
weak nuclear, 123
Force of gravity, 251
Fossil fuel
abandonment, 7
ancient sunshine, 2
geology, 2
photosynthesis, 1, 2
Fourier, 4
Fourier series, 118
Fourier transform, 83
Franklin, 124
Free space
impedance of, 37
permeability, 36, 63
permittivity, 33, 39
Frequency
carrier, 140
scaling, 258
spectrum, 140

G
Gallium arsenide, 203
Gallium nitride, 217
Gauss law, 52, 149
Generator
induction, 18
three phase, 196
Geometrical optics, 177
Geo-stationary, 23, 220
Gilbert, 123
Glaser, 208
Global warming, 19
Graphite
pyrolytic, 223
Greenhouse effect
population, 6
reforestation, 5, 6
Greenhouse gas, 21
Greenpeace, 7
Guided wave
cut-off, 49
dominant mode, 49
modes, 49
TE, 48
TM, 48
Gunn diode, 216, 217

H
Hansen-Woodyard, 105, 114
Headlight, 16
Heat loss, 15
Heaviside, 143
Heliostat
tracking, 17, 199
Helium, 258
Helmholtz equations, 57
Hemisphere, 10
Hertz, 125, 143
Hertz, Heinrich, 28
Hertzian dipole, 65
Holocene, 3
Hottel/Whillier/Bliss, 15
Huygen, 76, 80, 81, 129, 176, 177
Hyperbolic sub-reflector, 204

I
IMPATT diode, 216, 217
Impedance
characteristic, 71
surface, 255
Incoherent, 242
Industrial revolution, 6
Infrared, 28
Integro-differential equation, 249
Interference pattern, 80
International limit, 220
Inverse square law, 52, 53, 189
Ionisation, 222

Iso-power density, 222
Isotropic elements, 98
Isotropic radiator, 98
Isotropic sink, 87
Isotropic source, 87
Iteration
  convergence, 159
  downhill, 158
Iterative equation, 159
Iterative procedure, 158

J
Jacobean, 158

K
Kinetic theory
  classical, 251
King, R.W.P., 248
Kirchoff, 30, 78, 155
Kirchoff's law, 59
Klystron, 212, 227
  amplifier, 212
  buncher cavity, 213
  catcher cavity, 213
  collector, 213
  efficiency, 213
  gain, 213
  re-entrant cavity, 213
  reflector, 214
  reflex, 214
  space charge, 213
  velocity modulation, 213

L
Laminar flow, 184
Latitude, 9
Lattice
  atomic, 259
  ionic, 259, 260
Lattice vibration, 243
Lens
  convex, 204
  debulking, 204
  Fresnel, 204
  lighthouse, 204
Lens per cell, 205
Les Mees, 20
Levenberg, 159
Light
  frequencies, 242
  interference, 128
  photon, 126
  quanta, 124, 126
  quantum of, 126
  speed of, 28
Lightning conductor, 124
Linear array, 98
Lithography, 243
Local oscillator, 217
Lodge, 125
Longitude, 10
Lorentz condition, 59
Lorentz force, 33, 34
Lorentz law, 214
Luna ring, 210

M
Macroscopic fields, 28
Magnesium dichloride, 260
Magnetic
  focusing, 212
  stored energy, 136
  vector potential, 137, 141
Magnetic current, 77
Magnetic current source, 55
Magnetic field, 141
Magnetic flux
  quantum of, 136
Magnetic potential
  vector, 64
Magnetism, 123
Magnetostatics, 62
Magnetron, 21, 212, 214, 223
  anode block, 214
  coaxial geometry, 214
  cycloidal paths, 214
  injection signal, 226
  magnetic focusing, 214
  multi-cavity, 214
  oscillation, 215
  periodic waveguide, 215
  ring-shaped cavity, 215
  π-mode, 215
Magnifying glass, 13
Manufacturing
  low cost, 209
Mauna Loa, 6
Maxwell, James Clerk, 28
Maxwell's equations
  differential form, 34

Maxwell's equations (*cont.*)
  integral form, 34
  second order, 28, 33, 35, 37, 52, 54–56, 247, 249
Mead, 142
Mean free path, 253
Mercury, 258
Mesosphere, 21
Metal
  high conductivity, 244
Metal electrons, 243
Metal penetration, 243
Metaphysics, 122, 124
Microscopic
  distance, 134
  time interval, 134
Microstrip
  accessibility, 234
  embedded source, 236
  low cost, 234
  patch fields, 235
  patch radiation, 236
  substrate, 235
  technology, 234
  TEM mode, 235
Microstrip line, 50
Microwave
  antenna, 98, 218
  antenna size, 220
  array, 209, 218, 220
  array antenna, 218
  beam, 21, 98, 209, 220
  beam control, 224
  beam direction, 224
  circulator, 224
  collector/rectifier, 238
  conjugating circuit, 230
  devices, 217
  dipole, 237
  directional device, 224
  duplex, 224
  feed, 223
  ferrite, 224
  filter, 218, 231
  foot print, 219
  frequency, 98, 212
  guidance signal, 225
  local oscillator, 230, 232
  low power mixer, 234
  magic-T, 225
  main lobe, 220
  microstrip, 220
  mixer circuit, 230
  mono-pulse, 224
  planar array, 21
  power, 21, 98
  power conversion, 237
  power density, 219
  power generation, 209, 212
  safe level, 23, 219
  safe limit, 220
  semiconductor, 216
  split-beam, 224
  stripline, 220
  tube, 21, 212
  tube lifetime, 216
Microwave tube
  failure rate, 216
Minimisation technique, 157
Modular array, 18
Molecules, 34
Molten salt, 197
Moment method, 140, 249
Momentum
  accumulated, 252
  conservation, 127, 184, 185
  quantum of, 136
  space, 185
Moon, 210
Multi-cavity, 214
Multi-pactor, 222
Mutual coupling, 98, 234

N
Nanometre scale, 242
Nanoscale
  reduction factor, 253, 259
Nanoscale filament, 251
Nanoscience, 243
Nanotechnology, 243
Nantenna
  copper, 23, 242, 259
NASA, 208
Natural satellite, 210
Newman condition, 36
Newton, 157, 160
Newton-Raphson, 157
Niobium, 260

O
Onnes, 258
Optical

frequency, 255
Optical antenna, 23
Optics
bandgap, 243
curved reflector, 182
dielectric medium, 182
geometric theory of, 180
geometrical, 177, 179, 182
high concentration, 186
ideal concentrator, 184, 187
imaging, 176, 182
inhomogeneous medium, 182
large reflectors, 174
least action, 182
lenses, 173
light beam, 176
linear reflector, 183
mirrors, 173
optical path, 182, 183
optimum concentration, 187
parabolic bowl, 186
phase front, 178
phase-space, 185
photon, 122
photovoltaics, 122
physical, 178
plane wave spectrum, 177
plane waves, 177
power conservation, 180
ray concentration, 185
ray optics, 174
rays, 176
reflector, 16
refractive index, 182
symmetry, 176
theory of, 177
trough reflector, 176
ultra-violet, 16
wavefront, 177, 178
white light, 180
Orthogonal fields, 52
Oscillator
reference, 234, 236

P
Patch
coupling slot, 236
edge slots, 236
radiation pattern, 236
Patch array
specifications, 236
Periodic structure, 243
Permeability
anisotropic, 63
inhomogeneous, 63
Permittivity
complex, 56, 61
Phase
cumulative, 136
Phase locked, 223
Phase locked loop, 234
Phased array, 98
Phase velocity, 48
Phasor, 179
Phonon, 243
Phosphor, 161
Photodetector, 134, 243
Photoelectric effect, 125
Photon
at slit, 130
collector, 130
detector, 131
diffraction, 130
exchange of, 123
high energy, 127
particle, 123
probability vector, 130
scattering, 128
source, 130
stream, 130
vector rotation, 131
wave function, 122, 128
Photon
event probability, 131
Photonic devices, 122
Photosynthesis
carbon dioxide, 2
glucose, 2
thermodynamic, 2
water, 2
Photovoltaic
array, 154, 161
cell circuit, 154
cell model, 154
cells, 161
modules, 161
park, 20
simulation, 154
Photovoltaic effect, 153
Photovoltaic
calculator, 145
Cassegrain concentrator, 204
cell fabrication, 20

Photovoltaic (*cont.*)
  concentrated, 202
  heat resistance, 203
  infrared, 204
  lens focusing, 204
  mini-reflector, 202
  multi-junction cell, 203, 204
  operation, 243
  PV junction, 146
  radio, 145
  solar panel, 13
  space array, 211
  space array size, 218
  street signs, 145
  technology, 18, 145
  triple junction cell, 205
  watch, 9, 145
Physics
  particle, 125
  solid state, 146
Pilot signal, 228, 232
Pin ball, 259
Pinball machine
  analogy, 251, 253
Planar array, 98
Planck, 125, 136
Plane wave, 177
  spectrum, 81
Planet
  surface of, 209
Plasma, 217
Plasmons, 243
Platform
  orbiting, 207
Poisson equation, 151
Polarisation
  circular, 237
  horizontal, 40
  linear, 237
  parallel, 40
  perpendicular, 40
  transverse electric, 41
  transverse magnetic, 41
  vertical, 40, 41
Polynomial, 99
Potential barrier, 154
Potential functions
  auxiliary, 76, 77
  magnetic, 58, 77
Power
  radiated, 53
  solar, 4
Poynting, 177, 181
Poynting vector, 38
Probabilistic, 123
Proton, 28
Puimichel, 20

Q
Quanta, 125
Quantisation, 134
Quantum electrodynamics, 122
Quantum mechanics
  wavefunction, 4, 30, 52, 259

R
Radar, 217, 224
  cross-section, 229
  target enhancement, 229
Radiant heat, 187
Radiation
  black body, 3, 4, 125
  diffuse, 9
  efficiency, 246
  heat transfer, 189
  infrared, 4
  mechanism, 244
  pattern, 129
  photosynthesis, 2
  pressure, 126, 127
  probabilistic pattern, 132
  quanta, 125
  resistance, 70
  sink, 54
  source, 54
  spherical, 87
  time retardation, 61
  trapping, 4
Radiation resistance, 246
Radio waves, 123
Radius effect, 259
Read diode, 216
Receiver
  insulation, 17
  tower, 17
Rectangular waveguide
  TE_{10} mode, 49
Rectenna, 23, 209, 227, 242
  array, 238
  efficiency, 218, 237, 238
  filter, 237
  mismatch, 238

omni-directional, 238
Rectification, 242
Rectify, 23
Reflection coefficient, 44
Reflector
automatic control, 17
electroplated, 16
electrostatic, 214
parabolic, 17
parabolic dish, 17
trough, 17
Reflector antenna, 54
Relativistic speed, 126
Resistance
residual, 259
zero, 259
Resonant circuit, 244
Retarded potential, 61
Retro-directive array, 23
Retrodirectivity, 229
Ring resonator, 137, 138
Root
approximation, 158
Root location, 109

S
Satellite, 207, 218
antenna, 209
conversion efficiency, 218
drift, 238
economics, 208
electrical issues, 209
frequency, 212
low cost technology, 220
maintenance, 209
oscillation, 23
power, 212
power generator, 209, 218
rotation, 238
solar power, 208
Saturation current, 149
Scanned array, 105
Scattering
compton, 126
cylindrical waves, 129
electromagnetic wave, 134
incoherent, 134
particle collisions, 126
standing wave, 128
two-slit, 128
Schelkunoff, 109
Science
electrical, 28
general, 259
Science fiction, 208
Semiconductor, 217
antimony, 146
arsenic, 146
barrier voltage, 151
boron, 146
charge carriers, 146
DC current, 153
depletion layer, 147, 148
diffusion, 147, 152
diode, 146, 216
E-field, 153
EM excitation, 153
forward bias, 149
gallium, 146
gallium arsenide, 217
germanium, 217
holes, 146
impurity atoms, 147
indium, 146
junction current, 152
junction equations, 149
lattice, 146
N-type, 146
photons, 154
PN junction, 147
P-type, 146
quantum mechanics, 154
reverse bias, 149
saturation, 149
silicon, 217
thermal agitation, 147
voltage barrier, 148
wave action, 153
zinc sulphide, 217
Semi-conductor
diode, 23, 146
physics, 52
Short current element, 65
Signal
difference, 225
sum, 225
Signal injection, 224
Silicon
amorphous, 19
crystalline, 161
doping, 161
hybrid ised crystalline, 161
micro-crystalline, 19

Silicon (*cont.*)
mono-crystalline, 19
multi-crystalline, 19
polycrystalline, 161
ribon, 161
solar grade, 161
thin-film, 20
wafer fabrication, 19, 20, 146
Silver, 258
Skin depth, 248
Skin effect, 247
Slot
aperture antenna, 221
doughring, 222
EM solver, 222
inclined edge, 221
longitudinal shunt, 221
radiation pattern, 222
Slot array, 221
Snell's law, 44, 204, 229
Solar
acrylic mirrors, 195
array, 17, 161, 209
array fill-factor, 161
array tracking, 208
cell, 19
cell fabrication, 170
collector, 12
collector efficiency, 188
concentrated power, 9, 16, 173
concentration level, 196
concentration limit, 187
concentrator design, 189
concentrator temperature, 191
concentrator thermodynamics, 189
concentrator truncation, 188
constant, 7, 21
declination, 12
dish/Sterling, 202
electricity grid, 166
elevation, 12
engineering, 30
equinox, 12
farm, 17
flat plate converter, 14
fluid receiver, 183
flux, 7
focused flux, 202
Fresnel reflectors, 198
Fresnel system, 192
geometry, 9
heat absorption, 191
heated fluid, 13
Heliostat system, 192
high concentration, 201
incoherent light, 242
industry practice, 154
infra-red, 191
insolation, 9, 207
instruments, 7
irradiance, 7, 168
mercury arc valve, 166
micro-gas turbine, 201
mirror array, 199
module efficiency, 166
module inverter, 166
modules, 161
movable mirrors, 199
optics, 7
orbiting systems, 208
panel, 13, 20
parabolic dish system, 192
passive design, 9
photodiode, 163
photovoltaic cell, 146
PN junction, 243
power, 183, 218
power collection, 7, 28
power density, 4
power generation, 9
power point tracking, 166
ray absorption, 9
ray angles, 10
ray randomising, 188
receiver, 196
receiver tower, 199
satellite, 8
scattering, 9
silicon, 19
silicon wafer, 170
space array, 209
space collection, 21
space platform, 210
spectrum, 21, 23
state-of-the-art, 202
station, 207
sterling engine, 201
sunspot cycle, 8
system architectures, 192
temperature, 14, 191
terrestrial, 207
thermal, 13
thermal emittance, 196
thyristor, 166
tracking, 17, 207
tracking algorithm, 167

transmittance, 14
trough system, 192
two axis tracking, 198
working fluid, 196
Solar array
electrical circuit models, 163
fault detection, 169
installation, 169
monitoring, 168
optimization, 168
Solar cell
anti-reflective, 170
bifaciality, 170
conductance, 160
conductance method, 160
maximum power, 165
protection, 170
Solar concentrator
alignment methods, 205
Solar farm, 166, 169, 194
Solar flux
concentration, 174
focusing, 174
ray optics, 174
receiver, 174
Solar storage
liquid sodium, 199
molten salt, 199, 200
potassium nitrate, 199
sodium nitrate, 199
Solar system
annual efficiency, 194
capacity, 194
capacity factor, 194
concentration, 194
dish/Sterling, 202
energy storage, 197
heliostat field, 199
land use, 194
linear Fresnel, 197
modular, 202
parabolic dish, 201
parabolic trough, 195
peak efficiency, 194
relative cost, 194
Spain, 199, 200
thermal efficiency, 194
Solid-state
efficiency, 217
power, 217
Solid-state oscillator, 212
Solid state source, 216
Source
phase locked, 233, 236
solid-state, 236
stable, 223
Source function
electric, 78
magnetic, 57, 78
magnetic current, 91
Space
array modularity, 211
array radiation, 211
array thermodynamics, 211
arrays in, 207
assembly in, 208
platform, 207
semi-infinite, 128
technical issues, 209
technology, 207
thermal management, 211
transport to, 208
travel, 207
unbounded, 128
Space charge, 213, 243
Space orbit, 21
Space platform, 98
Spectral frequency, 38
Spectroscopy, 243
Spectrum
angular, 81
function, 81, 84, 87
plane wave, 81
visible, 21
Spectrum function, 100
Specularity factor, 254
Spherical waves, 178
Split beam, 225
Standing wave, 128
Stationary phase, 178
Statistics, 134
Stephan-Boltzmann, 190
Sterling engine, 18
Stokes' law, 34
Stratosphere, 21
Stratton, 122
Stripline, 50
Sunlight
concentration, 174
Superconduction
loop, 134, 258
Superconductor, 211

Superconductor (*cont.*)
high temperature, 260
mercury, 258, 259
Superposition, 80, 99
Surface current, 247
Surface impedance, 249
Synthetic oil, 196

T
Target acquisition, 98
Temperature
absolute zero, 259
critical, 259
high, 260
low, 260
maximum, 192
room, 259
transition, 14, 260
Thales of Miletus, 123
Thermal management, 209
Thermal power station, 194
Thermal storage, 197
Thermodynamics
conservation, 3
efficiency, 186
equilibrium, 5, 6
first law, 5, 180
greenhouse effect, 5
photosynthesis, 2
second law, 147, 190
Thin-film
hetero-junction, 20
poly-crystalline, 20
Time retardation, 61
Tin, 260
Tower, 17
Tracking
heliostat, 199
two axis, 198, 199
Tracking array, 98
Transform pair, 83, 84
Transistor, 216
geometries, 216
Transit time, 253
Transmission coefficient, 45
Transmission line
open-circuit, 244
standing wave, 244
TEM mode, 244
Transmission line match, 237
TRAPATT diode, 216, 217
Tschebyscheff function, 117, 118
Tunnel diode, 216, 217
Turbine, 187
Turbine/generator, 174

U
Ultraviolet, 28
Uncertainty principle, 122, 132
Unit circle, 107
Uranium, 7

V
Vacuum, 52, 212
Vacuum diode, 214
Vector notation, 30
Vector potential
magnetic, 247, 249
Vector-differential theorem, 34
Velocity modulation, 213
Visible spectrum, 21, 28

W
Wave diffraction
at slit, 128
gratings, 128
Wave equation
inhomogeneous, 57, 78
scalar, 36
source present, 60
vector, 57, 139
Wave penetration, 242
Wave spectrum
continuous, 134
Wave-front, 34
Waveguide
aluminium, 222
boundary condition, 221
circulator, 231
coaxial line, 50
complete modes, 140
cross-section, 222
directional feed, 223
frequency range, 222
hardware, 227
horn, 227
matched load, 138
microstrip line, 50

mode, 139, 140, 221, 222
orthogonal field, 221
parallel plate, 50
periodic, 260
phase shifter, 227, 228
power handling, 222
propagation coefficient, 140
quantised modes, 141
rectangular, 49, 140, 221
ring resonator, 138, 141
standing wave, 140
stripline, 46, 50, 139
T-junction, 227
tree, 227
Waveguide H-plane, 225
Waveguide slot array, 221
Winston, 184, 185

X
X-ray tube, 126
X-rays, 123, 126

Y
Yacht, 229
Young
two-slit experiment, 128
Yttrium, 260

Z
Zero gravity, 220
z-transform, 98

MIX
Papier aus verantwortungsvollen Quellen
Paper from responsible sources
FSC® C105338

If you have any concerns about our products,
you can contact us on
ProductSafety@springernature.com

In case Publisher is established outside the EU,
the EU authorized representative is:
Springer Nature Customer Service Center GmbH
Europaplatz 3, 69115 Heidelberg, Germany

Printed by Libri Plureos GmbH
in Hamburg, Germany